MW00787567

This book is dedicated to our wives, Anne Miller and Renee Miller

and

the legacy of plant knowledge and appreciation given us by John D. Freeman,
the utmost plant person, Trillium scholar,
and universal colleague to many.

Forest Plants of the Southeast and Their Wildlife Uses

REVISED EDITION

James H. Miller
USDA Forest Service, Southern Research Station
Auburn University

Karl V. Miller
D. B. Warnell School of Forest Resources
The University of Georgia

Featuring photography by Ted Bodner

The University of Georgia Press · Athens and London

in cooperation with the Southern Weed Science Society

Revised edition published in 2005 by the University of Georgia Press
Athens, Georgia 30602
www.ugapress.org

Printed and bound by Toppan Leefung
The paper in this book meets the guidelines for permanence
and durability of the Committee on Production Guidelines for
Book Longevity of the Council on Library Resources.
Most University of Georgia Press titles are available from popular ebook vendors

Printed in China
23 24 P 8 7

Library of Congress Cataloging-in-Publication Data

Miller, James H. (James Howard), 1944–
Forest plants of the Southeast and their wildlife uses /
James H. Miller, Karl V. Miller ;
featuring photography by Ted Bodner.—
Rev. ed.
x, 454 p. : col. ill., col. map ; 24 cm.
"In cooperation with the Southern Weed Science Society."
Includes bibliographical references (p. 428-432) and indexes.
ISBN 0-8203-2748-4 (pbk. : alk. paper)
1. Forest plants—Southern States—Identification.
2. Forest plants—Southern States—Pictorial works.
3. Animal–plant relationships—Southern States.
I. Miller, Karl V. II. Title.
QK125.M55 2005
581.7'3'0975—dc22 2004063744

British Library Cataloguing-in-Publication Data available

ISBN-13: 978-0-8203-2748-8

For more information about the Southern Weed Society, see www.swss.ws.
Income from this book goes to the Society's Endowment Fund
to support the Graduate Student Program.

CONTENTS

List of Figures

List of Genera

Grasses, Sedges, and Rushes

Woody Vines and Semiwoody Plants

Shrubs

Palms and Yucca

Cane

Cactus

Ferns

Ground Lichen

Acknowledgments

The purpose of this book is educational. Many have encouraged, sponsored, and contributed to this goal. The project was conceived and sustained by the Forest Plant Identification Subcommittee of the Southern Weed Science Society. Subcommittee members are the authors and Terry R. Clason, Cheryl A. Cobb, Andrew W. Ezell, Fred G. Fallis, William S. Garbett, John D. Gnegy, Dwight K. Lauer, Bobby Watkins, and Jimmie L. Yeiser, with executive board liaison by Shepard M. Zedaker. The Southern Weed Science Society provided a valuable institutional platform and financial support. The Society's leadership and members of the Forest Vegetation Management Section have been most supportive of this undertaking.

The majority of photography was by Ted Bodner, Auburn, Alabama, in collaboration with the first author. Ted Bodner captured plant beauty in detail, light, and composition, which conveys both reality and appreciation. Other valuable plant images were contributed by Dwight K. Lauer and John W. Everest, Auburn University; Tim R. Murphy, The University of Georgia; John R. Seiler, Virginia Tech University; Charles T. Bryson, Agricultural Research Service; Billy Craft, formerly with the Natural Resources Conservation Service; and Bruce A. Sorrie, private consultant.

Reviewers for botanical accuracy and completeness championed the bridge between the world of plants and inquisitive humans. Harold E. Grelen, author of several plant field guides while with the U.S. Forest Service, provided review comments, herbarium specimens, and photographs. Alvin R. Diamond, Jr., ecologist at Troy State University, shared much in ecological understanding, recommendations on species to include, and made a complete review of descriptions and photographic accuracy. David Bourgeois, botanist with Westvaco Corporation, gave advice on species inclusions and greatly improved botanical accuracy through his complete technical and editorial review. Suzanne Oberholster, former botanist with the U.S. Forest Service, gave thorough reviews on selected sections. Without these reviewers, their counsel on species lists, and their hours of scrutiny of detail, the descriptions would have been less than complete. Comprehensive reviews were made by John W. Everest, Auburn University, and Timothy B. Harrington, The University of Georgia. A. Sydney Johnson, retired professor at The University of Georgia, provided a thorough review of the wildlife use sections. Their knowledgeable comments greatly strengthened content and clarity, and for their counsel we are grateful.

Erwin B. Chambliss, U.S. Forest Service, Auburn, contributed to every facet of the book and gave tirelessly to the effort. Charles K. McMahon, U.S. Forest Service, Auburn, is recognized for the creative support and counsel he gave throughout the project. We are indebted to Betty R. Maxey, Juanita Crawford, and Carroll Garrett, U.S. Forest Service, Auburn, for their invaluable assistance with organizing botanical information.

Many plant subjects were cultured through the efforts of Robert H. Rush, Jr., curator of the Auburn University Arboretum, and Richard H. Martin, lands manager for Auburn University School of Forestry and Wildlife. The Auburn University Herbarium's regional collection was invaluable to the project.

This publication was made possible through the financial contributions of the Southern Weed Science Society, American Cyanamid, DowAgro Science, DuPont, and Monsanto.

This book would not have been possible without the support of the Southern Research Station of the U.S. Forest Service and the D.B. Warnell School of Forest Resources at The University of Georgia.

Introduction

Much is to be learned and understood about the rich flora of the Southeastern United States. This book is meant to be a guide to both prevalent and unique species that inhabit Southeastern forests, forest openings, margins, and right-of-ways. It is organized by plant genera. Descriptions of 180 genera are provided in sections for forbs, grasses and grass-likes, woody vines and semiwoody plants, shrubs, palms, cane, cactus, ferns, and ground lichen. Details for 330 species are provided. Most are common species found among the over 5,000 that reside within the temperate and subtropical areas of the Southeast. Others are included because they are valued wildlife plants, wetland plants, non-native invasives, or have aesthetic interest. Those species included are rarely the only species in a genus and more commonly are one of many. Identification of an unknown plant specimen to the species level may require the use of a botanical text containing a plant key, or wildflower guides, such as those listed in the reference section. Scientific and common names and botanical organization are according to USDA Natural Resources Conservation Service's Plants Database, http://plants .usda.gov. Scientific names are important because common names have not been given to many plants.

The physiographic provinces of the Southeast Forest Region are shown in Figure 1. Forests in the region range from central hardwood and high-elevation boreal forests in the north, to oak savanna forests on the western fringe, to swamp forests on the Atlantic Coast, and to tropical forests in the southern-most extremities of Florida. Dune and marsh communities occur along the coastal margins. Mixed pine-hardwood forests predominate throughout the mid-region with a similar structure that fosters a common flora. Arteries of bottomland hardwoods and riparian vegetation interlace throughout the region. This book focuses on plants that inhabit pine and pine-hardwood forests, upland and bottomland hardwood forests, and riparian habitat. Plants particular to the tropical parts of Florida and the higher elevations of the Southern Appalachian Mountains are not described, but are the focus of other books (see references).

Southeastern forests have for long periods been molded by strong natural and human influences. Much of the current forested area has been cleared and cultivated in the past, often followed by an extended period of open range grazing and repeated burning. Old field and natural forest succession followed, with episodes of increasingly intensified logging and decreases in burning frequency. With human migrations and occupations came introductions of non-native plants, which continue today. Non-native plants, often termed exotics, are included and denoted according to their origin. Non-native plants are increasingly replacing native flora.

Generalized descriptions of wildlife uses for many of the plants are presented to the extent understood. Complex communities of plants, animals, insects, and microorganisms inhabit the forests in this region. Intricate relations have evolved over millions of years. A need for increased knowledge of these linkages and interactions is becoming crucial. Accurate plant identification should aid progress in research, wiser use, management, and appreciation of our flora.

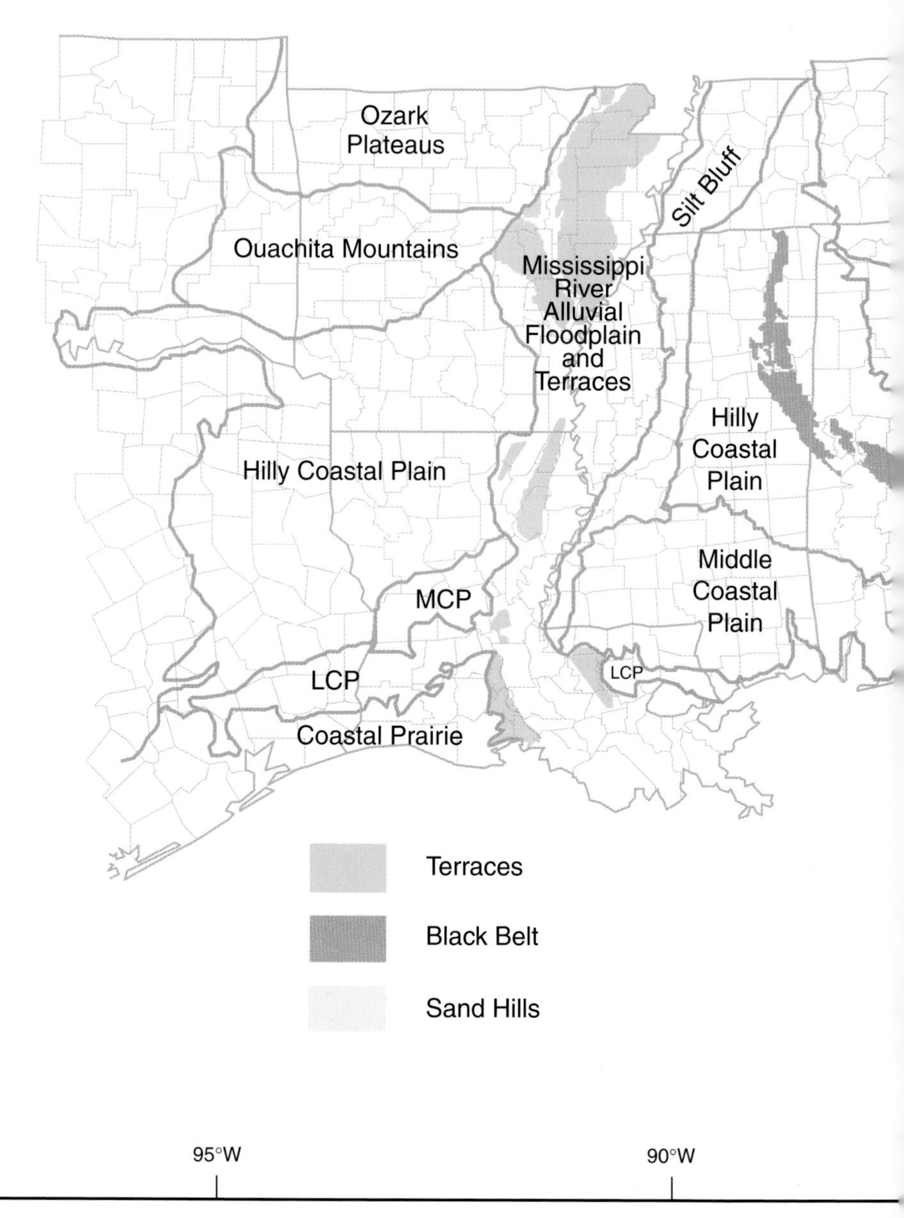

Figure 1. Physiographic provinces of the Southeastern United States.

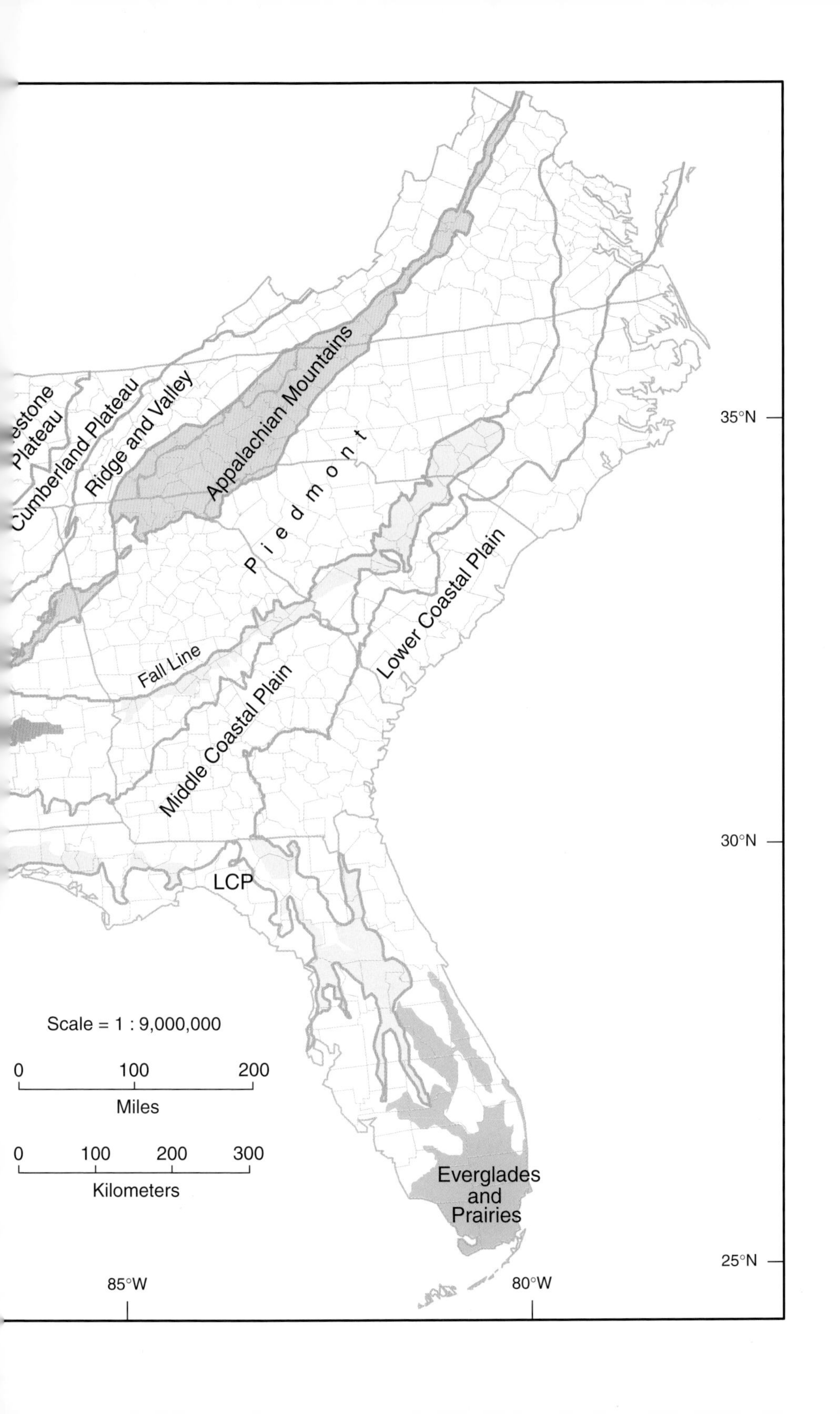

Plateau
Cumberland Plateau
Ridge and Valley
Appalachian Mountains
P i e d m o n t
Fall Line
Lower Coastal Plain
Middle Coastal Plain
LCP
Everglades and Prairies
Scale = 1 : 9,000,000
0
100
200
Miles
0
100
200
300
Kilometers
35°N
30°N
25°N
85°W
80°W

Format

The following is the page format outlining the information content:

Genus scientific name **Genus common name** **Family name**

Genus description. Growth form. Annual or perennial. Leaf, flower, fruit, and seed characteristics. Number of species in the SE.

Common Species
Species scientific name author **- common name** **Bayer international code**
Plant: Growth habit, annual or perennial, common size, branching and rooting habit, colonizing characteristics. Special features of importance are sometimes provided, such as: if poisonous, fall foliage color, flower or plant fragrance, and winter form.
Stem: Often includes thickness, details on branching, hairiness, color, and bark texture with progressive age for woody plants.
Leaves: Alternate or opposite or whorled, if evergreen, if compound, shape, common dimensions (length and width), margins, distinctive tips or bases, color and hairiness of surfaces, petiole length; if bracts, stipules or stipels present.
Flowers: Months of flowering within the region. Inflorescence type or flower arrangement, common size, flower type, petal description or numbers, color and size, sepal description, bract description.
Fruit and seeds: Months of mature seed dispersal within the region. Fruit type, common size, color and hairiness, bracts, seed description, common size.
Range: Native or exotic; country or region of origin if introduced. Delineation of range starting with the southwestern corner of range, if limited to certain physiographic provinces.
Ecology: If nitrogen fixer. Stand and vegetation types commonly inhabited. How species persists and spreads
Synonyms: Previously used scientific names, important varieties, other common names.

Other species in the SE: Scientific names for other species of the genus in the region, if less than 12 species and space permitting.

Wildlife: Plant uses by mammals, birds, and sometimes butterflies. Sub-regional differences of use.

Abbreviations

Plant Scientific Names. The following abbreviations are used in scientific names of species: **var.** means variety; **ssp**. means subspecies; and **X** means a cross between two species that are denoted within the accompanying parenthesis; and **ex** joins two authors involved with recognizing a species, which was named by the author listed second; **pro sp**. means a proposed species name.

Physiographic Provinces. The physiographic provinces are shown in Figure 1.

Mt means Southern Appalachian Mountains
Pd means Piedmont
Cp means Coastal Plains (Hilly, Middle, and Lower Coastal Plains)

Units of Measure. Metric units are used with the following abbreviations: **m** means meter, **cm** means centimeter, and **mm** means millimeter (see the cover flaps for a ruler).

Range. Frequently, **SE** is used to mean the Southeastern region. The two letter state abbreviations (postal codes) are used in the range descriptions, as follows:

AL Alabama
AZ Arizona
AR Arkansas
CA California
CO Colorado
CT Connecticut
DE Delaware
FL Florida
GA Georgia
ID Idaho
IL Illinois
IN Indiana
IA Iowa
KS Kansas
KY Kentucky
LA Louisiana
ME Maine
MD Maryland
MA Massachusetts
MI Michigan
MN Minnesota
MS Mississippi
MO Missouri
MT Montana
NE Nebraska
NV Nevada
NH New Hampshire
NJ New Jersey
NM New Mexico
NY New York
NC North Carolina
ND North Dakota
OH Ohio
OK Oklahoma
OR Oregon
PA Pennsylvania
RI Rhode Island
SC South Carolina
SD South Dakota
TN Tennessee
TX Texas
UT Utah
VT Vermont
VA Virginia
WA Washington
WV West Virginia
WI Wisconsin
WY Wyoming

Abbreviations for selected provinces of Canada are as follows: **Ont** means Ontario, **Que** means Quebec, and **BC** means British Columbia. The remaining provinces are used less frequently and are unabbreviated.

Sections of states and provinces are denoted by the following: **n** means north, **e** means east, **s** means south, and **w** means west. Likewise quadrant directions are abbreviated as **ne, se, sw,** and **nw**.

Forbs

Acalypha **Copperleaf or Threeseed mercury** **Euphorbiaceae**

Upright to spreading annual herbs, with densely hairy stems and alternate leaves, having two lateral veins from the base that parallel the midvein. Flowers in leaf axils or terminal clusters, with male and female flowers separate on the same plant, and both subtended by coarsely-toothed leafy bracts. Flower petals absent, bracts hide the tiny female flowers and the 3-seeded fruit, while male flowers extend beyond the bracts on spikelets. 5 species in the SE.

Common Species

Acalypha gracilens Gray **- slender copperleaf** **ACGR2**

Plant: Ascending annual, 20-80 cm tall, with many lateral branches from near the plant base, spreading outward and tips upward, from a distinct taproot. Plant gradually turns reddish-copper in late-summer or fall (thus the common name).
Stem: Slender, spreading, dense ascending or appressed hairy.
Leaves: Alternate, elliptic to elliptic-lanceolate, slender, 2-6 cm long and 0.5-2 cm wide, 2 distinct lateral veins parallel the leaf-margin to mid-leaf with other shorter veins, margins smooth on lower half and wavy to crenate above middle, finely hairy on both surfaces, green during summer turning copper to golden in late-summer, petioles of principal leaves 0.4-1.5 cm long, floral bracts near petioles can be mistaken for leaves.
Flowers: Jun-Dec. Axillary spikes, 0.5-2 cm long, within wedge-shaped leafy bracts having 5-13 triangular teeth, bracts occurring on female flowers at spike base, male flowers clustered along upper portion of spike, petals absent, only sepals.
Fruit and seeds: Jul-Feb. Capsule, 1-2 mm long, hairy, occurring in 3's within a leafy bract, splitting at maturity to release many seeds, reddish to black.
Range: Native. TX to FL and north to NH and west to WI, mostly Pd and Cp in the SE and infrequent in Mt.
Ecology: Early invader of site prepared forest plantations and right-of-ways. Frequent, scattered plants. Spreads by animal-dispersed seeds.
Synonyms: slender threeseed mercury.

Other species in the SE: *A. ostryifolia* Riddell, *A. rhomboidea* Raf., *A. setosa* A. Rich., *A. virginica* L.

Wildlife: Used by Northern Bobwhite and Mourning Dove along with other seed-eating songbirds. All species, particularly *A. gracilens* and *A. virginica* are common, but rarely abundant. Therefore, they are of moderate importance as seed producers or White-tailed Deer forage. However, slender copperleaf may become abundant on disturbed sites and is considered a high-use White-tailed Deer forage plant throughout the Southeast, especially during summer.

Acalypha Copperleaf or Threeseed mercury Euphorbiaceae

J. Miller (Oct)

Acalypha gracilens

T. Bodner (Jun)

Acalypha gracilens

T. Bodner (Sep)

Acalypha gracilens

Agalinis — False foxglove or Gerardia — Scrophulariaceae

Upright, slender-stemmed annual herbs (except for the perennial *A. tenuifolia*), with narrow, opposite leaves. Showy late-summer or fall flowers in racemes, pink to rose-purple usually with 2 yellow lines and purple spots in the throat, the corolla 5 lobed, with 2 upper and 3 lower. Seeds minute within a spherical capsule. Parasitic plants on the roots of grasses. 10 species in the SE.

Common Species

Agalinis purpurea (L.) Pennell **- purple gerardia** **AGPU5**

Plant: Upright annual, 40-120 cm tall, with many slender spreading branches from above mid-plant, very leafy, from a taproot with many laterals.
Stem: Finely slender, slightly hairy, green with light-green stripes, semi-circles under axils.
Leaves: Opposite, simple, sessile, linear to very narrowly lanceolate, often curled, 2-4 cm long and 0.5-2 mm wide, slightly hairy above, often turning purplish late season.
Flowers: Aug-Nov. Axillary in upper leaves and terminating branchlets forming indistinct racemes, corolla tube 5 lobed (2 upper and 3 lower), 2-4 cm long, lavender to purple with yellow throat and dark purple spots, 1 white slender stigma projecting and white anthers at right-angles forming a cross.
Fruit and seeds: Sep-Dec. Spherical capsule, 4-6 mm wide, splitting at maturity to release many minute seeds.
Range: Native. TX to FL and north to NH and west to MN and NE.
Ecology: Early invader of disturbed soils, but persists in grass cover as single scattered plants or colonies. Occurs in dry to moist sites and open to semi-shady habitats. Commonly inhabits open forests. Spreads by seeds.
Synonyms: *Gerardia purpurea* L., beach false foxglove.

Agalinis fasciculata (Ell.) Raf. **- gerardia** **AGFA2**

Similar to *Agalinis purpurea* except **leaves** in fascicles (tufts), linear, 0.5-1 mm wide; **flowers** (Aug-Oct) distinct racemes, hairs on petal margins about 1 mm long, spots inside throat lavender. **Range:** Native. MS to FL and north to TN and NC.
Ecology: Same as *A. purpurea*, except most common on roadsides and old fields and less in forested areas.
Synonyms: purple false foxglove.

Other species in the SE: *A. aphylla* (Nutt.) Raf., *A. auriculata* (Michx.) Blake, *A. gattingeri* (Small) Small, *A. linifolia* (Nutt.) Britt., *A. maritima* (Raf.) Raf., *A. obtusifolia* Raf., *A. setacea* (J.F. Gmel.) Raf., *A. tenuifolia* (Vahl) Raf.

Wildlife: *Agalinis purpurea and A. tenuifolia* are considered by some to be high preference White-tailed Deer forages, whereas *A. fasciculata* is a moderate to low preference forage. Other authors suggest purple gerardia is a poor quality forage. Gerardia are typical bee flowers with bee guides, especially bumblebees. *Agalinis purpurea* and *A. maritima* are the preferred food plants of the larval Dark Buckeye butterfly.

Agalinis **False foxglove or Gerardia** **Scrophulariaceae**

T. Bodner (Oct)

Agalinis fasciculata

J. Miller (Oct)

Agalinis purpurea

T. Bodner (Sep)

Agalinis fasciculata

Ambrosia Ragweed Asteraceae

Annual and perennial herbs, with the annual species comprising some of the most commonly-occurring plants in the SE. Leaves opposite on the lower plant and often alternate on the upper parts, usually lobed or dissected, margins serrate, and long petioles. Male and female flowers separate, with small female heads occurring below male heads in the axils of leaves or bracts, ray petals and plumed pappus absent. Nutlets (achenes) dark brown or black, spherical or nearly spherical, and beaked. 6 species in the SE.

Common Species

Ambrosia artemisiifolia L. - annual or common ragweed AMAR2

Plant: Common, erect, late-summer annual, 1-3 m tall, with densely-branched stems forming a bushy rounded top, from a shallow taproot. Foliage fragrant when crushed.
Stem: Upright, slender and hairy early, becoming stout and hairless, branched from base.
Leaves: Opposite basally and alternate upward, deeply bipinnately dissected, 4-10 cm long and 3-6 cm wide, segments mostly less than 1 cm wide.
Flowers: Aug-Nov. Terminal spike-like racemes, erect to drooping, 5-15 cm long, male flowers in tiny saucer-shaped heads, 1.5-2.5 mm wide, terminating the flower stalk above fewer female flowers, 2-4 mm wide, enclosed in foliar bracts.
Fruit and seeds: Sep-Dec. Beaked nutlet (achene), 3-4 mm long and 2-2.5 mm wide, yellowish-brown to reddish brown sometimes bluish near tip, central beak 1-2 mm long surrounded by 4-7 shorter projections, nutlet containing 1 ovoid seed.
Range: Native. Throughout US, most common in the Eastern and North Central States.
Ecology: Very common in the SE, early inhabitant of disturbed soils such as new forest plantations, declining in abundance after the second year. Found singly or in dense colonies. Spreads by bird-dispersed seeds. Viable seeds persist in the soil bank.

Ambrosia trifida L. - giant ragweed AMTR

Similar to *Ambrosia artemisiifolia* except **plant** larger annual, 1-6 m tall, branching; **stem** angled, up to 3-10 cm wide, striate green, often hairless below and spreading hairy above; **leaves** opposite, palmately 3-5 lobed and upper ones unlobed, 7-30 cm long, margins serrate, surfaces scurfy hairy, petioled; **flowers** (Sep-Dec) male flowers abundant in slender terminal racemes, flowers 3-5 mm wide, 5-6 lobed, stalk 2-7 mm long, female flowers in tufted fascicles (4-5) in upper leaf axils, 6-13 mm long and 5-10 mm wide, ascending sharp projections toward the apex, hairy; **fruit and seeds** (Oct-Dec) beaked woody nutlet (achene), 5-10 mm long and 2-3 mm wide, black, obovoid with several ribs, each rib ending in a short spine, nutlet containing 1 black seed, 4-5 mm long, finely pitted.
Range: Native. TX to c FL and north to Que and west to CO and BC, mainly Pd, Mt, and Mississippi River Valley in the SE.
Ecology: Early inhabitant of disturbed lands near rivers and streams. Common species adjacent to the Mississippi River on alluvial terraces and in young forests on moist soils. Usually on wetter soils than common ragweed. Often occurring in colonies. Spreads by seeds, persistent in soil bank.
Synonyms: horse-cane, richweed, great ragweed.

Other species in the SE: *A. bidentata* Michx., *A. confertiflora* DC., *A. hispida* Pursh, *A. psilostachya* DC.

Wildlife: Common ragweed ranks among the most important seed and cover-producing plants for Northern Bobwhite. Seeds also are consumed readily by numerous songbirds including Mourning Dove, American Goldfinch, Dark-eyed Junco, and species of sparrows. It also is a common preferred White-tailed Deer browse during spring and summer. Giant ragweed is a moderate to poor White-tailed Deer browse, but seeds are little used by birds.

Ambrosia Ragweed Asteraceae

T. Bodner (Sep)

Ambrosia artemisiifolia

T. Bodner (May)

Ambrosia artemisiifolia

T. Bodner (Jun)

Ambrosia trifida

T. Bodner (Oct)

Ambrosia artemisiifolia

Apocynum **Indianhemp** **Apocynaceae**

Perennial herbs, having stout erect stems with branching above mid-plant, thick milky sap when injured, from woody rootcrown with horizontal roots. Leaves opposite, ovate to lanceolate, with entire margins and tipped by a tiny point (mucronate). Minute flowers in terminal round-topped clusters (cymes), corolla white, pink, or greenish, and 5 lobed. Fruit paired slender dry pods, splitting along sutures to release many plumed seeds. 3 species in the SE.

Common Species

Apocynum cannabinum L. **- Indianhemp** **APCA**

Plant: Erect, stout-stemmed and a somewhat fleshy perennial, 30-150 cm tall, with milky sap, one to few stems and branched above mid-plant, from a woody base and horizontal roots. Often in small colonies from rootsprouts. Poisonous to cattle.
Stem: Deep red, round in cross-section, hairless, opposite branched above.
Leaves: Opposite and evenly spaced, often erect or ascending, oblong-lanceolate to lanceolate to ovate, 4-12 cm long and 1.5-6 cm wide, rounded bases and tips, tip with tiny point, dark green and hairless with whitish midvein and lateral veins above, whitish green and sparsely hairy (to hairless) beneath, margins entire and smooth, petioles 2-7 mm, reddish.
Flowers: May-Aug. Terminal round-topped branched clusters (cymes), 3-6 cm wide, exceeded by lateral branches, corolla white to greenish, urn-shaped, 3-5 mm long and 2-3 mm wide, 5 lobed.
Fruit and seeds: Sep-Oct. Dangling dry pods, paired, 10-15 cm long and 1-1.5 cm wide, splitting on a side to release many seeds, 4-6 mm long, thin and flat, topped by a tuft of soft milky hairs.
Range: Native. Throughout US, s Canada, and n Mexico.
Ecology: Occurs on dry to moist sites and open to semi-shady habitats. Commonly occurring in small groupings in open forests, new forest plantations, pastures, and old fields, and along right-of-ways and forest edges. Persists by woody rootstocks, colonizes by rootsprouting, and spreads by wind-dispersed seeds.
Synonyms: hemp dogbane.

Other species in the SE: *A. androsaemifolium* L., *A. xfloribundum* Greene (pro sp.) [*androsaemifolium* x *cannabium*].

Wildlife: No wildlife use reported, although various species of Indianhemp are frequented by butterflies such as the Spicebush Swallowtail.

Apocynum **Indianhemp** **Apocynaceae**

T. Bodner (Sep)

Apocynum cannabinum

J. Miller (May)

Apocynum cannabinum

J. Miller (May)

Apocynum cannabinum

Asclepias **Milkweed** **Asclepiadaceae**

Perennial herbs, having stout erect stems often with milky sap, spreading or reclining, mainly occurring as single plants. Leaves opposite, sometimes whorled or alternate. Flowers often in showy, terminal clusters (umbels), each flower with 5 downward-curved petals and a 5-lobed petal crown (corona), the lobes consisting of a hood and horn. Fruit a fleshy pod becoming dry papery and splitting along a side to release many plumed seeds. Usually poisonous to livestock. Over 20 species in the SE.

Common Species

Asclepias tuberosa L. **- butterfly milkweed** **ASTU**

Plant: Upright, stout-stemmed perennial, 30-80 cm tall, sap not milky, branched from the base and infrequently branched from upper nodes, from a woody rootstock and occurring as individual scattered plants.
Stem: Branching from base, green and hairy.
Leaves: Alternate and numerous, elliptic to obovate or oblanceolate and some hastate, becoming linear upward, 4-10 cm long and 3-25 mm wide, hairy (especially beneath), margins entire and some rolled under.
Flowers: May-Aug. Showy terminal hemispherical- or flat-topped clusters (umbels), 2-5 cm broad, corolla in a petal crown (corona), yellow to orange or red, with hoods yellow or red, horns shorter than hoods.
Fruit and seeds: Jun-Nov. Erect boat-shaped pod, fleshy turning dry-papery, 8-15 cm long and 1-1.5 cm wide, splitting on a side to release many plume-tufted seeds, 1-2 mm long.
Range: Native. TX to FL and north to ME and west to MN and AZ.
Ecology: Occurs on dry to moist sites and open to semi-shady habitats. Commonly occurring singly in open forests, new forest plantations, pastures, and old fields, and along right-of-ways and forest edges. Persists by woody rootstocks and spreads by wind-dispersed seeds.
Synonyms: pleurisy-root.

Asclepias variegata L. **- white milkweed** **ASVA**

Similar to *Asclepias tuberosa* except **plant** 20-100 cm tall; **stem** smooth or hairy; **leaves** opposite, 2-5 pairs, ovate to nearly lanceolate, 5-14 cm long and 3-7 cm wide and smaller upward, smooth or slightly hairy; **flowers** (Apr-Jun) umbels 3-6 cm broad, corolla bright white with reddish-purple bands between hood and horns; **fruit and seeds** (Jun-Oct) same as above. **Range:** Native. TX to FL and north to CT and west to MO and OK.
Synonyms: redring milkweed.

Wildlife: Butterfly milkweed is a poor White-tailed Deer forage, whereas *A. variegata* is reportedly of moderate preference. Milkweed seeds are consumed occasionally by Northern Bobwhite. All milkweeds are excellent producers of nectar that is attractive to numerous butterflies and other insects.

Asclepias **Milkweed** **Asclepiadaceae**

T. Bodner (Jun)

Asclepias variegata

T. Bodner (Jun)

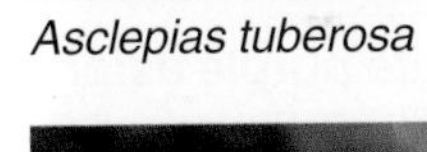

Asclepias tuberosa

T. Bodner (Jun)

Asclepias variegata

T. Bodner (Jun)

Asclepias tuberosa

Aster **Aster** **Asteraceae**

Perennial herbs, rarely annuals, that usually bloom in late summer and fall. Numerous small aster-type flower heads, yellow or red centers, surrounded by whorls of white, blue, or pink to purple ray flowers, enclosed within overlapping whorls of involucral bracts. Leaves numerous and alternate, often narrowly lanceolate to linear, and occurring as rosettes in spring. Seed in a bristled nutlet. Dried plants persist standing through winter. 66 species in the SE.

Common Species

Aster paternus Cronq. **- white-topped aster** **SEAS3**

Plant: Upright perennial, 20-60 cm tall, 1-several stems from a leafy rosette and a branching rootcrown with fibrous roots.

Stem: Moderately stout for asters, hairless basally and scurfy-hairy upward.

Leaves: Basal rosette and alternate on stems, broadly oblanceolate to elliptic with tapering bases, basal leaves 2-10 cm long and 1-5 cm wide and smaller upward, becoming sessile, all minutely hairy on margins and sometimes over surfaces, tips acute.

Flowers: Jun-Jul. Small heads in a flat-topped terminal corymb, both ray (outer) and disk (inner) flowers white, only 5-15 ray petals, involucral bracts white, with spreading green tips, in unequal rows.

Fruit and seeds: Aug-Feb. Narrow-tapered nutlet (achene), 1.6-2 mm long, silver hairy, topped with brownish-white bristles, 4-5 mm long, nutlet containing 1 seed.

Range: Native. AL to FL and north to s Canada and southwest to KY.

Ecology: Persists in new forests after harvesting because of rootcrowns. Inhabits open dry forests, right-of-ways, and roadsides. Spreads by wind-dispersed seeds.

Synonyms: *Sericocarpus asteroides* (L.) B.S.P., toothed whitetop aster.

Aster patens Ait. **- late purple aster** **SYPAP2**

Plant: Upright perennial, 20-150 cm tall, 1-3 widely branching stems arising from a slender rhizome having fibrous roots.

Stem: Slender, ascending, bristling short hairy, often bushy with many alternate branches from upper leaf axils.

Leaves: Alternate, lacking petioles and clasping around stem with a cordate base, broadly ovate to oblong, 3-15 cm long and 1-4.5 cm wide and smaller upward, margins entire, rough with stiff hairs on both surfaces.

Flowers: Sep-Nov. Flower heads at branch ends in panicles, blue to purplish or rarely pink rays (outer) with yellow to purple or red disks (centers), ray petals 15-25 and heads large for asters, 2-3 cm wide.

Fruit and seeds: Nov-Dec. Flat-ellipsoid nutlet (achene), 2.8-3.3 mm long, tan, ribbed, topped with tan to white bristles, 5-7 mm long, nutlet containing 1 seed.

Range: Native. TX to n FL and north to NH and west to MO and KS, rare in Lower Cp.

Ecology: Early invader of disturbed places such as new forest plantations and right-of-ways, and old fields. Occurs mostly on dry sites and open habitat. Scattered in occurrence or forming loose colonies. Colonizes by rhizomes and spreads by wind-dispersed seeds.

Synonyms: *Symphyotrichum patens* (Ait.) Nesom var. *patens,* spreading aster, skydrop aster.

Aster Aster Asteraceae

T. Bodner (Jun)

Aster paternus

T. Bodner (Oct)

Aster patens

Aster **Aster** **Asteraceae**

Aster pilosus Willd. - **white heath aster** **SYPIP3**

Plant: Erect perennial, 0.6-1.5 m tall, 1 to several stems from base, with or without short rhizomes having fibrous roots.
Stem: Stout at base, long hairs often in lines on the upper stem, many alternate branches appearing in late summer.
Leaves: Basal and alternate on stem, elliptic to linear with tapering base and acute tip, 2-10 cm long and 0.5-1 cm wide and much smaller and more numerous upward, margins entire to slightly toothed (crenate), many long hairs to hairless, basal leaves if persistent, oblanceolate with petioles.
Flowers: Sep-Nov. Flower heads along branches in panicles, ray petals (outer petals) white or rarely pink, 20-30, with yellow to red disks (centers), heads 2-3 cm in diameter, enclosed in an urn-shaped involucre with rows of unequal spreading bracts having tapering wedge-shaped tips, white-margins, and green midribs.
Fruit and seeds: Nov-Jan. Tapered nutlet (achene) 0.7-1.3 mm long, tan, topped with whitish bristles, 4-5 mm long, nutlet containing 1 seed.
Range: Native. LA to n FL and north to s Canada and west to WI and IA.
Ecology: Early invader of disturbed places such as new forest plantations, right-of- ways, and fields. Often abundant in old fields. Found as scattered plants or small colonies. Spreads mainly by wind-dispersed seeds.
Synonyms: *Symphyotrichum pilosum* (Willd.) Nesom var. *pilosum,* frost aster, hairy white oldfield aster. 4 varieties in the SE.

Aster dumosus L. - **bushy aster** **SYDUD2**

Similar to *Aster pilosus* except **plant** mostly leafy branched flower stalks, bushy in late summer, 0.3-1 m tall, from creeping rhizomes or swollen roots; **stem** hairless below with short-hairy upward; **leaves** rigid, scurfy above and hairless below, 2-12 cm long and 0.3-1 cm wide and smaller and numerous upward, margins often rolled under and with scattered serrations; **flowers** (Aug-Oct) on long spreading stalks (peduncles), 2-15 cm long, having many small, rigid, leaf-like bracts, ray petals 15-30, white to pale lavender to bluish, involucral bracts unequal with short, green, and V-shaped tips. **Range:** Native. TX to n FL and north to s Canada and west to WI and IA, common to Cp in the SE.
Ecology: Occurs on moist to dry sites and open to semi-shady habitats. Commonly inhabits roadsides. Colonizes by rhizomes and spreads by wind-dispersed seeds.
Synonyms: *Symphyotrichum dumosum* (L.) Nesom var. *dumosum,* rice button aster. 6 varieties in the SE.

Wildlife: Although common and often abundant throughout the Southeast, asters are of relatively minor importance to wildlife. The basal rosettes are eaten by White-tailed Deer during winter and early spring and the leaves and stems are consumed occasionally from spring through fall. Wild Turkey infrequently consume the flowers and seed heads.

Aster **Aster** **Asteraceae**

J. Miller (Oct)

Aster pilosus

T. Bodner (Sep)

Aster pilosus

J. Miller (Oct)

Aster dumosus

J. Miller (Oct)

Aster dumosus

Bidens **Beggarticks** **Asteraceae**

Annual, biennial, or perennial herbs, with long-slender nutlets (achenes), 3-4 angled, having barbed awns that stick to fur or clothing (thus the common name). Leaves opposite (in pairs) and simple or highly dissected. Stems upright or trailing with the ends erect, rarely hairy, and solid piths. Ray flowers (outer petals) few, yellow or white (rarely pink), with disk flowers (centers) yellow. Involucral bracts enclosing flowers in 2 rows, the outer green and the inner yellowish to brownish. Dried plants remain standing into winter. 13 species in the SE.

Common Species

Bidens bipinnata L. **- Spanish needles** **BIBI7**

Plant: Upright annual, 0.2-1.7 m tall, with scant foliage on spreading branches from mid-plant, tipped by spiny tufted seeds, from a taproot.
Stem: Slender, square and smooth, green with dark-green or maroon lines.
Leaves: Opposite, mostly 2-3 times pinnately dissected (thus the sci. name), 4-20 cm long, dull green and veins evident as ridges above, short appressed hairy along margins and main veins beneath, veins dark-green lines on the lower surface, petioles short appressed hairy and grooved on upper surface.
Flowers: Jul-Oct. Heads occur singly at the ends of long nearly-leafless stems in flat-topped corymbs, ray petals (outer) absent or short and yellow, disks (centers) yellow to yellowish-green, involucral bracts in 2 rows, outer 7-10, not leafy and shorter than inner ones.
Fruit and seeds: Aug-Feb. Thin and angled nutlet (achene), 8-18 mm long and less than 1 mm wide, smooth or sparsely hairy, green becoming black, topped with 3-4 barbed awns, nutlet containing 1 seed.
Range: Native. Throughout Eastern US.
Ecology: Early invader of disturbed places, often in sandy soils and moist or wet places. Forms colonies, especially along trails. Spreads by seeds attaching to fur or clothing.

Other species in the SE: *B. aristosa* (Michx.) Britt., *B. bientoides* (Nutt.) Britt., *B. cernua* L., *B. connata* Muhl. ex Willd., *B. coronata* (L.) Britt., *B. discoidea* (Torr. & Gray) Britt., *B. frondosa* L., *B. laevis* (L.) B.S.P., *B. mitis* (Michx.) Sherff, *B. pilosa* L., *B. tripartita* L., *B. vulgata* Greene.

Wildlife: Because the plants are rarely abundant, the seeds of beggar-ticks are of minor importance to wildlife. However, they are consumed readily by Northern Bobwhite and other seed-eating songbirds. When occurring near swamps and flooded bottomlands, the seeds are consumed frequently by Wood Ducks. Beggar-ticks are a low to moderate preference White-tailed Deer forage.

Bidens Beggarticks Asteraceae

J. Miller (Oct)

Bidens bipinnata

T. Bodner (Sep)

Bidens bipinnata

T. Bodner (Sep)

Bidens bipinnata

Blephilia **Woodmint** **Lamiaceae**

Upright perennial leafy-stemmed herbs (mints), with 2-6 dense whorls of flowers (head-like) encircling square hairy stems at intervals, forming distinctive terminal spikes. Flowers irregular, 2-lipped, white to lavender to bluish, appearing around the head. Petioled leaves opposite, elliptic to lanceolate, with serrate margins. Nutlets ellipsoid to oblong, smooth and shiny, 1 per flower. 3 species in the SE.

Blephilia ciliata (L.) Benth. **- horsemint** **BLCI**

Plant: Ascending perennial mint, 40-80 cm tall, few stems from the base and rarely branching above, from woody rootcrowns. Plants remain standing during winter.
Stem: Erect or leaning, squarish or angled in cross-section, short-gray hairs often with scattered longer hairs.
Leaves: Opposite, lanceolate to elliptic, 3-6 cm long and 1-3 cm wide, often curving downward, shallowly serrate, tips pointed to rounded, gray hairy above and whitish downy beneath, petioles 5-15 mm long and those immediately below flowers less than 5 mm long. Leaves odorless even though a mint.
Flowers: May-Jul. Terminal whorls or heads, 3-5, spaced at intervals along a stalk forming a distinctive spike, each whorl subtended by rows of grayish leafy bracts with hairy margins and sharp-pointed tips, corolla irregular, 2-lipped, bluish white to lavender with purple dots, hairy, appearing around the heads, densely crowded calyx tubes forming the head, each with pointed lobes.
Fruit and seeds: Aug-Feb. Nutlet (achene), about 1 mm long, ellipsoid to ovoid, black, smooth and lustrous, nutlet containing 1 seed. Nutlets shakened from dried calyx tubes by wind, water, or trampling.
Range: Native. TX to GA and north to VT and west to WI, MO, and AR, mainly Pd and upper Cp in the SE.
Ecology: Occurs mainly in small groups in dry forests openings and along forest edges, occupying more moist and shaded sites in MO and KY than in rest of the SE. Persists by rootcrowns and spreads by animal- and water-dispersed seeds.
Synonyms: downy pagoda-plant.

Blephilia hirsuta (Pursh) Benth. **- woodmint** **BLHI**

Similar to *Blephilia ciliata* except **plant** taller, 80-130 cm tall, much branched; **stem** densely hairy (or sparsely hairy), hairs 1-2 mm long; **leaves** broadly lanceolate to ovate, 4-8 cm long and 2-4 cm wide, margins serrate, petioles 1-2 cm long with those immediately below flowers greater than 1 cm long, peppermint odor when crushed; **flowers** (Jun-Sep) white with purple dots. **Range:** Native. AR to n GA and north to Que and west to MN and MO, mainly Mt in the SE.
Ecology: Occurs in groups in moist shady forests in the mountains.
Synonyms: hairy pagoda-plant.

Other species in the SE: *B. subnuda* R. Simmers & Kral.

Wildlife: No wildlife value reported.

Blephilia **Woodmint** **Lamiaceae**

J. Miller (Jun)

Blephilia ciliata

J. Miller (Jun)

Blephilia ciliata

Centrosema **Butterfly pea** **Fabaceae**

Twining or trailing, perennial leguminous vines, slender stemmed with 3-leaflet leaves on long petioles, minutely hairy throughout. Flowers circular, violet to lavender in mid-summer with many leafy bracts beneath. Extremely long, flattened legume (pods). 3 species in the SE, only spurred butterfly pea is widespread.

Common Species

Centrosema virginianum (L.) Benth. **- spurred butterfly pea** **CEVI2**

Plant: Trailing or climbing perennial vine, up to 1.5 m long, usually twining through the vegetation being rough hairy throughout, from a persistent woody rootcrown. Very similar in appearance to *Clitoria mariana*, also called butterfly pea, another viney legume having scoop-shaped flowers, short pods, and less hairy or hairless.

Stem: Slender vine, twining and curvy in parts and often twining on itself, branching infrequently at leaf axils, minutely rough hairy.

Leaves: Alternate, 3 leaflet, leaflets 2-7 cm long and 0.5-3 cm wide, ovate to lanceolate, margins entire, bases rounded, tips long tapering, dark green and rough hairy above and lighter below with veins even paler green, petioles 1-5 cm long, 2 stipules at leaf axils persistent, 1.5-4 mm long, ovate-lanceolate to lanceolate.

Flowers: Jun-Aug. Pea-type but upside down and flattened, almost circular, 2-4 cm wide, blue-violet to lavender with a white center and white back, 1-3 flowers in axillary clusters, numerous bracts along flower stalks, stalks 2-4 cm long, partially hiding the calyx, bracts about 1 cm long.

Fruit and seeds: Jul-Nov. Legume (pod) very long, slender and flat, 7-14 cm long (including long beak) and 3-4 mm wide, pod halves twisting when opening, 10-20 seeds per pod, cylindric, 2-3 mm long, brown to gray with black mottling. Flowers and pods present together.

Range: Native. TX to FL and north to NJ and west to TN and AR.

Ecology: Nitrogen fixer. Later invader of open, moist or alluvial woodlands, forest openings, and margins of pocosins and swamps. Less common on sandhills, old fields, and roadsides. Found singly or in patches. Persists by rootcrowns and spreads by animal-dispersed seeds.

Other species in the SE: *C. arenicola* (Small) F.J. Herm. [only in FL], *C. sagittatum* (Willd.) Brandeg. *ex* Riley [only Alachua Co., FL].

Wildlife: The seeds of spurred butterfly pea are of moderate importance for Northern Bobwhite in the Piedmont and Coastal Plain. They likely are also consumed by seed-eating songbirds, although no use is reported. *Centrosema* are moderate to high preference White-tailed Deer forages.

Centrosema **Butterfly pea** **Fabaceae**

T. Bodner (Aug)

Centrosema virginianum

T. Bodner (Aug)

Centrosema virginianum

Chamaecrista **Partridge pea** **Fabaceae**

Annual or perennial legumes, with pinnately compound leaves (tiny leaflets in paired rows) having nectar glands and persistent stipules at the petiole base. Showy yellow, irregular flowers. Flattened legume (pods) that split and twist to release several to many seeds. 7 species in the SE and 5 only in FL.

Common Species

Chamaecrista fasciculata (Michx.) Greene **- showy partridge pea** **CHFA2**

Plant: Upright annual legume, 0.3-1.2 m tall, very leafy with few to many alternate branches above mid-plant, from a taproot. Often growing in small to extensive colonies.

Stem: Moderately stout, ascending, brown and hairy, occasionally partially green and hairless.

Leaves: Alternate, pinnately compound (small leaflets in rows of 6-18 pairs), 3.5-6 cm long, sensitive at times (folding slowly when touched), leaflets 0.5-2 cm long and 2-6 mm wide, with distinct tips, each leafstalk has a saucer-shaped nectar gland near base and paired persistent stipules at base, stipules with striate veins.

Flowers: May-Sep. Bright-yellow irregular flowers, 2.5-3.5 cm wide, fascicles of 1-6 in upper leaf axils, petals 5, 1-2 cm long, stamens 10, 10-13 mm long.

Fruit: Nov-Dec. Legume (pod), oblong and flat, 3-5.5 cm long and 3-5 mm wide, hairy or smooth, splitting at maturity and springing open to release 4-20 seeds, flattened oval or somewhat triangular, 3.5-5 mm long, blackish and pitted. Flowers and immature pods often present together.

Range: Native. TX to s FL and north to ME and west to Ont, ND, and NE.

Ecology: Nitrogen fixer. Frequent and abundant in 1-2 year old forest plantations and infrequently-mowed roadsides, recently abandoned fields, and open forests. Extensive colonies along firelines, roadside ditches, and logging landings. Establishment improved with burning. Spreads by bird-dispersed seeds.

Synonyms: *Cassia fasciculata* Michx., sleeping plant.

Chamaecrista nictitans (L.) Moench **- partridge pea** **CHNI2**

Similar to *Chamaecrista fasciculata* except **leaves** smaller, 2-5 cm long; **flowers** smaller, 0.8-1 cm wide, never hairy. **Range:** Native. Same as above with several varieties.

Synonyms: *Cassia nictitans* L., small partridge pea, wild sensitive plant.

Other species in the SE: *C. deeringiana* Small & Pennell. Only found in FL: *C. lineata* (Sw.) Greene, *C. pilosa* (L.) Greene, *C. rotundifolia* (Pers.) Greene, *C. serpens* (L.) Greene.

Wildlife: The seeds of *C. fasciculata* and *C. nictitans* are an important component in the winter diet of Northern Bobwhite, particularly in the Piedmont and Coastal Plain. In the Coastal Plain, these two species often are promoted for Northern Bobwhite by planting, prescribed fire, or soil disturbance. Partridge peas are a moderately preferred White-tailed Deer browse. They also are larval food plants of the Little Yellow, Orange Sulphur, Cloudless Giant Sulphur, and Sleepy Orange butterflies.

Chamaecrista **Partridge pea** **Fabaceae**

J. Miller (Sep)

Chamaecrista fasciculata

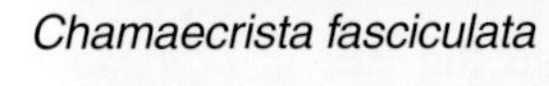

J. Miller (Sep)

Chamaecrista fasciculata

T. Bodner (Aug)

Chamaecrista nictitans

T. Bodner (Aug)

Chamaecrista fasciculata

Chimaphila **Wintergreen** **Pyrolaceae**

Low evergreen, perennial herbs or subshrubs, often in small colonies from litter-covered stems or creeping rhizomes. Leaves in terminal whorls, thick with sharp-toothed margins. Flowers terminal on stalks, single or in clusters, petals 5, white to pink. Spherical capsules with 5 cells, splitting to release many seeds. 2 species in the SE.

Common Species

Chimaphila maculata (L.) Pursh **- spotted wintergreen** **CHMA3**

Plant: Small evergreen herb or subshrub, 10-30 cm tall, upright branches from a horizontal main stem or rhizome usually covered by pine needle litter, often forming small colonies under shady forests.

Stem: Fleshy, whitish to grayish, infrequently branched.

Leaves: Opposite or whorled, 2-3, evergreen and thick, lanceolate with rounded base, 2-6 cm long and 0.5-2.5 cm wide, margins sharply dentate, dark green with whitish streaks along veins above and pale green beneath.

Flowers: Apr-Jun. Terminal stalked clusters of 1-5 flowers extending above the top leaf whorl, petals 5, white, flowers 12-18 mm wide, sepals 5.

Fruit and seeds: Jul-Oct. Spherical capsule, erect, 4-5 mm long, within persistent sepals, splitting to release many seeds, 0.5-0.6 mm long, light brown.

Range: Native. AL to GA and north to WV and west to KY in the SE (also north to MA and west to MI).

Ecology: Inhabits ground layer of upland pine and hardwood forests, being more frequent in pine forests and pine plantations. Colonizes by underground stems and spreads by seeds.

Synonyms: pipsissewa, striped prince's pine.

Other species in the SE: *C. umbellata* (L.) W. Bart. [only in VA, WV, NC and GA - pipsissewa or prince's pine].

Wildlife: Wintergreens are consumed in low quantities by Ruffed Grouse during winter.

Chimaphila **Wintergreen** **Pyrolaceae**

J. Miller (Sep)

Chimaphila maculata

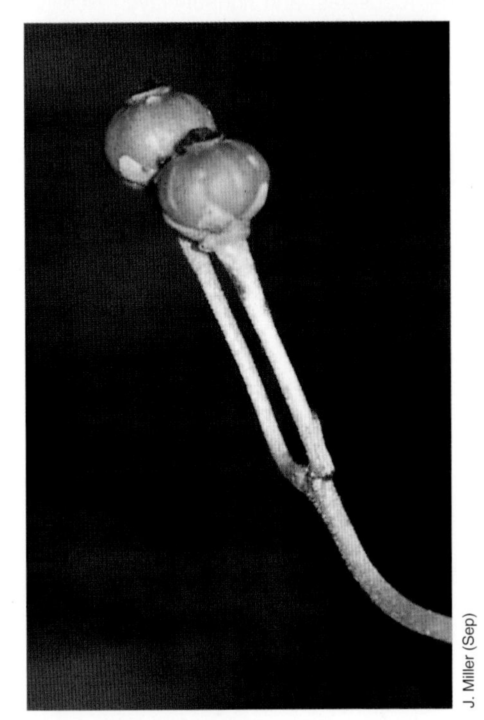

J. Miller (Sep)

Chimaphila maculata

J. Miller (Sep)

Chimaphila maculata

Chrysopsis Goldenaster Asteraceae

Perennial and annual herbs with alternate leaves, usually densely to finely hairy. Flower heads with numerous yellow ray petals and yellow disks, disks flat or convex, enclosed in urn-shaped involucre with overlapping rows of green-white bracts. Nutlets (achenes) striate and topped with both elongated bristles and shorter coarse bristles or scales. Recent taxonomic studies have separated the genera of golden asters into *Pityopsis* and *Heterotheca*, and added *Chrysopsis*. 8 species in the SE.

Common Species

Chrysopsis mariana (L.) Ell. **- Maryland goldenaster** **CHMA14**

Plant: Upright perennial herb, 20-80 cm tall, with large basal leaves in early season and later, long stems with small-leaves, from a short woody rhizome or branching rootstock, forming colonies.

Stem: Moderately stout, green becoming brown to purplish, shaggy long hairy when young, later becoming hairless with gland-tipped hairs (sticky) on the uppermost parts (requires hand lens).

Leaves: Basal leaves oblanceolate to obovate with tapering petioles, up to 18 cm long and 3.5 cm wide, with irregular teeth and indentations, stem leaves alternate and smaller, evenly spaced on stem and smaller upward, lanceolate to oblong, and sessile.

Flowers: Jun-Oct. Heads few to many in curve-topped terminal panicles, yellow rays and disks, rays petals 13-21, about 1 cm long, disks 1-2 cm wide, sepals and flower stalks sticky with glandular hairs, involucre 7-10 mm high, urn-shaped being constricted around the collar.

Fruit and seeds: Sep-Dec. Flat-obovoid nutlet (achene), 2-3 mm long, reddish to brown, topped with tan bristles, 6-7 mm long, nutlet containing 1 seed.

Range: Native. LA to FL and north to s Ont and southwest to KY.

Ecology: Inhabits new forests, forest openings, roadsides, and right-of-ways. Persists and colonizes by rhizomes and spreads longer distances by wind-dispersed seeds released in early winter.

Synonyms: *C. mariana* var. *macradenia* Fern., *Heterotheca mariana* (L.) Shinners

Other species in the SE: *C. godfreyi* Semple, *C. gossypina* (Michx.) Ell., *C. latisquamea* Pollard, *C. lineanifolia* Semple, *C. pilosa* Nutt., *C. scabrella* Torr. & Gray, *C. subulata* Small.

Wildlife: *Chrysopsis pilosa* is consumed in minor amounts by White-tailed Deer during the summer months. No other wildlife value reported.

Chrysopsis **Goldenaster** **Asteraceae**

J. Miller (Oct)

Chrysopsis mariana

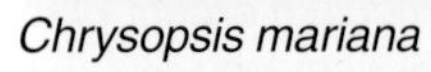

J. Miller (Oct)

Chrysopsis mariana

J. Miller (Oct)

Chrysopsis mariana

J. Miller (Oct)

Chrysopsis mariana

Cirsium **Thistle** **Asteraceae**

Spiny, biennial or perennial herbs with toothed leaves in basal rosette or alternate on fleshy stems. Bristly-appearing disk flowers of purple, lavender, pink, red, white, or yellow. Seeds in achenes with plume-like bristles for wind dispersal. 13 species in the SE.

Common Species

Cirsium horridulum Michx. **- yellow thistle** **CIHO2**

Plant: Upright biennial, 0.2-1.5 m tall, first appearing as a densely spiny-leafed rosette followed by an elongated stem with spiny leaves, from a fleshy root.
Stem: Stout, long hairy, leafy, infrequently branched.
Leaves: Basal rosette and alternate on stem, very spiny, lanceolate to oblanceolate, 10-30 cm long and 3-10 cm wide, margins dentate, often arching downward on stem.
Flowers: Mar-Jun (and later when mowed). Large heads in terminal racemes, reddish-purple or yellow and rarely white, large spiny bracts beneath each flower in whorls, with smaller inner ones.
Fruit and seeds: May-Jul. Flat-obovoid nutlet (achene), tan, 4-6 mm long, topped with white bristles, 3-4 cm long, nutlet containing 1 seed.
Range: Native. Coastal states TX to FL and north to ME, rarely Mt.
Ecology: Inhabits open and disturbed places including new forest plantations and roadsides, especially sandy soils and along salt and fresh marshes. Occurs as scattered plants. Spreads by wind- and bird-dispersed seeds.
Synonyms: *C. spinosissimus* (L.) Scop., *Carduus spinosissimus* Walt., bristly bull, field thistle, horrible thistle.

Cirsium lecontei Torr. & Gray **- Le Conte's thistle** **CILE2**

Similar to *Cirsium horridulum* except **plant** only to 1 m tall with 1-3 long branches; **stem** white hairy; **leaves** 1.5-3 cm long and 2-4 cm wide, margins lobed or dentate-serrate and spiny, lower leaves white hairy when young, upper surfaces with scattered long hairs or hairless; **flowers** (Jun-Aug) terminal on branches, purple with involucre 3-4 cm long and 1.5-3 cm broad, many spines on bracts, 0.5-2 cm long. **Range:** LA to FL and north to NC, only in Cp.
Ecology: Occurs in moist to wet pine forests and savannas.
Synonyms: *Carduus lecontei* (Torr. & Gray) Pollard.

Other species in the SE: *C. altissimum* (L.) Hill, *C. arvense* (L.) Scop., *C. carolinianum* (Walt.) Fern. & Schub., *C. discolor* (Muhl. ex Willd.) Spreng., *C. engelmannii* Rydb., *C. muticum* Michx., *C. nuttallii* DC., *C. pumilum* (Nutt.) Spreng., *C. repandum* Michx., *C. virginianum* (L.) Michx., *C. vulgare* (Savi) Ten.

Wildlife: Thistle seeds are the preferred fall and winter food of the American Goldfinch. Seeds also are consumed by other songbirds including Pine Siskin, Purple Finch, and Carolina Chickadee. Flowers are a nectar source for the Ruby-throated Hummingbird and numerous species of butterflies, especially the swallowtails.

Cirsium **Thistle** **Asteraceae**

T. Bodner (May)

Cirsium horridulum

J. Miller (Apr)

Cirsium horridulum

B. Craft (Jun)

Cirsium lecontei

Clitoria **Butterfly pea or Pigeonwings** **Fabaceae**

Trailing perennial legumes, having slender stems with 3-leaflet leaves. Flowers showy, light blue to lavender, blooming throughout summer and early fall, having a tiny pair of bracts beneath. Fruit a legume (pod), long with a beaked tip, swollen at the seeds, splitting to release several to many seeds. 1 species widespread in the SE and 2 only in s FL and TX.

Common Species

Clitoria mariana L. **- butterfly pea** **CLMA4**

Plant: Perennial, initially erect then trailing and sprawling, up to 1 m long, with many stems from a persistent woody rootcrown. Very similar in appearance to *Centrosema virginianum*, also called butterfly pea, another viney legume having circular flowers, very long flattened pods, and hairy throughout.

Stem: Slender, green vines, straight (not usually twining or curvy), fine ascending hairy or hairless, branching infrequent.

Leaves: Alternate, 3 leaflet, leaflets ovate to lanceolate, 2-5 cm long and 0.7-3 cm wide, margins entire, tip tapering but rounded at end with tiny point (mucronate), dark green above and whitish green beneath, stipules usually persistent at leaf axils, 2-4 mm long, leaflet petioles slightly swollen.

Flowers: Jun-Aug. Axillary clusters, 1-3 flowers, pea-type but upside down, outer petals open scoop-shaped, 3-4 cm long and 2-4 cm wide enclosing smaller clasping petals (1-2 cm long), light blue to lavender with purple and sometimes white spots in the center, notched at the apex, small white keel petals beneath, stalks (peduncles) 0.5-4 cm long and stems (pedicels) 4-10 mm long, subtended by small striate bracts, 1-3 mm long.

Fruit and seeds: Jul-Oct. Legume (pod), oblong and beaked, 3-6 cm long and 5-10 mm wide, swollen at seeds, smooth, pod halves twisting when opening, releasing 2-6 seeds, blackish, 2-3 mm long, sticky and adherent.

Range: Native. AZ to FL and north to NY and west to IA.

Ecology: Nitrogen fixer. Occurs in open dry forests and roadside edges as scattered plants or small patches. Persists by rootcrowns and spreads by seeds.

Synonyms: Atlantic pigeonwings.

Other species in the SE: *C. fragrans* Small [only in FL], *C. ternatea* L. [only in TX and FL].

Wildlife: Seeds are consumed infrequently by Northern Bobwhite and songbirds, perhaps due to the sticky seed coat. Butterfly pea is a low to moderate preference White-tailed Deer browse. It also is a host plant for the larval Golden-banded Skipper and Lilac-banded Longtail butterflies.

Clitoria **Butterfly pea or Pigeonwings** **Fabaceae**

T. Bodner (Jun)

Clitoria mariana

T. Bodner (Jun)

Clitoria mariana

Cnidoscolus **Spurge-nettle** **Euphorbiaceae**

Upright herbs, often with bristly-stinging hairs. Leaves lobed or divided and entire. Flowers of white sepals (no petals), male and female flowers separate on same plant (monecious). Fruit a 3-lobed spiny capsule. 2 species in the SE.

Common Species

Cnidoscolus stimulosus (Michx.) Engelm. & Gray **- bullnettle** **CNST**

Plant: Erect or reclining perennial with white stinging hairs, 15-30 cm tall (to 70 cm tall), unbranched to numerous ascending branches with white watery sap, from a persistent fleshy taproot.

Stem: Stout, V-branched, light green, bristly stinging hairy.

Leaves: Alternate, deeply 3- to 5-lobed and lobes toothed (maple leaf like), blade 6-20 cm long and wide, veins whitish, stinging hairy beneath veins and minutely on margins, petioles about equaling blade length.

Flowers: Late Mar-Aug. Terminal clusters or terminal on axillary branches of larger plants, flowers white and 5-lobed (calyx, no corolla), 2-4 cm across, female flowers usually in cluster center with outer male flowers.

Fruit and seeds: May-Sep. Upright spherical to cylindrical pods, 9-12 cm long and 8-10 cm wide, covered by stinging hairs, 3 chambered, splitting to release 1 seed per chamber, seed dark brown, 8-9 mm long and 4-5 mm wide.

Range: Native. LA to FL and north to VA , mainly in Cp (rare in Pd).

Ecology: Persists in sandy dry forests of all ages, right-of-ways, old fields, and sand dunes. The stems and leaves have stinging hairs that can produce a painful rash. Persists by rootcrowns and spreads by animal-dispersed seeds.

Synonyms: treadsoftly, mala mujer, finger rot.

Other species in the SE: *C. texanus* (Muell.-Arg.) Small [TX, OK, AR, and LA].

Wildlife: Spurge-nettle seeds are an infrequent food of the Northern Bobwhite and other songbirds.

Cnidoscolus **Spurge-nettle** **Euphorbiaceae**

T. Bodner (May)

Cnidoscolus stimulosus

T. Bodner (Jul)

Cnidoscolus stimulosus

J. Miller (May)

Cnidoscolus stimulosus

Collinsonia **Horsebalm or Stoneroot** **Lamiaceae**

Erect perennial herbs with thick, hard, tuberous roots and slightly quadrangular stems. Leaves opposite, ovate to obovate, with coarsely serrate margins. Flowers on an upright terminal stalk, yellow or white, both corolla and calyx are 2-lipped. Often fragrant. Fruit a 4-parted nut, splitting at maturity into 1-seeded nutlets. Mostly found in dense forests. 4 species in the SE.

Common Species

Collinsonia serotina Walt. **- southern horsebalm** **COSE11**

Plant: Upright perennial, unbranched or slightly branched, 40-150 cm tall, from a thick, hard, rhizome-like tuber. Plant parts with fragrance of licorice or lemon.
Stem: Stout, rigid, short fine hairs.
Leaves: Opposite, 6 or more, ovate to elliptic, thick, largest 15-25 cm long and 7-17 cm wide, with short stiff hairs or long soft hairs beneath, licorice fragrance when crushed, petioles 4-13 cm long.
Flowers: Jul-Sep. Along an upright terminal stalk, 10-30 cm long, yellowish-white petals 1.2-1.5 cm long, calyx 3.2-7.5 mm long (greater than half the length of the corolla), both irregular with 2 lips, stamens 2-4. Lemony fragrance.
Fruit and seeds: Aug-Oct. Spherical compound nut, 1.9-2.4 mm wide, dark brown, smooth, splitting at maturity into 4 separate nutlets, each containing 1 seed.
Range: Native. LA to n FL and north to NC, mainly in Cp and rarely in Pd.
Ecology: Occurs in sandy but moist forests of all ages, preferring shady to semi-shady habitats.
Synonyms: *C. canadensis* L. var. *punctata* (Ell.) Gray, *C. punctata* Ell., *Hypogon verticillata* Raf., *Micheliella anisata* (Sims) Briq., Blue Ridge horsebalm.

Collinsonia canadensis L. **- horsebalm** **COCA4**

Similar to *Collinsonia serotina* except **leaves** hairless or stiff hairy beneath; **flowers** calyx (sepals) 2.3-4.3 mm long (less than half length corolla), 2 extended stamens. **Range:** Native. MS to n FL and north to MD and west to KY, mainly in Mt and Pd, rarely in Cp.
Ecology: Later successional plant in moist forests and may persist briefly in forest openings and open forests.
Synonyms: richweed, citronella horsebalm.

Other species in the SE: *C. tuberosa* Michx., *C. verticillata* Baldw.

Wildlife: *Collinsonia* seeds are a minor component of the diet of the Northern Bobwhite and likely other seed-eating songbirds.

Collinsonia **Horsebalm or Stoneroot** **Lamiaceae**

J. Miller (Oct)

Collinsonia serotina

Commelina **Dayflower** **Commelinaceae**

Erect or reclining, annual or perennial herbs, lily-like, with alternate, parallel-veined leaves that surround a grass-like stem with their basal sheaths. Irregular blue flowers, enclosed in a spathe (a leaf-like hood or envelope), with or without a small white petal. Fruit a capsule with few seeds. Plants root at nodes of reclining stems. 6 species in the SE.

Common Species

Commelina communis L. **- Asiatic dayflower** **COCO3**

Plant: Upright annual herb, becoming reclining and branched by mid-summer, 15-60 cm long, resembling a grass, rooting at nodes and occasionally mat-forming in late summer.
Stem: Grass-like, hairless, green, branching at rooted nodes.
Leaves: Alternate, parallel veined, oblanceolate to ovate, 3-15 cm long and 1-3.5 cm wide, smooth and hairless, margins entire, base a sheath surrounding the stem.
Flowers: Apr-Nov. Solitary at the top of a stalk, irregular with 2 upper and larger blue petals, 1-1.5 cm long, and a smaller inconspicuous white petal extending below, anthers 6 (4 yellow tipped), flower enclosed within a green, folded, boat- or heart-shaped spathe (leaf-like envelope), 1.7-2.5 cm long and 0.8-1.5 cm wide.
Fruit and seeds: Apr-Nov. Capsule, 2-3 chambered (1 chamber empty), splitting to release 4 seeds, brown or reddish, 2-4 mm long.
Range: Native to Asia. Throughout Eastern US.
Ecology: Common in moist, open habitat having low vegetation, such as new forest plantations and open forests, and along forest edges and right-of-ways. Spreads by animal-dispersed seeds.

Other species in the SE: *C. benghalensis* L., *C. caroliniana* Walt., *C. diffusa* Burnm. f. [common dayflower], *C. erecta* L., *C. virginica* L.

Wildlife: Seeds of the dayflowers are consumed in moderate to low quantities by Mourning Dove and Northern Bobwhite. *Commelina virginica* and *C. caroliniana* have been listed as moderate preference White-tailed Deer forages in Mississippi.

Commelina **Dayflower** **Commelinaceae**

T. Bodner (Jul)

Commelina communis

J. Miller (Sep)

Commelina communis

Conyza **Horseweed** **Asteraceae**

Annual herbs with tall leafy stems (except for short and branching *C. ramosissima*) having sessile narrow leaves arranged alternately or spiraled. Numerous inconspicuous flowers on axillary branches in the upper plant with short white or lavender petals, yellow centers, and involucral bracts in rows beneath. Seeds in plumed nutlets (achenes). Dried plants remain standing through winter. 4 species in the SE.

Common Species

Conyza canadensis (L.) Cronq. **- horseweed** **COCA5**

Plant: Upright, summer or winter annual, 1-2 m tall in late-summer, very leafy and unbranched except for floral stalks in upper leaf axils, from shallow fibrous roots.

Stem: Leafy, round in cross-section, scattered coarse hairs when young becoming hairless, green-brown striations.

Leaves: Alternately spiraling, numerous and crowded along stems, linear to oblanceolate, 3-10 cm long and 1-15 mm wide and smaller upward, sessile, margins entire or sometimes toothed having coarse hairs near base. Lower leaves shedding upward during late-summer.

Flowers: Jul-Nov. Numerous small heads along stalks from upper leaf axils forming panicles, white or lavender ray petals (outer), 1-2 mm long, yellow disks (centers), rows of green midveined involucral bracts, involucre 3-4 mm long.

Fruit and seeds: Aug-Jan. Bristle-tufted nutlet (achene), yellowish, 1-2 mm long and tapering from base to bristly apex, hairy with ribbed margins, bristles white to tan, 2-3 mm long, nutlet containing 1 seed.

Range: Naturalized from Europe. Throughout US and also occurs in Canada, Mexico, and Central and South America.

Ecology: Early invader of disturbed places such as new forest plantations, cultivated and abandoned fields, right-of-ways, and pastures. Common on open sites the year after burning, often in extensive and dense colonies, and decreasing the following year and thereafter. Spreads by wind-dispersed seeds (probably persistent in soil).

Synonyms: *Erigeron canadensis* L., *Leptilon canadense* (L.) Britt., mares-tail, hogweed, butterweed, Canadian horseweed.

Other species in the SE: *C. bonariensis* (L.) Cronq., *C. floribunda* Kunth, *C. ramosissima* Cronq.

Wildlife: Generally a poor wildlife plant, horseweed is used occasionally by White-tailed Deer in the spring and summer, particularly on poor quality or overpopulated deer ranges.

Conyza Horseweed Asteraceae

T. Bodner (Jun)

Conyza canadensis

T. Bodner (May)

Conyza canadensis

T. Bodner (Oct)

Conyza canadensis

T. Bodner (Oct)

Conyza canadensis

Coreopsis Coreopsis or Tickseed Asteraceae

Annual or perennial herbs with showy, flat-disk flowers with ray petals often having 4-5 toothed tips, yellow (sometimes with a red-brown base), less often lavender or pink to white. Involucral bracts beneath flower heads of 2 types, the outer row is leaf-like and the inner row is membrane-like, all are joined at base. Leaves opposite or less commonly alternate, simple to deeply lobed or 3-leaflet. Seeds in flat and curved nutlets (achenes), winged or wingless, resembling ticks (thus the common name). 18 species in the SE.

Common Species

Coreopsis major Walt. - greater tickseed COMA6

Plant: Upright to leaning perennial, 0.5-1 m tall, arising in tufts and increasingly opposite branched, from long slender rhizomes.
Stem: Moderately stout at base with very slender flower stalks, initially erect increasingly leaning in summer to fall, opposite branched from all nodes, lower branches the longest.
Leaves: Appearing as whorl of 6 leaves at each node, actually 2 opposite 3-leaflet leaves (divided to the base) without petioles, leaflets lance-shaped to ovate, 3-8 cm long and 1-3 cm wide, margins entire to slightly wavy, usually dark green with only the midvein obvious, surfaces short-soft hairy.
Flowers: May-Aug. Flat-topped clusters of flat disk heads terminal on branches, 6-10 yellow ray petals (outer), 5-toothed tips or not, yellow or red disks (centers), flower heads 5-9 cm wide, involucral bracts in two rows, greenish outer row, narrower and about as long as brownish inner row, both 2-10 mm long and rounded tipped.
Fruit and seeds: Sep-Mar. Flat-curved nutlet (achene), oblong, 5-6 mm long and 1.3-1.8 mm wide, blackish on convex side, with narrow dark-brown wings along both sides, ovate in outline and resembling a tick when laying flat with convex side up.
Range: Native. LA to n FL and north to PA and southwest to KY.
Ecology: Occurs often as scattered plants or loose groupings in dry open forests, old fields, and along forest edges. Present in all stages of forest succession in semi-shady openings. Persists and colonizes by rhizomes and spreads by animal-dispersed seeds.
Synonyms: *C. major* var. *major, C. major* var. *stellata* (Nutt.) B.L. Robins., wood tickseed.

Coreopsis tripteris L. - threeleaf or tall tickseed COTR4

Plant: Upright single-stemmed perennial, 1-3 m tall, from a stout rhizome with a taproot.
Stem: Stout base becoming slender above, branching below mid-plant, short hairy becoming hairless.
Leaves: Opposite, deeply lobed with 3 divergent segments and becoming simple in upper plant, lobed segments narrowly lanceolate or narrowly elliptic, 4-12 cm long and 0.5-3 cm wide, petioles absent to 4 cm long.
Flowers: Jun-Sep. Heads several or many on long stalks, 2-8 cm long, yellow to golden ray petals, 1-2.5 cm long, tips 3-5 toothed, disks yellow becoming purplish or deep red, outer involucral bracts lanceolate, 3-6 mm long, inner bracts oval, to 1 cm long.
Fruit and seeds: Oct-Jan. Bristly nutlet (achene), 4-7 mm long and 1.8-2.3 mm wide, dark brown, topped with erect bristles and sometimes with 2 short barbed awns.
Range: Native. LA to GA and north to s Canada and west to WI and KS.
Ecology: Present in all stages of forest succession. Occurs in openings as scattered plants, along right-of-ways, and in old fields, mostly wet to moist habitats such as stream sides and ditches. Spreads by seeds attaching to fur or clothing.

Wildlife: Members of this genus generally are considered poor White-tailed Deer forages, although *C. major* is listed as a preferred browse species in Mississippi during spring and summer. Seeds likely are used to a some extent by songbirds and rodents.

Coreopsis — Coreopsis or Tickseed — Asteraceae

T. Bodner (Jun)

Coreopsis tripteris

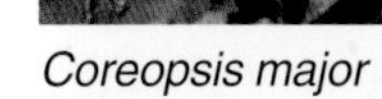

J. Miller (May)

Coreopsis major

T. Bodner (Jun)

Coreopsis tripteris

T. Bodner (Aug)

Coreopsis major

Croptilon Scratchdaisy Asteraceae

(formerly *Haplopappus*)

Annual and biennial herbs, with many terminal small aster-type flowers, bright-yellow and enclosed within rows of whitish-green bracts forming a cylinder. Leaves alternate and simple. Nutlets 4- to 5-angled or striate with brownish bristly plumes. 2 species in the SE.

Common Species

Croptilon divaricatum (Nutt.) Raf. **- goldenweed** **CRDI17**

Plant: Upright, hairy-stemmed annual, 20-100 cm tall, unbranched below and generally much branched above mid-plant, from a taproot.
Stem: Slender, rough or spreading hairy, sometimes glandular hairy (slightly sticky), green to purplish brown.
Leaves: Alternate, well spaced along stems (not crowded), linear to oblanceolate, 3-10 cm long and 3-7 mm wide and smaller upward (the lowest sometimes larger), margins entire or with few spiny teeth on lower leaves, sessile.
Flowers: Aug-Nov. Numerous small heads along branched stalks from upper leaf axils (panicles or corymbs), bright-yellow ray petals (7-11), 4-6 mm long, yellow disks (centers), less than 1 cm wide, involucral bracts greenish-white in rows forming a cylinder or tube, 4-6 mm long.
Fruit and seeds: Oct-Dec. Flat-ellipsoid nutlet (achene), 2-2.5 mm long, short appressed hairy, topped with brownish barbed bristles, 4-5 mm long, nutlet containing 1 seed.
Range: Native. TX to FL and north to VA and west to KS and OK, mainly Cp.
Ecology: Early invader of disturbed places such as new forest plantations and old fields, and along forest edges and right-of-ways. Most common on sandy soils. Spreads by wind-dispersed seeds.
Synonyms: *Haplopappus divaricatus* (Nutt.) Gray, slender scratch daisy.

Other species in the SE: C. *hookerianum* (Torr. & Gray) House.

Wildlife: Goldenweed is a low to moderate preference White-tailed Deer forage. No other wildlife use reported.

Croptilon Scratchdaisy Asteraceae

J. Miller (Sep)

Croptilon divaricatum

T. Bodner (Oct)

Croptilon divaricatum

T. Bodner (Sep)

Croptilon divaricatum

T. Bodner (Sep)

Croptilon divaricatum

Crotalaria **Rattlebox** **Fabaceae**

Annual or perennial legumes, with leaves of 1 to 7 leaflets, and yellow to orange pea-type flowers. Few to many flowers along a terminal or axillary stalk (raceme). Fruit a legume (pod), appearing inflated, ovoid to oblong, green at first and turning black at maturity, splitting to release 3-20 spherical seeds. 4 native species and 9 introduced species in the SE.

Common Species

Crotalaria spectabilis Roth **- rattlebox or showy crotalaria** **CRSP2**

Plant: Upright summer annual, 0.5-2 m tall, from a woody-like taproot, often occurring in small colonies.

Stem: Stout, green or purplish, waxy and hairless, sometimes ribbed.

Leaves: Alternate, simple, paddle shaped (obovate), 5-15 cm long and 1.5-2.5 cm wide, short petioles, smooth above and densely hairy beneath, persistent flap-like stipules at leaf axils, 5-8 mm long and 4-6 mm wide.

Flowers: Jul-Oct. Many showy, pea-type, yellow flowers along few to several terminal stalks (racemes), 10-50 cm long, petals 5, 1.7-2.5 cm long, calyx 5 lobed, stalks with a persistent bract at base, 5-12 mm long and 5-9 mm wide.

Fruit and seeds: Aug-Oct. Legume (pod), cylindric with hair-like tip, 3-5 cm long, appearing inflated, initially green maturing to black, splitting on top side along suture to release few seeds, kidney shaped, 4-5 mm wide, black to brown at maturity.

Range: Native of India and introduced widely in the SE. TX to FL and north to VA and west to s MO and OK, mainly Cp and Pd.

Ecology: Nitrogen fixer. Common roadside plant, either planted in patches or natural as scattered plants. Invades adjacent harvested forests and usually disappears until further disturbance. Spreads by seeds.

Synonyms: *C. retzii* A. Hitchcock.

Other species in the SE: *C. incana* L., *C. lanceolata* E. Mey., *C. ochroleuca* G. Don, *C. pallida* Ait., *C. pumila* Ortega, *C. purshii* DC., *C. retusa* L., *C. rotundifolia* Walt. *ex* J.F. Gmel., *C. sagittalis* L., *C. verrucosa* L., *C. virgulata* Klotzsch, *C. zanzibarica* Benth.

Wildlife: *Crotalaria* seeds (not showy crotalaria) are consumed infrequently by Northern Bobwhite. Showy crotalaria was introduced as a soil builder and a forage crop. However, it contains pyrrolizidine alkaloids and the entire plant is considered toxic to livestock and humans.

Crotalaria **Rattlebox** **Fabaceae**

T. Bodner (Oct)

Crotalaria spectabilis

Croton **Croton** **Euphorbiaceae**

Erect annual and perennial herbs (rarely shrubs), densely covered with star-shaped hairs having a velvety white or brownish appearance. White-velvety male and female flowers separate on the same plant (monoecious), both in terminal clusters along short stalks (racemes), petals inconspicuous or absent. Leaves and branches alternate, except leaves may appear whorled and branches in 3's at shortened internodes in the upper plant. Fruit a capsule, 3-chambered and 1-3 seeded. Seeds glossy, ovoid or oblong. 7 species in the SE with 2 varieties of *Croton capitatus* and 7 varieties of *Croton glandulosus.*

Common Species

Croton capitatus **Michx.- woolly croton or hogwort** **CRCA6**

Plant: Upright annual, to 1.5 m tall, white- or brown-velvety appearing especially at terminals, sparsely branched at base and above mid-plant, from a taproot (not fragrant).
Stem: Stout, green and densely brown to orange hairy (less on lower stem), alternate ascending branches that can be in 3's in upper plant.
Leaves: Alternate, lanceolate (or heart-shaped near base), 4-15 cm long and 1.5-6 cm wide, margins entire with minor teeth, velvety whitish hairy above and white or golden-brownish hairy beneath, petioles 2-9 cm long.
Flowers: Jul-Oct. Terminal or axillary spikes, 1-3 cm long, in upper plant, petals absent and bracts appearing woolly due to dense hairs, male flowers above female flowers.
Fruit and seeds: Jul-Nov. Hard-coated capsule, nearly spherical, 6-9 mm wide, splitting at maturity ejecting 3 seeds, ovoid and flattened on a side, 3-5 mm long, glossy brown.
Range: Native. TX to FL and north to s NY and west to e CO.
Ecology: Early invader of disturbed soils and overgrazed pastures. Decreases on new tree plantations after the first year, but persisting up to 5 years. Occurs as scattered individual plants, especially on drier soils and slopes of Pd and Cp. Spreads by gravity- and bird-dispersed seeds.
Synonyms: goatweed, doveweed.

Croton glandulosus **var.** ***septentrionalis*** Muell.-Arg. **- tropic croton** **CRGL2**

Similar to *Croton capitatus* var. *capitatus* except **plant** only to 60 cm tall, rough-hairy and glandular, freely branched, taproot spicily fragrant; **stem** green with light-green stripes, rough brownish hairy (not woolly), branches from upper leaf whorl; **leaves** lanceolate to oblong, 2-9 cm long and 0.5-4 cm wide, margins toothed, brownish hairy beneath, petioles 0.5-4 cm long with 2 saucer-shaped glands just beneath the leaf margin base; **flowers** (May-Oct) spikes about 1 cm long, but not woolly only slightly white hairy; **fruit and seeds** (Jun-Nov) solitary or small clusters at base of upper leaves, capsule 4-5 mm wide, seeds 3-3.5 mm long, glossy, grayish tan mottled with black. **Range:** Native. TX to FL and north to MD and west to se NE and e KS.
Synonyms: vente conmigo.

Other species in the SE: *C. alabamensis* E.A. Sm. *ex* Chapman, *C. argyranthemus* Michx., *C. lindheimerianus* Scheele, *C. monanthogynus* Michx., *C. punctatus* Jacq.

Wildlife: Woolly croton, and likely other species, are toxic to livestock and not preferred as White-tailed Deer forages, although moderate summer use of *C. grandulosus* and *C. monanthogynus* has been reported from Oklahoma. *Croton capitatus* and *C. grandulosus* are important food of the Mourning and Ground Doves, whereas *C. monanthogynus* is taken less frequently. Northern Bobwhite and some songbirds use croton seeds during summer. Crotons are preferred larval food plants of Goatweed and Gray Hairstreak butterflies.

Croton Croton Euphorbiaceae

Croton glandulosus var. *septentrionalis*

Croton capitatus

Croton glandulosus var. *septentrionalis*

Croton capitatus

Desmodium **Ticktrefoil or Desmodium** **Fabaceae**

Mostly upright (with 2 trailing species) perennial legumes (1 annual species), having alternate, 3-leaflet leaves. Leaflets broadly ovate to linear with stipules at the base of each leaf petiole (some deciduous) and stipels at base of each leaflet (mostly persistent). The similar *Lespedeza* spp. have no stipels at the base of leaflets. Flowers are pea-type (irregular), less than 1 cm long, pink and purple (some white), scattered loosely to tightly along terminal or axillary stalks. Fruit a flat legume (pod), covered with bristled and hooked hairs, with flattened seeds in distinctive segments. Jointed pod segments break apart to become the "sticktights" and "beggarlice" that stick to clothes and fur by the hook-ended hairs. About 29 species in the SE, with some distinctions blurred.

Common Species

Desmodium obtusum (Muhl. *ex* Willd.) DC. **- stiff ticktrefoil** **DEOB5**

Plant: Upright perennial legume, 0.5-1.5 m tall, with multiple stems, from a thick woody root up to 20 cm long.

Stem: Slender, progressively leaning from summer to winter, sparsely to densely hooked-ended hairy and straight hairy throughout, dark green to brown.

Leaves: Alternate, 3 leaflet with the terminal the longest, leaflets elliptic-ovate to lanceolate, 4-6 cm long and 2-3 cm wide, margins entire, hook-ended hairy (clinging to clothes), petioles 3-12 mm long.

Flowers: Jul-Oct. Terminal panicles, irregular pea-type, petals pink to purple, 4-6 mm long, stalks hook-ended hairy (requires hand lens).

Fruit and seeds: Oct-Feb. Flat legume (pod) with 1-3 rounded segments, each 3-5 mm long, hook-ended hairy attaching to fur and clothing.

Range: Native. Throughout Eastern US.

Ecology: Nitrogen fixer. Inhabits open forests with the perennial roots remaining after timber harvest to occur in new forests as scattered plants or small colonies. Common along forest edges and right-of-ways. Persisting by woody rootcrowns and spreading by animal-dispersed seeds attaching to fur.

Synonyms: *D. rigidum* (Ell.) DC., panicled ticktrefoil.

Desmodium laevigatum (Nutt.) DC. **- smooth ticktrefoil** **DELA2**

Similar to *Desmodium obtusum* except **plant** lacking hairs (glabrous) or only scattered hairy; **leaves** with petioles 2-6 cm long, terminal leaflet ovate to elliptic-ovate, 4-6 cm long and 2-4 cm wide, blunt tipped often minutely indented, darker green and smooth above and paler green and hairy only on midvein beneath; **flowers** (Sep-Oct) on long slender branching stalks, petals pink to lavender to purple, 8-10 mm long, sepals densely short hairy, flower stems 10-20 mm long; **fruit and seeds** (Nov-Feb) pod with 3-5 segments, each convex or straight above and angled or rounded below, hook-ended hairy for attaching to fur.

Synonyms: smooth tickclover.

Desmodium Ticktrefoil or Desmodium Fabaceae

T. Bodner (Sep)

Desmodium obtusum

J. Miller (Jul)

Desmodium obtusum

T. Bodner (May)

Desmodium laevigatum

T. Bodner (Sep)

Desmodium obtusum

Desmodium **Ticktrefoil or Desmodium** **Fabaceae**

Desmodium strictum (Pursh) DC. **- pinebarren ticktrefoil** **DEST2**

Similar to *Desmodium obtusum* except **plant** 40-70 cm tall; **stem** hook-ended hairy; **leaves** with leaflets linear to narrowly oblong, 2-4 cm long and 5 mm or less wide, petioles 0.5-1.5 cm long; **flowers** pink to purple, stalks inconspicuous; **fruit** a flat legume (pod) with 1-2 segments, each slightly concave above and rounded below. **Range:** Native. LA to FL and north to VA, mainly Cp and Pd.
Ecology: Occurs in dry forests, including sandhills, and along right-of-ways.
Synonyms: beggarlice, linear-leaf tickclover.

Desmodium rotundifolium DC. **- prostrate ticktrefoil** **DERO3**

Plant: Perennial trailing legume, often partially covered with leaf litter, 0.5-1.5 m long, from a branching woody root.
Stem: Trailing, slender, long hairy in the middle.
Leaves: Alternate, 3 leaflet, leaflets almost round to broadly oval, 1-4 cm long and wide, both surfaces long hairy, petioles 1-2 cm long and long hairy, stipules persistent, ovate, clasping stem, 5-10 mm long.
Flowers: Jul-Oct. Axillary and occasionally terminal erect panicles or racemes, stalks stout, petals purple, 8-10 mm long, sepals sparsely hairy.
Fruit and seeds: Oct-Dec. Flat legume (pod) with 3-6 segments, each distinct with both margins indented, each 4.5-6.2 mm long.
Range: Native. TX to FL and north to s Canada and west to MI, mainly Pd and Mt in the SE.
Ecology: Nitrogen fixer. Occurs commonly in dry forests, open to shady habitats, including new forest plantations, and along forest edges and right-of-ways. Sometimes planted for wildlife openings. Persists by woody rootcrowns and spreads by animal-dispersed seeds.
Synonyms: dollarleaf, common trailing desmodium.

Wildlife: Desmodiums are among the most important species in the diet of the Northern Bobwhite. In some areas of the Southeast, Ruffed Grouse and Wild Turkey consume considerable quantities of the seeds, particularly during fall and winter. Most species of desmodium are considered high preference White-tailed Deer forages and are used heavily during summer. Desmodiums are a preferred host of the larval Hoary Edge butterfly.

Desmodium Ticktrefoil or Desmodium  Fabaceae

Desmodium strictum

T. Bodner (May)

Desmodium rotundifolium

B. Sorrie (Jun)

Diodia **Buttonweed** **Rubiaceae**

Annual and perennial herbs, low branching, erect to spreading and often mat forming. Leaves opposite and sessile, with fringed stipules that contain flowers and fruit. Small 4-petaled flowers, white to pink, solitary, and sessile. Fruit with 2 seeds. 2 species in the SE.

Common Species

Diodia teres Walt. **- poorjoe** **DITE2**

Plant: Annual, usually spreading, but occasionally erect, often in mats, 10-40 cm tall, from shallow roots with a slender taproot.

Stem: Simple or branched, slender and nearly round in cross-section, often reclining, up to 80 cm long, minutely hairy.

Leaves: Opposite, simple and sessile, linear or narrowly lanceolate, 2-3.5 cm long and 3-6 mm wide, rough hairy, stipules at leaf base with bristles.

Flowers: Jun-Nov. Solitary and paired in upper leaf axils, white to pink, petals 4, 1-2 mm long, hairless, sepals 4, 2-4 mm long.

Fruit and seeds: Jul-Jan. Top-shaped (obovoid) tiny nut, 4-5 mm long, hairy, with 4 short green calyx teeth at top, splitting at maturity into 2 (or rarely 3) 1-seeded nutlets.

Range: Native. TX to FL and north to PA and west to WI and NE.

Ecology: Occurs on dry sites and open to semi-shady habitats, including limestone glades. Early invader of new pine plantations, declining in abundance within the first 3 years. Common plant along forest edges and right-of-ways, and in pastures and abandoned fields. Spreads by seeds apparently persistent in the soil.

Synonyms: *Diodella teres* (Walt.) Small, *Diodia teres* var. *setifera* Fern. & Grisc., rough buttonweed.

Diodia virginiana L. **- Virginia buttonweed** **DIVI3**

Similar to *Diodia teres* except **plant** larger perennial from a woody rootcrown, also forming dense colonies; **leaves** larger, 3-6 cm long and 4-12 mm wide; **flowers** with petals 2-4 mm long, sepals 2, 4-6 mm long. **Range:** Native. TX to FL and north to NJ and west to IL and MO.

Ecology: Occurs in wet areas such as savannas, ditches, and margins of streams, marshes and ponds. Persists by rootcrowns and spreads by seeds.

Wildlife: *Diodia teres* seeds are consumed in low to moderate amounts by Wild Turkey and Northern Bobwhite, and perhaps by seed-eating songbirds. It is considered a low preference White-tailed Deer forage, whereas the less abundant *D. virginiana* may be more preferred.

Diodia **Buttonweed** **Rubiaceae**

T. Bodner (Jun)

Diodia teres

T. Bodner (Jul)

Diodia virginiana

Duchesnea Mock-strawberry or Duchesnea Rosaceae

Low hairy perennial herb, spreading by stolons, having alternate 3-leaflet leaves with long petioles. Flowers rose-type, yellow with 5 petals, resembling wild strawberry (*Fragaria* spp.) which has white flowers. Fruit strawberry-like, but dry and tasteless. 1 naturalized species in the SE.

Common Species

Duchesnea indica (Andr.) Focke **- Indian strawberry** **DUIN**

Plant: Low-growing perennial, to 15 cm tall, spreading by stolons along the ground surface and rooting at nodes.

Stem: Absent, but runners may be initially erect and long-petioled leaves upright from stolon nodes.

Leaves: Alternate and rosettes, 3 leaflets, leaflets ovate to elliptic, 2-6 cm long and 1-4 cm wide, margins crenate to doubly crenate, petioles 8-20 cm long.

Flowers: Feb-Nov. Solitary from leaf axils, yellow, 5 petals, calyx 5 lobed, lanceolate, special wide bracts 3-5 toothed near apex, stamens 20 in 3 lengths, pistils more than 100.

Fruit and seeds: Apr-Nov. Strawberry-like, green turning red at maturity, with many tiny red nutlets on the surface, not poisonous, but not sweet.

Range: Introduced from Asia and widely naturalized. TX to FL and north to NY and west to MO (also CA, OR, and WA).

Ecology: Occurs in moist open forests and new forest plantations, along forest margins and right-of-ways, also in pastures and lawns. Persists by rootcrowns, colonizes by stolons, and spreads by animal-dispersed seeds.

Synonyms: *Fragaria indica* (Andr.) Focke, Indian mock-strawberry, mock strawberry, snakeberry.

Wildlife: No wildlife value reported and likely of little consequence as a wildlife forage.

Duchesnea **Mock-strawberry or Duchesnea** **Rosaceae**

T. Bodner (Jul)

Duchesnea indica

Elephantopus Elephantsfoot Asteraceae

Perennial herbs with large leaves in basal rosettes, often flat on the soil surface, in an imaginary shape of an elephant's footprint (thus the common name), from a woody rootcrown. Terminal tight clusters of pink or purple aster-type flowers subtended by 3 ovate leaf-like bracts on 1-5 branching stems. Tapered nutlets topped with 5 barbed bristles. 4 species in the SE.

Common Species

Elephantopus tomentosus L. - **elephantsfoot** ELTO2

Plant: Perennial herb having a basal rosette of hairy leaves on or near the soil surface, with forked branching flower stalks, 10-60 cm tall, from a woody rootcrown.
Stem: Slender, with few Y-forking ascending branches, slightly rough hairy.
Leaves: Basal rosette and alternate on stem, basal leaves oblanceolate to obovate, 8-25 cm long and 3-10 cm wide, margins crenate becoming serrate at the tapering base, stem leaves ovate, sessile, 2-4 cm long and 1-3 cm wide, both densely velvet hairy on both surfaces and often wrinkled at veins.
Flowers: Jul-Nov. Terminal tight clusters of sessile tubular heads on forking slender stalks, pink to purple (rarely white), petals linear, 1-1.5 cm long, calyx hairy and glandular, subtended by 3 foliar bracts, ovate, 1-3 cm long, green and long hairy.
Fruit and seeds: Aug-Dec. Tapered nutlet (achene), 4-5 mm long, ribbed, hairy, topped with 5 barbed bristles, 6-8 mm long, nutlet containing 1 seed.
Range: Native. TX to FL and north to MD and west to KY, mainly Cp and Pd in the SE.
Ecology: Occurs commonly in new and open forests from bottomlands to sandhills. Inhabits areas with low ground vegetation. Persists by rootcrowns and spreads by seeds.
Synonyms: *E. nudicaulis* Poir., devil's-grandmother, tobaccoweed, hairy elephantsfoot.

Elephantopus carolinianus Raeusch. - **Carolina elephantsfoot** ELCA3

Similar to *Elephantopus tomentosus* except **plant** 20-100 cm tall with leafy stems and basal leaves mostly absent; **stem** leafy and much branched with long hairs below. **Range and ecology**, occurring in the same range but on wetter sites, such as along margins of streams, swamps, ditches, bottomland hardwood forests, and moist hillsides.
Synonyms: leafy elephantsfoot.

Other species in the SE: *E. elatus* Bertol., *E. nudatus* Gray.

Wildlife: The flowering heads of *Elephantopus* are occasionally consumed by White-tailed Deer, however, these species should be considered very low preference forages.

Elephantopus **Elephantsfoot** **Asteraceae**

T. Bodner (Sep)

Elephantopus carolinianus

T. Bodner (Aug)

Elephantopus tomentosus

T. Bodner (Sep)

Elephantopus carolinianus

J. Miller (Sep)

Elephantopus tomentosus

Erechtites **Fireweed** **Asteraceae**

Erect large annual herb, light green, with stout stems having many lanceolate leaves. Occurs in colonies after fire and soil disturbance. Flowers heads nonshowy in many-branched terminal panicles. White plumed seeds blowing in late fall and early winter. 1 species in the SE.

Common Species

Erechtites hieracifolia (L.) Raf. *ex* DC. **- American burnweed** **ERHI2**

Plant: Large erect annual, to 3 m tall, leafy and light green, from a fibrous root. May produce a rank odor when cut or crushed. Often occurs in dense colonies after fire.

Stem: Single to several, slightly hairy or not, fleshy, grooved, solid with white pith, large plants branched above mid-plant.

Leaves: Alternately spiraling, numerous along stem, lower leaves oblanceolate and upper leaves lanceolate to elliptic, 5-20 cm long and 0.5-7 cm wide and smaller upward, margins sharply serrate to irregularly lobed (variable in form), midvein white, lower leaves with petioles and upper leaves clasping stem.

Flowers: Jul-Nov. Numerous upright heads in panicles, involucral sheath cylindric with a swollen base, pale green to purplish, 1-2 cm long and 4-8 mm broad, petals barely visible, cream to pinkish.

Fruit and seeds: Sep-Dec. Tapered nutlet (achene), 2-3 mm long, brown, ribbed, topped by soft and bright-white bristles, 1-1.4 cm long, nutlet containing 1 seed.

Range: Native. TX to FL and north to Newfoundland and west to Ont.

Ecology: Occurs singly or in dense colonies after fire or soil disturbance on open sites, such as new forest plantations, declining in abundance the following years. Infrequent plants can remain in the understory of plantations on moist to wet sites. Persistent on mowed right-of-ways. Spreads by wind-dispersed seeds, apparently persistent in the soil for long periods.

Synonyms: eastern fireweed, pilewort.

Wildlife: Fireweed is a low preference White-tailed Deer forage. The small seeds are sometimes consumed by songbirds such as the Indigo Bunting, American Goldfinch, and others.

Erechtites **Fireweed** **Asteraceae**

T. Bodner (Sep)

Erechtites hieracifolia

T. Bodner (Sep)

Erechtites hieracifolia

J. Miller (Sep)

Erechtites hieracifolia

T. Bodner (May)

Erechtites hieracifolia

Erigeron **Fleabane** **Asteraceae**

Annual or perennial herbs, having upright stems and alternate leaves (sometimes all basal). Heads in corymbs or panicles with numerous ray flowers, white to lavender to blue, and yellow disk flowers. Seeds within plumed nutlets. Daisy fleabanes, although resembling, differ from asters in that their flower stalks have no leaves and they start blooming earlier in spring and summer than most asters. 8 species in the SE.

Common Species

Erigeron strigosus Muhl. *ex* Willd. **- prairie fleabane** **ERST3**

Plant: Upright annual or rarely biennial, 0.3-1m tall, sparsely branching and sparsely leafy, from shallow fibrous roots.

Stem: Slender, 1-few branched, short appressed or scattered hairy, green with lighter stripes becoming brown.

Leaves: Alternate with early basal rosette, basal leaves oblanceolate to elliptic, to 15 cm long (petiole included) and 3 cm wide, margins entire to irregular toothed, base tapering, stem leaves linear to oblanceolate, 2-10 cm long and 2-18 mm wide, sessile, margins entire to wavy, appressed hairy throughout.

Flowers: Apr-Oct. Terminal heads on slender stalks, white to lavender ray petals, usually 40-60, 3-6 mm long, yellow disks, 6-12 mm wide, involucral bracts in single row, 2-4 mm long.

Fruit and seeds: Jun-Dec. Tapered nutlet (achene), 0.6-1 mm long, yellowish-tan, topped with whitish bristles, 1.5-2 mm long, nutlet containing 1 seed.

Range: Native. Throughout US and s Canada.

Ecology: Early invader of new forest plantations and common on right-of-ways, abandoned fields, pastures, and waste places. Spreads by wind-dispersed seeds.

Synonyms: daisy fleabane, rough fleabane, white-top.

Erigeron annuus (L.) Pers. **- daisy fleabane** **ERAN**

Similar to *Erigeron strigosus* except **plant** up to 1.5 m tall and more leafy; **stem** scattered long hairy near base (not appressed hairy); **leaves** basal when present, 5-55 mm wide, stem leaves lanceolate, 3-12 cm long; **flowers** larger, with white to pale lavender ray petals up to 10 mm long.

Synonyms: annual fleabane.

Other species in the SE: *E. philadelphicus* L., *E. procumbens* (Houst. *ex* P. Mill.) Nesom, *E. pulchellus* Michx., *E. quercifolius* Lam., *E. tenuis* Torr. & Gray, *E. vernus* (L.) Torr. & Gray

Wildlife: Most of the fleabanes are low to moderate White-tailed Deer forages and generally only consumed in poor range conditions. However, *E. strigosus* has been reported as a preferred forage in a Mississippi study.

Erigeron Fleabane Asteraceae

T. Bodner (Jun)

Erigeron annuus

J. Miller (Jun)

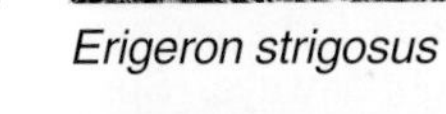

Erigeron strigosus

T. Bodner (May)

Erigeron annuus

T. Bodner (May)

Erigeron strigosus

Eupatorium **Thoroughwort or Boneset** **Asteraceae**

Upright perennial herbs (rarely annuals), often growing in clumps, from a rhizome or rootcrown. Leaves opposite, whorled, or rarely alternate, toothed or coarsely to finely dissected. Flower heads numerous and small in corymbs or panicles, usually white, but also pink, blue, and purple. Most often flowering in late summer. Seeds in bristle-plumed nutlets (achenes). Thoroughworts, along with asters and goldenrods, are the most common composites in the SE. 35 species in the SE.

Common Species

Eupatorium capillifolium (Lam.) Small **- dogfennel** **EUCA5**

Plant: Commonly-occurring, tall, erect annual or short-lived perennial, 0.5-3 m tall, single to several stems from base and often branched above mid-plant, from a thick woody rootcrown having fibrous roots.

Stem: Stout at base, green and downy hairy above becoming reddish purple and smooth below.

Leaves: Alternate or often opposite at stem base, deeply dissected into fine hair-like segments less than 0.5 mm wide and 2-10 cm long, green, feathery in appearance, strong odor when crushed.

Flowers: Sep-Nov. Elongated much-branching conical panicle, 30-80 cm long and 20-50 cm wide, often nodding, flowers 3-6 per head, petals greenish-white and sometimes with purple lobes, 1.5-3 mm long.

Fruit and seeds: Nov-Mar. Slender nutlet (achene), 1-1.6 mm long, dark-brown, topped with white bristles, 2-3 mm long, nutlet containing 1 seed.

Range: Native. TX to s FL and north to MA and TN and west to AR, infrequent in Mt.

Ecology: Occurs on dry to wet sites and open habitat, as single plants or small colonies. One of the most common forbs on new forest plantations, also inhabiting old fields, pastures, and mowed right-of-ways. Spreads by wind-dispersed seeds and regrows from woody base. Weak allelopath to young pines and probably other plants.

Synonyms: summer cedar, cypressweed.

Eupatorium compositifolium Walt. **- yankeeweed** **EUCO7**

Similar to *Eupatorium capillifolium* except **plant** perennial, 0.5-1.5 m tall, rarely taller, forming loose tufts from rhizomes; **leaves** segments 1-2.5 mm wide; **flower** heads numerous, panicles 10-45 cm long and 10-30 cm wide, rarely nodding, petals white with greenish to purplish bracts; **fruit and seeds** slender nutlet, 1.3-1.8 mm long. **Range:** Native. TX to s FL and north to NC and west to AR, mainly Cp and Pd in the SE.

Eupatorium **Thoroughwort or Boneset** **Asteraceae**

T. Bodner (Oct)

Eupatorium compositifolium

J. Miller (Jun)

Eupatorium capillifolium

J. Miller (Oct)

Eupatorium compositifolium

J. Miller (Oct)

Eupatorium capillifolium

Eupatorium **Thoroughwort or Boneset** **Asteraceae**

Eupatorium leucolepis (DC.) Torr. & Gary **- justiceweed** **EULE**
Similar to *Eupatorium capillifolium* except **plant** stiffly erect to leaning, 0.4-1 m tall, from a rootcrown or short rhizome; **leaves** thick, opposite, linear-oblong, 3-5 cm long and 3-10 mm wide and smallest upward, 2 faint lateral veins from the base paralleling the margin, sessile, tapering tipped, rough hairy with glandular hairs; **flowers** (Jul-Oct) dense, terminal, flat-topped corymb, petals white, 3-4 mm long, involucral bracts whitish hairy and exceeding the petals, with exserted white pistils; **fruit and seeds** (Nov-Feb) slender nutlet, 2-3 mm long, topped with white bristles. **Range:** Native. LA to FL and north to VA, mainly Cp.
Ecology: Most abundant on sandy soils, especially sandhills, and peat soils. Commonly occurs in open forests, right-of-ways, and forest edges. Persists and colonizes by rhizomes and spreads by abundant wind-dispersed seeds.

Eupatorium serotinum Michx. **- late boneset** **EUSE2**
Plant: Upright perennial, 0.4-2 m tall, several stems, from a woody rootcrown.
Stem: Stout, short hairy, branched in upper plant, often dark purplish, solid.
Leaves: Opposite, lanceolate to elliptic with tapering tips, 5-15 cm long and 2-8 cm wide, margins coarsely serrate, midvein and 2 lateral veins prominent (3-nerved), sparsely hairy on both surfaces, petioles 1-4 cm long, purplish.
Flowers: Aug-Oct. Terminal clusters of heads, 15-40 cm wide, petals white and upper involucral bract margins white hairy, bracts in 3 series, the outer half as long as the inner, broadly rounded tips.
Fruit and seeds: Dec-Feb. Nutlet (achene) 1.7-2.8 mm long, topped with white bristles, 3-3.5 mm long, nutlet containing 1 seed.
Range: Native. TX to FL and north to NY and west to IL and KS.
Ecology: Occurs on dry to wet sites and open to partially-shady habitats, mainly as scattered single plants. Most common on moist sites such as bottomlands. Common early plant in new forest plantations and present on right-of-ways and cultivated fields. Present on floodplains and margins of tidal marshes. Persists by woody rootcrown and spreads by wind-dispersed seeds.
Synonyms: eupatorium, late-flowering thoroughwort.

Eupatorium — Thoroughwort or Boneset — Asteraceae

T. Bodner (Aug)

Eupatorium serotinum

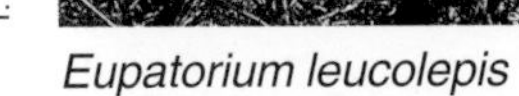

T. Bodner (Aug)

Eupatorium leucolepis

T. Bodner (Sep)

Eupatorium serotinum

T. Bodner (May)

Eupatorium serotinum

Eupatorium **Thoroughwort or Boneset** **Asteraceae**

Eupatorium aromaticum L. **- aromatic eupatorium** **AGARA**

Similar to *Eupatorium serotinum* except **plant** 0.3-0.8 m tall, 1-few stems; **stem** minutely hairy; **leaves** opposite, narrowly to broadly triangular to ovate to somewhat heart-shaped, relatively thick, 3-7 cm long and 2-5 cm wide and smallest upward, margins crenate to serrate-crenate, only short hairy on midvein beneath, petioles 0.2-2 cm long; **flowers** (Aug-Oct) petals white, hairy on tips, involucral bracts, 4.5-6 mm long, rarely white; **fruit and seeds** (Dec-Feb) nutlets 2.1-3.3 mm long, topped with white bristles. **Range:** Native. East LA to n FL and north to MA and west to OH, all physiographic provinces in the SE.
Ecology: Common scattered plant in dry open forests and forest plantations. Occurring even in shady habitat.
Synonyms: recently classified as *Ageratina aromatica* (L.) Spach., lesser snakeroot.

Eupatorium coelestinum L. **- blue mistflower** **COCO13**

Similar to *Eupatorium serotinum* except **plant** 0.3-1 m tall, forms small colonies from slender rhizomes; **leaves** opposite, ovate to lance-ovate, 4-8 cm long and 2-6 cm wide, margins crenate to crenate-serrate, sparsely appressed hairy and often glandular hairy; **flowers** (Aug-Nov) petals purplish to violet to blue, rarely white, and bract margins not white; **fruit and seeds** nutlet about 1.5 mm long, resinous glandular, topped with scant white bristles, 3-4 mm long. **Range:** Native. TX to n FL and north to PA and west to IL and KS, mainly in Cp and Pd of the SE.
Ecology: Occurs on moist sites and semi-shady habitat. Present in bottomland forest openings, right-of-ways, wet meadows, and ditches.
Synonyms: *Conoclinium coelestinum* (L.) DC., mistflower, blue boneset, ageratum, purple eupatorium.

Eupatorium fistulosum Barratt **- joe-pye-weed** **EUFI**

Plant: Erect and robust perennial, 2-3 m tall, from a persistent rootcrown.
Stem: Upright, stout, single or few from base, purplish and white waxy appearing, hairless.
Leaves: Mostly in whorls of 4-7, elliptic to lanceolate, 8-30 cm long and 2.5-15 cm wide, often arching downward, margins finely crenate to crenate-serrate, tapering tips and bases, fine hairy beneath or not, longest petioles 2-7 cm long.
Flowers: Jul-Oct. Large terminal divergently-branched panicle-like corymb, 10-50 cm long and 10-30 cm wide, petals bright pink-purple, 3.5-5 mm long.
Fruit and seeds: Oct-Dec. Nutlet (achene) 3.2-4.5 mm long, topped with tan bristles, nutlet containing 1 seed.
Range: Native. TX to c FL and north to ME and west to IA, mainly Pd and Mt, and infrequent in Cp of the SE.
Ecology: Occurs as scattered plants in open forests and old fields, and along forest edges and right-of-ways. Persists by rootcrowns and spreads by wind-dispersed seeds.
Synonyms: queen-of-the-meadow, trumpetweed.

Wildlife: Dogfennel has very little wildlife importance other than cover. Other species in this genus have been reported as moderate preference browse plants. Joe-pye-weed flowers are attractive to many of the larger butterflies such as the Monarch, sulphurs, swallowtails, and others.

Eupatorium Thoroughwort or Boneset Asteraceae

T. Bodner (Oct)

Eupatorium coelestinum

J. Miller (Sep)

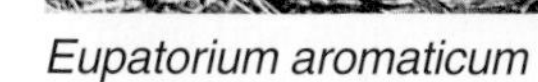

Eupatorium aromaticum

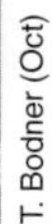

T. Bodner (Oct)

Eupatorium coelestinum

T. Bodner (Jul)

Eupatorium fistulosum

Euphorbia **Spurge** **Euphorbiaceae**

Annual and perennial herbs and small shrubs, with white latex sap. Leaves usually opposite, but sometimes alternate or whorled, and simple or compound. Stalked flowers, with 4-5 petal-like lobes, often white, subtended by whorls of bracts. Capsule extending from the flower with 3 chambers and 1 seed per chamber. 20 species in the SE.

Common Species

Euphorbia pubentissima Michx. **- false flowering spurge** **EUPU7**

Plant: Upright perennial, 10-80 cm tall, 1-several branching stems with milky sap, from a stout rootcrown.

Stem: Very slender, progressively more branched during the growing season, branching widely in whorls (or opposite), green to limegreen above becoming sandy to reddish or purple basally, fine scattered hairy (requires hand lens) or hairless.

Leaves: Alternate below, whorled (or paired) at base of inflorescence, elliptic to oblanceolate, 1-4 cm long and 0.8-2 cm wide, dark green and smooth with whitish midvein above and light green with protruding midvein beneath, margins entire and often rolled under, sessile or short petioled, somewhat drooping downward.

Flowers: Mar-Sep. Terminal clusters on long thin stalks (0.2-5 cm long) in flat-topped arrangement, sepals 5 (no real petals), white (to greenish to pink), 1-5 mm long, tiny yellow centers turning to stalked seed capsules.

Fruit and seeds: Apr-Nov. Smooth capsule, 3 lobed, 2-3 mm long, often drooping from flowers on stems 3-4 mm long, splitting at maturity releasing 3 seeds, 2.2-2.5 mm long, light gray to reddish brown.

Range: Native. TX to FL north to NY and west to MN and NE.

Ecology: The most common species of the spurge family in the SE and very variable. Occurs on moist to dry sites and open habitat. Persists after disturbance in forests as scattered plants and occurs in open forests and along forested right-of-ways. Persists by rootcrowns and spreads by animal-dispersed seeds. Poisonous to livestock.

Synonyms: *E. corollata* var. *corollata* L., *E. corollata* var. *zinniiflora* (Small) Ahles, tramp's spurge, prairie spurge.

Wildlife: Seeds of several of the spurges, including flowering spurge, are frequently eaten by Mourning Dove and Wild Turkey, and occasionally by Northern Bobwhite. Seed-eating songbirds likely consume seeds from various spurges. White-tailed Deer also eat minor amounts, particularly *E. pubentissima*.

Euphorbia **Spurge** **Euphorbiaceae**

T. Bodner (May)

Euphorbia pubentissima

T. Bodner (May)

Euphorbia pubentissima

J. Miller (Sep)

Euphorbia pubentissima

Euthamia Goldentop Asteraceae

Upright perennial branched herbs, with alternate leaves, narrow to linear, having silvery resinous glands. Leaves crowded along the upper stem. Numerous small golden-yellow heads in flat-topped densely-branched clusters. Flowers with overlapping involucre bracts being shiny yellowish with greenish tips. Nutlets (achenes) small, oblong, and hairy with end bristles. *Euthamia*, with its yellow flat-topped flower clusters, was previously included within *Solidago* (goldenrod), having conical panicles or racemes. 4 species in the SE, with numerous varieties.

Common Species

Euthamia tenuifolia (Pursh) Nutt. **- slender goldentop** **EUTE7**

Plant: Ascending to nodding perennial, 40-100 cm tall, from a long-sender rhizome having few lateral roots. Often forming dense colonies.

Stem: Slender, scattered fine hairy becoming hairless, 4-angled becoming round basally, branched near ground and much above mid-plant, alternately spiraling, upper stem light green due to minute silvery glands (requires hand lens).

Leaves: Alternate, crowded along the upper stem, long linear, 2-7 cm long and 1-3 mm wide and smaller upward, sessile, tapering tipped, often light green due to minute silvery surface glands becoming glandless, tufts of small leaves often in upper axils.

Flowers: Sep-Nov. Terminal on densely branched flat- to curve-topped corymbs or panicles, 100's of tiny cylindric flowers, golden-yellow, 3-4 mm tall, involucre sheath golden-green to silvery-yellow with greenish pointed tips, 1-2 mm tall.

Fruit and seeds: Dec-Mar. Oblong nutlet (achene), 0.7-1 mm long, topped with white bristles, 3-3.3 mm long, nutlet containing 1 seed.

Range: Native. MS to FL and north to NJ, mainly in the Cp.

Ecology: Occurs as small dense colonies in new forest plantations, open forests, and along forest margins and right-of-ways, especially in Cp and lower Pd. Dry sites and open to semi-shady habitats. Persists by rhizomes and spreads by wind-dispersed seeds.

Synonyms: *Solidago microcephala* (Greene) Bush, *S. minor* (Michx.) Fern., flat-topped goldenrod, small-headed goldenrod.

Other species in the SE: *E. graminifolia* (L.) Nutt. [formerly *Solidago graminifolia* (L.) Salisb.], *E. gymnospermoides* Greene, *E. leptocephala* (Torr. & Gray) Greene [formerly *Solidago leptocephala* Torr. & Gray].

Wildlife: As with the *Solidago* species, White-tailed Deer, Ruffed Grouse, Eastern Cottontail, and Wild Turkey occasionally consume the basal rosettes of *Euthamia* during winter. Seeds are consumed in minor amounts by various songbirds, particularly the American Goldfinch.

Euthamia **Goldentop** **Asteraceae**

T. Bodner (Oct)

Euthamia tenuifolia

J. Miller (Aug)

Euthamia tenuifolia

Fragaria **Strawberry** **Rosaceae**

Low-growing perennial herbs, with tufts of erect petioled 3-leaflet leaves, from short rhizomes, and spreading by stolons. Rose-type flowers with 5 white petals, 5 sepals, 20 stamen, and numerous pistils. Fruit a strawberry, juicy and red at maturity. Several subspecies, varieties, and crosses cultivated in the SE.

Common Species

Fragaria virginiana Duchesne **- wild strawberry** **FRVI**

Plant: Stemless erect-leaved perennial, less than 20 cm tall, from rhizomes and with stolons running along the soil surface.

Stem: Short and not apparent.

Leaves: Alternate, 3 leaflets, leaflets elliptic to obovate, 3-10 cm long, coarsely toothed, lateral veins parallel at about a 30 degree angle from the midvein.

Flowers: Mar-Jun. Terminal stalked clusters, stalks appressed hairy, petals 5, white, 6-15 mm long, sepals 5, 4-8 mm long, subtended by 5 leafy bracts that remain on the fruit.

Fruit and seeds: May-Jun. Strawberry, less than 2.5 cm long and 1-1.5 cm wide, red nutlets embedded in the fruit, the nutlets of similar *Duchesnea indica* (Indian mock-strawberry) set atop the fruit.

Range: Native. OK to FL and north to s Canada and west to MN.

Ecology: Occurs on moist to dry sites and semi-shady habitats as scattered plants or small colonies. Occurs in new forest plantations, along forest margins and right-of-ways, and in old fields. Persists after disturbance by rhizomes, colonizes by stolons, and spreads by animal-dispersed seeds.

Synonyms: Several varieties cultivated, Virginia strawberry.

Wildlife: Strawberry fruits are available for a short period during late spring and early summer. They are consumed by numerous songbirds, particularly the American Robin and Cedar Waxwing. The leaves are eaten by a variety of wildlife species including Ruffed Grouse, White-tailed Deer, Eastern Cottontail, and other small mammals.

Fragaria **Strawberry** **Rosaceae**

T. Bodner (Apr)

Fragaria virginiana

Gaillardia Blanketflower Asteraceae

Annual or short-lived perennial herbs, frequently shrub-like with woody bases, from a taproot. Flower heads spherical or hemispherical with purple or yellow centers, yellow or red ray petals when present (frequently deciduous), having involucral bracts in 2-3 series, green to purple, and reflexed into a skirt. Leaves alternate or all basal, petioled below and sessile upward, both leaves and stem short hairy. Seeds within hairy nutlets (achenes). 2 species in the SE.

Common Species

Gaillardia aestivalis Foug. **- lanceleaf blanketflower** **GAAE**

Plant: Upright annual or short-lived perennial, 10-60 cm tall, much branched and shrub-like, from a taproot.

Stem: Slender, round, alternate branched, fine soft hairy.

Leaves: Alternate, oblanceolate below and linear to oblong above, 1.5-7 cm long and 2.5-14 mm wide and smaller upper, petioled below and sessile above, lower deciduous in late-summer.

Flowers: Jul-Nov. Spherical heads on long stalks from upper leaf axils, stalks with few bracts or without bracts, heads 1-2 cm wide, purple on spherical disks (sometimes yellow), ray petals red, deciduous in late-summer, 6-15, 1-2 cm long, sepals curve under to be inconspicuous.

Fruit and seeds: Oct-Nov. Nutlet (achene), 2-3 mm long, densely long hairy, with membranous scales at apex, nutlet containing 1 seed.

Range: Native. TX to FL and north to NC and west to KS.

Ecology: Occurs often in small colonies. Spreads along forest roads and into new plantations by wind-dispersed seeds.

Synonyms: gaillardia.

Other species in the SE: *G. pulchella* Foug. [turf and crop weed on sandy sites].

Wildlife: No wildlife value reported.

Gaillardia Blanketflower Asteraceae

J. Miller (Oct)

Gaillardia aestivalis

T. Bodner (Sep)

Gaillardia aestivalis

J. Miller (Oct)

Gaillardia aestivalis

J. Miller (Sep)

Gaillardia aestivalis

Galactia Milkpea Fabaceae

Perennial viney legume, with trailing or twining stems, except the erect and partially-woody, *G. erecta*. Leaves usually with 3 leaflets or 4 to 6 leaflets (*G. erecta*) or 5 to 9 leaflets (*G. elliottii*), having distinct leafstalks with minute stipules and bracts (stipels). Flowers at leaf axils, singly or several in racemes. Pea-type flowers with red, purple, pink, or white petals. Calyx 4-lobed and 2-lipped. Fruit a legume (pod), oblong and turgid, splitting at maturity with twisting halves releasing few to many seeds. 11 species in the SE, only 4 are taxonomically distinct, the others are considered a complex.

Common Species

Galactia volubilis (L.) Britt. **- downy milkpea** **GAVO**

Plant: Twining or trailing perennial vine, 1-2 m long, from woody tuber.

Stem: Slender, twining on itself or on vegetation, densely (to sparsely) hairy, hairs to 0.5 mm long.

Leaves: Alternate, 3 leaflet, leaflets ovate to oblong-lanceolate, 1.5-4.5 cm long and 1-3 cm wide, rounded base and tip, tip often minutely pointed becoming indented, sparsely hairy or hairless above and downy hairy beneath, petiole 8-10 mm long, tiny stipules at base or not.

Flowers: Jun-Sep. Widely-spaced in axillary racemes, 1-3 stalked flowers, pea-type, pink to purple, 5-8 mm long, calyx 4 lobed, 2-3.5 mm long, pedicels 1-4 mm long, subtended by triangular bracts, 0.4-1.2 mm long.

Fruit and seeds: Jul-Dec. Legume (pod), 2-5 cm long and 4-5 mm broad, flattened and swollen at seeds, short spreading hairy, pointed tipped, calyx persistent at base, splitting at maturity to release 2-8 seeds, mottled, 2-3 mm long.

Range: Native. TX to FL and north to NY and west to IN and KS.

Ecology: Nitrogen fixer. Occurs on moist to dry sites and in open to semi-shady habitats, as scattered plants or small patches in open areas. Inhabits forest plantations, open forests, margins, and right-of-ways. Persists by woody tubers and spreads by animal-dispersed seeds.

Synonyms: *Galactia macreei* M.A. Curtis, *G. mississippiensis* (Vail) Rydb., *G. volubilis* var. *mississippiensis* Vail.

Wildlife: In the Southeast, milkpea seeds are an important component of the diet of the Northern Bobwhite, often taken in much higher quantities than would be expected based on their low abundance. The seeds likely are also consumed by other songbirds and small mammals. The plants occasionally are browsed by White-tailed Deer.

Galactia Milkpea Fabaceae

T. Bodner (Sep)

Galactia volubilis

Galax **Galax or Galaxy** **Diapensiaceae**

Evergreen, low-growing perennial herb from rhizomes, forming continuous ground cover or patches, common in Blue Ridge Mountains. Leaves thick and rounded with upright long petioles. Small white flowers on erect spike-like racemes. Capsules crowded atop the upper part of an erect stalk. 1 species in the SE.

Common Species

Galax urceolata (Poir.) Brummitt **- galax** **GAUR2**

Plant: Low growing, evergreen perennial, 10-30 cm tall, tufts of long-petioled leaves forming colonies, from scaly creeping rhizomes having fibrous red roots. Colonies may emit a skunk-like odor.

Stem: Short and not apparent, at ground line with overlapping bracts.

Leaves: Alternate, evergreen, ovate to round (orbicular) with cordate base, 3-10 cm long and wide, margins dentate-serrate, petioles 3-25 cm long, dark green and shiny in summer often turning maroon to pink in fall.

Flowers: May-Jul. Erect slender spike,10-30 cm tall, minute flowers crowded around and along the upper part, petals 5, white, 4-6 mm long, flowers open from bottom to top, scentless.

Fruit and seeds: Aug-Nov. Capsules along upper stalk, ovoid, 2.7-3 mm long, grainy surface, 3-celled, splitting to release several lustrous seeds, reddish brown, 1-1.2 mm long.

Range: Native. AL to GA north to VA and west to KY, common in Mt and local in the Pd and Cp valleys.

Ecology: Persistent ground cover in mountain forests and occurring near streams at lower elevations. Persists and colonizes by rhizomes and spreads by seeds.

Synonyms: *G. aphylla* L., galaxy, wandflower, colt's foot, beetleweed, skunk weed.

Wildlife: Galax is winter browse of moderate importance for White-tailed Deer in the southern Appalachian Mountains, particularly in years of poor acorn production. No other wildlife value reported.

Galax Galax or Galaxy Diapensiaceae

T. Bodner (May)

Galax urceolata

T. Bodner (Nov)

Galax urceolata

Galium **Bedstraw or Cleavers** **Rubiaceae**

Spreading, weak-stemmed annual or perennial herbs, often rough surfaced, with leaves in whorls having entire margins. Flowers in simple to profuse cymes or one to several flowers per upper leaf axil. Corolla tube with 3-5 lobes (petals), projecting outward at right angles, usually white but also greenish, yellowish, and maroon (*G. latifolium*), and sepals absent. Fruit dry or sometimes fleshy, non-splitting, and round to kidney shaped. Roots often red or orange. 15 species in the SE.

Common Species

Galium aparine L. **- catchweed bedstraw** **GAAP2**

Plant: Spreading annual, stems trailing and branched, 20-80 cm long, often forming mats.
Stem: Angled, rough textured with short recurved bristles, often pale green.
Leaves: Whorled, 6-8 per node, linear to narrowly oblanceolate, 2-4.5 cm long and 2-6 mm wide, tips sharp-pointed, surfaces rough, margins hooked bristly.
Flowers: Mar-May. Single or in terminal and axillary clusters, corolla white, 2-3 mm wide, 3-5 lobed, no sepals, 4 short stamens, 2-celled ovary that becomes paired fruit.
Fruit and seeds: Mar-May. Dry hook-bristled fruit, paired on stalks, spherical or kidney-shaped, 2-3 mm long and 4-5 mm wide, green turning brown or black, seed remaining in fruit.
Range: Naturalized from Europe. Throughout US, infrequent in Lower Cp.
Ecology: Common spring plant of new forest plantations, forest margins, right-of-ways, fields, and disturbed sites. Spreads by seeds attaching to animals.
Synonyms: goose-grass, spring-cleavers, stickywilly.

Galium pilosum Ait. **- hairy bedstraw** **GAPI2**

Plant: Erect or ascending perennial herb, 20-80 cm tall or long, infrequently branched, from a persistent rootcrown with taproot and reddish lateral roots.
Stem: Slender, square between nodes, rough when young, with hooked hairs on angles, becoming smooth.
Leaves: Four per node, elliptic to ovate, 1-2.5 cm long and 6-12 mm wide, tips sharp-pointed, rough on top and smooth glandular beneath.
Flowers: May-Aug. Multiple-branched flower clusters from terminal and upper axils on stalks 1-5 cm long, corolla white to greenish or maroon.
Fruit and seeds: Jul-Nov. Dry hook-bristled fruit, spherical, 2-3 mm long and 3-4 mm wide, green maturing brown or black, containing 1 seed.
Range: Native. Throughout Eastern US.
Ecology: Common summer plant of new forest plantations, forest margins, right-of- ways, fields, and disturbed sites. Scattered plant in young plantations, even in heavy shade. Spreads by seed attaching to animals.

Galium hispidulum Michx. **- coastal or fleshyfruit bedstraw** **GAHI**

Similar to *Galium pilosum* except **plant** smaller, 10-30 cm tall, erect or spreading, often evergreen; **stem** smooth, hairy only at nodes; **leaves** also 4 per node, thick, 7-15 mm long and 3-6 mm wide, sparsely erect hairy, sunken glands beneath; **flowers** corolla white; **fruit and seeds** fruit fleshy, spherical blue to whitish-blue, smooth and hairless, 2-3.5 mm long and 4-5 mm wide. **Range:** Native. LA to n FL and north to NJ, mainly in Cp.
Ecology: Occurs in open and shady pine forests, pastures and right-of-ways.

Wildlife: The bedstraws are poor to moderately preferred summer forages of White-tailed Deer. No other wildlife use reported.

Galium **Bedstraw or Cleavers** **Rubiaceae**

J. Miller (Oct)

Galium hispidulum

T. Bodner (Mar)

Galium aparine

J. Miller (Oct)

Galium hispidulum

Gamochaeta **Everlasting** **Asteraceae**

Winter and summer annual and biennial herbs, with leafy rosettes yielding erect often branching stems. Rosette and alternate stem leaves with entire margins and woolly hairy, at least on the lower surfaces. Flower heads along stalks, inconspicuous petals, yellow to whitish, within more conspicuous brownish to pinkish bracts, in a loose to tight cylinder. Nutlets small with bristle plumes. 4 species in the SE, all until recently listed as varieties of *Gnaphalium purpureum.*

Gamochaeta purpurea (L.) Cabrera **- purple cudweed** **GAPU3**

Plant: Upright annual or biennial with a leafy flower stalk, 10-40 cm tall, from a basal rosette, often found in small groups. Persists standing dry over winter.
Stem: Ascending, one to few from basal rosette, seldom branching, white woolly hairy.
Leaves: Initially a basal rosette and alternate on the lower stem, oblanceolate to spatula-shaped, 3-10 cm long and 1-2 cm wide and smaller upward, sessile, margins entire, tip rounded with a tiny sharp point (mucronate), white woolly hairy beneath and less so above.
Flowers: Mar-Jul. Scattered small heads in a terminal spike, leafy bracts subtend each stemless flower cluster, flowers mainly involucral bracts, 3-4 mm long, light-brown initially tinged with pink-purple (thus common name).
Fruit and seeds: Jun-Sep. Nutlet (achene), 0.5-0.7 mm long, yellowish to reddish, resin dotted, nutlet containing 1 seed.
Range: Native. Throughout US.
Ecology: Commonly occurring after disturbance in new forests, open forests, and along forest margins, on dry to moist sites. Spreads by wind- and gravity-dispersed seeds. Seeds germinating after falling, producing several ages of plants at a site.
Synonyms: *Gnaphalium purpureum* L., rabbittobacco, everlasting, catsfoot, spoonleaf purple everlasting.

Other species in the SE: *G. americana* (P. Mill.) Weddell [formerly *Gnaphalium purpureum* L. var. *americanum* (P. Mill.) Klatt], *G. falcata* (Lam.) Cabrera [formerly *Gnaphalium purpureum* L. var. *falcatum* (Lam.) Torr. & Gray], *G. pensylvanica* (Willd.) Cabrera [formerly *Gnaphalium purpureum* L. var. *spathulatum* (Lam.) Baker], *G. sphacilata* (Kunth) Cabrera [formerly *Gnaphalium sphacilatum* Kunth].

Wildlife: *Gamochaeta* are low preference White-tailed Deer forages; primary use is winter foraging on the basal rosettes. Some use by Eastern Cottontail has been reported.

J. Miller (Jul)

Gamochaeta purpurea

T. Bodner (Sep)

Gamochaeta purpurea

Geranium **Geranium or Cranesbill** **Geraniaceae**

Winter annual or perennial herbs, from rhizomes, with leaves palmately or radially lobed or cleft, basal and also opposite on stems. Flowering commencing in spring, pink to purplish-pink, in pairs and paired clusters, stalked and subtended by leafy bracts; petals 5, sepals 5, and stamens 10. Fruit a tubular capsule, within and extending from leafy bracts, the capsule terminated by a slender beak resembling a tiny cranesbill (thus the common name). The beak splits into 5 sections that coil upward when ripe, with a dry fruit attached at each tip before dispersal. 6 species in the SE.

Common Species

Geranium carolinianum L. **- Carolina geranium** **GECA2**

Plant: Winter annual or biennial, diffusely-branched, sprawling to ascending, 20-70 cm tall, from fibrous roots with a shallow taproot.

Stem: Much branched from the base, moderately stout, greenish-pink to red and densely hairy.

Leaves: Alternate basally and opposite above, round to oval in outline, 2-6 cm long and wide, palmately cleft into 5-9 finger-like lobes or toothed segments, hairy on both surfaces, petioles 3-12 cm long, hairy.

Flowers: Feb-Jun (and sporadically later). Terminal compact clusters (cymes) of 2-several flowers at tips of stems and branches, petals 5, whitish pink to pale purple, 4-6 mm long, tip notched, sepals 5, hairy, tip pointed, subtended by a whorl of small leaves.

Fruit and seeds: May-Jul. Tubular capsule with pointed beak, 8-12 mm long, green and hairy becoming dry at maturity and splitting at base to spring up into 5 curving segments each tipped with a kidney-shaped seed, 2 mm long, hairy for attaching to fur.

Range: Native. Throughout US.

Ecology: Early invader of disturbed places such as new forest plantations, right-of- ways, and abandoned fields. Decreases with shading. Spreads by animal-dispersed seeds.

Synonyms: wild geranium, stork's-bill, cranesbill.

Other species in the SE: *G. columbinum* L., *G. dissectum* L., *G. maculatum* L., *G. molle* L., *G. pusillum* L.

Wildlife: Carolina geranium is a common open-field plant and is a preferred winter forage of White-tailed Deer in the Southeast, averaging 19 percent crude protein in the vegetative state. Seeds are consumed by the Northern Bobwhite and Mourning Dove, and by songbirds and small mammal species.

Geranium **Geranium or Cranesbill** **Geraniaceae**

T. Bodner (Mar)

Geranium carolinianum

T. Bodner (May)

Geranium carolinianum

J. Miller (Apr)

Geranium carolinianum

Gnaphalium Cudweed Asteraceae

Woolly white-haired annual and biennial herbs (other regions perennials), having erect stems with alternate leaves, and most with leafy basal rosettes. Stems, lower leaf surfaces, and rows of flower bracts often densely white hairy to woolly. Whitish to yellow flower heads enclosed within whitish bracts, in terminal branching panicles or corymbs. Nutlets (achenes) oblong, 0.5-0.8 mm long, hairless, often with resinous glands. 5 species in the SE.

Common Species

Gnaphalium obtusifolium L. - rabbittobacco PSOBO

Plant: Erect, densely white-hairy, annual or biennial, 30-100 cm tall, 1 to several stems from a basal rosette, from fibrous roots. Emits fragrance of pepper or tobacco when rubbed or crushed (thus common name, rabbit tobacco). Plants remain standing during the winter and following spring (thus other common name, everlasting).
Stem: Slender, very white hairy becoming less hairy and slightly glandular below.
Leaves: Alternate, numerous and crowded on the lower stem, gray-green and almost hairless above and densely matted white hairy beneath, narrowly lanceolate to oblanceolate, 2.5-8 cm long and 4-12 mm wide, margins wavy, midvein obvious, sessile, lower leaves turning dark brown or black on upper surface in late summer to fall. Dried leaves remain during winter, twisted to reveal whitish undersides.
Flowers: Aug-Oct. Terminal on few branches in the upper plant (panicles or corymbs), cylindric, disk flowers whitish to pale yellow, bracts white woolly hairy, involucral sheath 6-8 mm long, remaining dry and spreading on the plant during winter, silvery brown.
Fruit and seeds: Sep-Oct. Nutlet (achene), 0.6-0.8 mm long, reddish brown, smooth, released with and without plumes, nutlet containing 1 seed.
Range: Native. Occurs throughout Eastern US, common on sandy Cp soils.
Ecology: Early invader of disturbed places with scattered plants in new forest plantations and old fields, and less frequently on mowed right-of-ways. Spreads by wind-dispersed seeds.
Synonyms: *Pseudognaphalium obtusifolium* (L.) Hillard & Burtt ssp. *obtusifolium,* fragrant cudweed, everlasting, sweet everlasting, catsfoot.

Gnaphalium helleri Britt. - Heller's cudweed PSHE4

Similar to *Gnaphalium obtusifolium* except **plant** less common and averaging smaller, 20-90 cm tall, fragrance less but also of pepper or tobacco; **stem** green and densely glandular hairy (greasy feeling); **leaves** slightly clasping stem, green and glandular hairy above, matted white hairy beneath; **flowers** becoming very white-tan woolly, ovoid.
Ecology: The two species often grow on the same sites and mature at similar times, although green everlasting may be later.
Synonyms: *Pseudognaphalium helleri* (Britt.) A. Anderb., green everlasting.

Other species in the SE: *G. stramineum* Kunth, *G. uliginosum* L., *G. viscosum* Kunth

Wildlife: Cudweeds are poor White-tailed Deer forages and have little value to other wildlife species.

Gnaphalium **Cudweed** **Asteraceae**

T. Bodner (Oct)

Gnaphalium obtusifolium

T. Bodner (Aug)

Gnaphalium obtusifolium

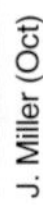

J. Miller (Oct)

Gnaphalium obtusifolium

T. Bodner (Sep)

Gnaphalium helleri

Helenium Sneezeweed Asteraceae

Erect annual or perennial herbs, with basal or alternate stem leaves, sessile and grading into the stem. Numerous flower heads terminal on stalks, reflexed yellow or whitish petals, yellow or reddish centers (disks). Involucral bracts in 2-3 series, inner ones being shorter. Nutlets (achenes) 4-5 angled with hairs on angles and scales at top. Most species are poisonous to livestock but rarely eaten. 9 species in the SE.

Common Species

Helenium amarum (Raf.) H. Rock **- bitter sneezeweed** **HEAM**

Plant: Upright annual herb, 0.2-1 m tall, branched near base with numerous branches at mid-plant, from a taproot with laterals along the entire length.
Stem: Stout at base and slender above, hairless and glandular (oily feel), light green above to purplish to brownish basally.
Leaves: Alternate, thread-like (linear) and fascicled, 1-8 cm long and 1-2 mm wide, green and glandular hairy (sticky), basal leaves dried or absent at flowering.
Flowers: May-Nov. Heads several to many on slender bractless stalks (3-10 cm long) extending above the leafy parts forming a flat-topped corymb, ray petals yellow, 5-10, 5-12 mm long, wedge-shaped, 3-toothed tipped, spreading to drooping, centers (disks) yellow in a dome, involucral bracts green and reflexed, 1.5-2.5 cm wide. Flowers fragrant.
Fruit and seeds: Jun-Nov. Non-plumed nutlet (achene), 4-5 angled, topped by awn-shaped scales, tips and bases hairy, nutlet containing 1 seed, brown, 1-1.5 mm, angled.
Range: Native. TX to FL and north to MA and west to IL and NE.
Ecology: Common inhabitant of highway right-of-ways and forest road edges, sustained by mowing and grading. Early invader of new forest plantations. Also common in the mid-SE on disturbed lands and pastures, but does not persist with cultivation. Poisonous to livestock. Spreads by seeds, but not wind dispersed, possibly animal dispersed.
Synonyms: *Helenium tenuifolium* Nutt., Spanishdaisy, bitterweed, yellowdicks.

Helenium autumnale L. **- fall sneezeweed** **HEAU**

Plant: Clump-forming perennial, 0.5-1.5 m tall, with 1 to several stems, on wet-moist soils.
Stem: Ascending, winged or wing-angled stems, short scattered hairy. Toxic to livestock.
Leaves: Alternate, elliptic to lanceolate to oblanceolate, 6-15 cm long and 1-3 cm wide and smaller upward, margins serrate to entire, base tapering to join wings on stem, all covered in sunken glands (requires hand lens), lower leaves deciduous at flowering.
Flowers: Sep-Nov. Heads several to numerous, each 1-2 cm broad and each on a leafy, slender stalk 2-6 cm long, collectively forming a flat- to rounded-topped cluster (corymb), petals yellow, 12-20, wedge-shaped and 3-lobed tipped, 16-25 mm long, spreading to drooping (reflexed), disk flowers yellow, in a dome, bracts green and reflexed.
Fruit and seeds: Oct-Jan. Nutlet (achene) about 1 mm long, topped by scales.
Range: Native. Throughout most of US and adjacent Canada.
Ecology: Inhabits stream and river banks and margins, and other moist places. Persists by rootcrowns and spreads by water-dispersed seeds.
Synonyms: autumn sneezeweed, common sneezweed.

Other species in the SE: *H. brevifolium* (Nutt.) Wood, *H. campestre* Small [only in AR], *H. drummondii* H. Rock [TX, LA, FL], *H. flexuosum* Raf. [e US], *H. pinnatifidum* (Nutt.) Rydb. [FL, SC, NC], *H. quadridentatum* Labill. [TX, LA, MS, AL], *H. vernale* Walt. [LA, FL, SC, NC].

Wildlife: Bitter sneezeweed has no wildlife value. Others, including *H. autumnale*, generally are low preference White-tailed Deer browse and questionable in importance to other wildlife species.

Helenium **Sneezeweed** **Asteraceae**

T. Bodner (Jun)

Helenium amarum

J. Miller (Oct)

Helenium autumnale

Helianthus **Sunflower** **Asteraceae**

Erect or ascending herbs, mainly perennials and a few annuals, with flowers in heads, having yellow entire petals (rays) and dark yellow or purplish centers (disks). Among the disk flowers are husky awl-shaped scales, often used in identification, being shed with the nutlets (achenes). Leaves often basal and stem leaves opposite, sometimes alternate above, entire or toothed, and 3-veined, but sometimes obscurely. Involucral bracts beneath the flower being leafy, often pointed, and in several series. 35 species in the SE.

Common Species

Helianthus angustifolius L. **- swamp sunflower** **HEAN2**

Plant: Erect becoming ascending, tall perennial, 1-2 m tall, 1-few branched from the base and infrequently branched above, from a short erect rootcrown with fibrous roots.
Stem: Moderately stout at base becoming slender and wiry below flowers, finely hairy being slightly rough to almost hairless, green to tan to brown.
Leaves: Alternately spiraling above and opposite at base, numerous, thick, long linear, 5-15 cm long and 0.2-1.5 cm wide and smaller upward, sessile, dark green and very rough hairy with wide whitish-green midvein beneath, margins entire and often slightly rolled under, mildly aromatic when crushed, small leaves often in axils, basal rosette in winter.
Flowers: Jul-Nov. Heads on long stalks (10-30 cm long) with yellow ray petals, 10-15, each 2-4 cm long, centers (disks) dark purplish red (rarely yellow), 1.5-2 cm wide, slightly domed, involucral bracts in several rows, green, lanceolate, 6-12 mm long, persistent.
Fruit and seeds: Sep-Mar. Dull-black nutlet (achene), 2-4 mm long, containing 1 seed.
Range: Native. TX to FL and north to NY and west to KY and MO.
Ecology: The most common sunflower in mid-SE, occurring in new forest plantations, open forests, old fields, and along forest margins and right-of-ways, on both lowland and upland sites (contrary to common name). Persists by rootcrowns and spreads by seeds.

Helianthus tuberosus L. **- Jerusalem artichoke** **HETU**

Similar to *Helianthus angustifolius* except **plant** taller, 1-3 m tall, from tuber-bearing rhizome (thus sci. name); **stem** stout, 2-3 cm wide at base, rough hairy; **leaves** opposite to alternate above, lanceolate to ovate, 10-25 cm long and 4-12 cm wide, rough hairy above and short hairy beneath, margins serrate, tapering to more or less a winged petiole, 2-8 mm long; **flowers** (Jul-Oct) with yellow ray petals, 3-5 cm long, dark yellow centers (disk), 1.5-2.5 cm wide, on stout stalks, involucral bracts broadly to narrowly lanceolate and tips recurving; **fruit and seeds** nutlet 5-6 mm long, light and dark brown mottled, with 1 seed.
Range: Native to western US (where cultivated for its edible tuber). Naturalized in Eastern US, except only now migrating into Gulf Coastal Plain of the SE.
Ecology: Persists after disturbance to occur in new forest plantations, forest edges, and roadsides, preferring moist habitats and lowland sites. Spreads by animal-dispersed seeds.
Synonyms: *H. tomentosus* Michx.

Helianthus divaricatus L. **- woodland sunflower** **HEDI2**

Similar to *Helianthus tuberosus* except **plant** only to 1.5 m tall; **stems** smooth; **leaves** opposite, narrowly lanceolate, 5-18 cm long and 1-5 cm wide, short non-winged petioles to 5 mm long; **flower** (Jun-Aug) petals 1.5-3 cm long; **fruit and seeds** nutlet 4-4.5 mm long.
Range: Eastern US as far south as c FL, MS, and LA.

Wildlife: Sunflowers generally are fair wildlife plants. White-tailed Deer commonly forage on the flower heads and new vegetative growth in mid- to late-summer. Seeds are consumed by Northern Bobwhite, Mourning Dove, and numerous small mammals and songbirds such as various sparrows, American Goldfinch, Carolina Chickadee, and Dark-eyed Junco.

Helianthus Sunflower Asteraceae

J. Miller (Jun)

Helianthus tuberosus

J. Miller (Jun)

Helianthus tuberosus

T. Bodner (Aug)

Helianthus tuberosus

T. Bodner (Oct)

Helianthus angustifolius

T. Bodner (May)

Helianthus divaricatus

T. Bodner (Jun)

Helianthus divaricatus

Heterotheca **False Goldenaster** **Asteraceae**

Erect annual or biennial herb, with yellow-orange flower heads in late summer. Stems and small leaves long hairy and dotted with resinous glands. Leaves alternate, simple, rough, margins dentate to serrate, lower ones petioled and deciduous, and upper ones sessile to clasping. Seeds in plumed nutlets (achenes). 2 species in the SE.

Common Species

Heterotheca subaxillaris (Lam.) Britt.&Rusby **- camphorweed** **HESU3**

Plant: Upright annual or biennial, 0.2-2 m tall, finely branched in the upper plant, from a taproot. Plant with fragrance of camphor (thus common name).

Stem: Stout, green to purple to brown with lighter stripes, scattered long white hairy, hollow pith.

Leaves: Alternate or basal, variable, simple, lanceolate to ovate or oblong, 2-9 cm long and 1-3 cm wide and smallest upward, margins dentate to serrate to entire, rough hairy both surfaces, lower ones petioled and deciduous, upper ones sessile to clasping.

Flowers: Jul-Nov. Flower heads numerous, occurring on branches with small leaves from upper leaf axils in flat- to round-top corymbs or panicles, yellow-orange rays and disks, ray petals 20-45, 5-10 mm long, disks 0.8-1.5 cm wide, involucre with rows of linear bracts, flower buds spherical.

Fruit and seeds: Aug-Dec. Nutlets (achenes) of 2 types, ray nutlets without hairs or topped with bristles, disk nutlets with hairs and a double-row of bristles, both nutlets obovoid, 2.3-3 mm long, containing 1 seed.

Range: Native. AZ to FL and north to DE and west to KS.

Ecology: Early invader of disturbed forests, sandy fields, and right-of-ways. A frequent scattered roadside plant in the Cp and Pd. Seed probably dispersed by wind and animals.

Synonyms: *Heterotheca scabra* (Pursh) DC., golden aster, camphor plant.

Other species in the SE: *H. camporum* (Greene) Skinners.

Wildlife: Camphorweed seeds are likely consumed by some songbirds such as the Indigo Bunting and others. Little or no other wildlife value reported.

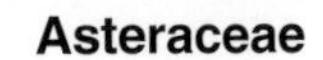

Heterotheca False Goldenaster Asteraceae

J. Miller (Oct)

Heterotheca subaxillaris

J. Miller (Oct)

Heterotheca subaxillaris

J. Miller (Oct)

Heterotheca subaxillaris

Hibiscus **Rosemallow or Hibiscus** **Malvaceae**

Annual or perennial herbs or shrubs, with showy white, cream, or yellow flowers, often with purple-red centers, in upper leaf axils. Hibiscus flowers are known for their 5 broad petals and a column of fused stamens around a pistil that extends from the corolla. Flower and fruit subtended with 7-15 linear bracts. Leaves unlobed to palmately lobed or dissected, often with long petioles and short stipules. Fruit a capsule with 5 sections and several to many seeds. 10 species in the SE.

Common Species

Hibiscus aculeatus Walter. **- comfortroot** **HIAC**

Plant: Spreading or ascending perennial, up to 1 m tall, from a rootcrown.
Stem: Stout becoming slender upward, short rough hairy.
Leaves: Alternate, palmately 3-5 lobed (V-shaped sinuses), 3-9 cm long and 4-10 cm wide, margins coarsely and irregularly toothed, both surfaces rough hairy, petioles 2-10 cm long.
Flowers: Jun-Oct. Solitary in upper leaf axils, petals 5, each 5-6 cm long, cream colored turning to yellow and finally fading to pink with crimson base, an extended crimson stamen column topped by a pistil with a 5-parted stigma, calyx lobes triangular to lanceolate, 8-12 mm long, sharp tipped, with an outer whorl of bracts, 8-10, linear, 1-2 cm long, usually with cleft apex, stiff hairy.
Fruit and seeds: Aug-Nov. Beaked spherical capsule, 1.7-2 cm long, stiff white hairy, opening into 5 slice-shaped sections, each with sharp beaks, sections splitting at maturity to release several seeds, rounded, 3.5-4 mm long, light-brown with whitish spots.
Range: Native. LA to FL and SC, mainly in Cp.
Ecology: Occurs as scattered plants in new forest plantations, open older forests, and along right-of-ways and roadside ditches. Persists by rootcrowns and spreads by seeds.

Hibiscus moscheutos L. **- crimsoneyed rosemallow or wild cotton** **HIMO**

Similar to *Hibiscus aculeatus* except **plant** multi-stemmed, upright, to 2 m tall; **leaves** dark green or grayish above and velvety hairy and whitish to grayish beneath, ovate-elliptic to lanceolate, 3-5 lobed or not, 8-24 cm long and 4-10 cm wide, bases rounded or cordate, tips long tapering, margins serrate to crenate. **Range:** Native. Throughout Eastern US.
Ecology: Occurs along marsh margins and other wet habitat, such as ditches.

Hibiscus laevis All. **- halberdleaf rosemallow** **HILA2**

Similar to *Hibiscus aculeatus* except **plant** upright with few to many stems arising from a rootcrown; **stem** hairless; **leaves** ovate to ovate-lanceolate, 7-15 cm long, usually with side lobes at base, margins serrate, green on both surfaces, petioles 4-15 cm; **flowers** (Jun-Sep) solitary in upper leaf axils, petals 6-7 cm long, reddish pink to white with red base, sepals widely triangular, 8-10 mm long, bracts below sepals linear, 1.7-2.5 cm long; **fruit and seeds** (Aug-Oct) capsule 2.3-3.5 cm long, hairless, seed ovoid, 3-3.2 mm long, dense reddish-brown hairy. **Range:** Native. Throughout SE, mainly Cp and Pd.
Ecology: Occurs on wet to moist sites and open to semi-shady habitats. Present along rivers and streams, and margins of marshes and ditches. Persists by rootcrowns and spreads by water-dispersed seeds.
Synonyms: *H. militaris* Cav.

Other species in the SE: *H. coccineus* Walt., *H. grandiflorus* Michx., *H. syriacus* L., *H. trionum* L. Only in FL: *H. acetosella* Welw. *ex* Hiern., *H. bifurcatus* Cav.

Wildlife: The various species of hibiscus are poor quality wildlife food plants.

Hibiscus Rosemallow or Hibiscus Malvaceae

J. Miller (Sep)

Hibiscus aculeatus

J. Miller (Sep)

Hibiscus aculeatus

B. Craft (Jun)

Hibiscus laevis

D. Lauer (Oct)

Hibiscus moscheutos

Hieracium **Hawkweed** **Asteraceae**

Perennial erect herbs, with latex-like white sap, erect stems commonly with larger basal leaves and smaller alternate stem leaves, from short rhizomes or rootcrown. Flowers in aster-type heads, yellow to red-orange. Nutlets black, cylindric and ribbed, bristle plumed. 10 species in the SE.

Common Species

Hieracium gronovii L. **- gronovius hawkweed** **HIGR3**

Plant: Upright hairy perennial, 0.3-1.5 m tall, commonly single stemmed with mainly basal leaves, from a rootcrown or short rhizome with fibrous roots.

Stem: Slender to stout, dense long hairy basally and decreasing in density upward.

Leaves: Mostly basal and alternate on lower stem, leaves absent on upper stem, broadly oblanceolate to obovate, 4-20 cm long and 1.2-5 cm wide and much smaller upward, hairy on both surfaces, margins slightly toothed or entire, base wedge-shaped or tapering becoming sessile, somewhat clasping upward.

Flowers: Jul-Nov. Heads at ends of numerous branched stalks in a panicle, petals yellow, slightly extending from a tight cylindric involucre, 6-8 mm long and 3-5 mm wide, light green.

Fruit and seeds: Aug-Dec. Nutlet (achene), 2.5-4 mm long, topped with tan bristles, 4-6 mm long, nutlet containing 1 seed.

Range: Native. TX to FL and north to MA and west to MI and KS.

Ecology: Persists after disturbance to occur in new forest plantations, especially dry sandy sites. Occurs along forest edges, right-of-ways, and in pastures. Persists and colonizes by rhizomes and spreads by animal- and wind-dispersed seeds.

Synonyms: hairy hawkweed, queendevil.

Other species in the SE: *H. aurantiacum* L. [orange hawkweed or devil's paint- brush], *H. caespitosum* Dumort. [formerly *H. pratense* Tausch], *H. longipilum* Torr. [hairy hawkweed], *H.* x *marianum* Willd. (pro sp.) [*gronovii* x *venosum*], *H. paniculatum* L., *H. pilosella* L. [mouse-ear], *H. piloselloides* Vill. [formerly *H. florentinum* All, king devil], *H. scabrum* Michx., *H. venosum* L. [rattlesnake-weed].

Wildlife: White-tailed Deer reportedly eat the seed heads of hawkweeds during summer and the leaves during winter, at least in minor quantities. Seed heads also are consumed in minor amounts by Ruffed Grouse and Wild Turkey.

Hieracium **Hawkweed** **Asteraceae**

T. Bodner (Aug)

Hieracium gronovii

T. Bodner (Aug)

Hieracium gronovii

Hypericum **St. Johnswort or Pineweed** **Clusiaceae**

Annual and perennial herbs and shrubs (see *Hypericum* in shrub section), with smooth opposite, whorled, or fascicled leaves of medium to small size, covered with tiny black dots (requires hand lens). Flowers in upper leaf axils, forming a flat-topped cyme, petals only 4 or 5, yellow (some pink). Fruit an ovoid capsule that splits lengthwise to release numerous cylindrical or oblong seeds. 30 species in the SE.

Common Species

Hypericum gentianoides (L.) B.S.P. **- pineweed** **HYGE**

Plant: Small upright annual herb, 10-40 cm tall, densely branched with scale-like inconspicuous leaves, from a taproot. Commonly occurs in colonies on new forest plantations. Resembles a pine seedling (thus the common name). Plant toxic to livestock.
Stem: Slender, smooth, many green and wing-angled lateral branches from a lustrous brown main stem.
Leaves: Opposite, tiny and scale-like, 1-5 mm long, appressed and ascending.
Flowers: Jul-Oct. Tiny, solitary, terminal and alternate on branches, petals 5, yellow, 1.5-4 mm long and 4-8 mm wide.
Fruit and seeds: Aug-Dec. Slender capsules, cone-shaped, 4-6 mm long, splitting lengthwise to release numerous seeds, 0.4-0.6 mm long, yellow to brown.
Range: Native. TX to FL and north to MA and west to Ont and MN.
Ecology: Common primary invader of disturbed soils, especially sandy soils, such as new forest plantations, forest roadsides, and right-of-ways. Forms loose to dense small colonies. Occurs on moist to dry sites and open habitat. Spreads by seeds, apparently remaining viable in soil for long periods.
Synonyms: *Sarothra gentianoides* L., orangegrass.

Wildlife: Hypericums generally are poor White-tailed Deer browse plants. Seeds of some species (not *H. gentianoides*) are consumed in small quantities by Northern Bobwhite and seed-eating songbirds.

Hypericum St. Johnswort or Pineweed Clusiaceae

T. Bodner (Aug)

Hypericum gentianoides

T. Bodner (Aug)

Hypericum gentianoides

Ipomoea **Morning-glory** **Convolvulaceae**

Annual or perennial herbaceous vines (or shrub, *I. carnea*), with alternate leaves. Flowers in leaf axils, funnelform with petals joined, corolla margin entire or only shallowly lobed, blooming mainly in the morning (thus the common name). Flowers with only one pistil. Fruit a spherical 2-4 valved capsule with 2-6 seeds. 16 species in the SE.

Common Species

Ipomoea pandurata (L.) G.F.W. Mey. **- bigroot morning-glory** **IPPA**

Plant: Trailing or slightly twining, perennial vine, 1-3 m long, from a large tuberous root (thus the common names).
Stem: Slender to stout, twining, hairless and smooth, purplish, branched at leaf axils.
Leaves: Alternate, heart- to arrowhead-shaped, 4-8 cm long and 3-8 cm wide, margins entire to wavy, hairy beneath; margin, midvein, and petiole often purplish.
Flowers: May-Jun. Clusters of 1-5 on short stalks in leaf axils, funnel-shaped, 6-8 cm long and wide, white with lavender inner throat, sepals 12-15 mm long.
Fruit and seeds: Jul-Oct. Ovoid capsule, about 1 cm long, brown, splitting at maturity to release many seeds, orange-slice shaped, hairy on angles.
Range: Native. TX to FL and north to CT and west to Ont, MI, and NE.
Ecology: Commonly occurs as scattered plants along right-of-ways and present in new forest plantations. Persists by large deep tuber and spreads by seeds.
Synonyms: man of the earth, wild potato vine, man-root.

Ipomoea purpurea (L.) Roth **- tall morning-glory** **IPPU2**

Plant: Trailing or climbing annual vine, 2 m or more long, infrequently branched, from a taproot.
Stem: Slender to stout, twining and climbing, hairy.
Leaves: Alternate, heart-shaped with distinct tip, 4-12 cm long or wide, margins entire, petioles 3-6 cm long, hairy.
Flowers: Jul-Sep. Clusters of 1-5 on long stalks in leaf axils, funnel-shaped, blue to purple (to white or variegated), 4-6 cm long and wide, sepals lance-elliptic, 8-15 mm long, hairy.
Fruit and seeds: Aug-Oct. Spherical capsule with hair beak, about 1 cm wide, brown, splitting at maturity to release many seeds, orange-slice shaped, dark brown to black, 4.5 mm long.
Range: Introduced from tropical America. TX to FL and north to ME and west to Ont, WI, and NE.
Ecology: Invader of disturbed places, such as new forest plantations, forest openings, and right-of-ways. Found as scattered plants. Spreads by animal-dispersed seeds.
Synonyms: *Convolvulus purpureus* L., common morning-glory.

Ipomoea hederacea Jacq. **- ivyleaf morning-glory** **IPHE**

Similar to *Ipomoea purpurea* except **leaves** 3-cleft or lobed (or 5-cleft or entire); **flowers** (Jul-Nov) clusters of 1-3, blue to rose-purple with white inner throat, 2.5-4.5 cm long and wide, sepals tapering to long recurved tips, 12-24 mm long. **Range:** Introduced from tropical America. AZ to FL and north to Ont and west to ND.

T. Bodner (Jul)

Ipomoea purpurea

T. Bodner (Jun)

Ipomoea pandurata

B. Craft (Jul)

Ipomoea hederacea

Ipomoea **Morning-glory** **Convolvulaceae**

Ipomoea cordatotriloba Dennst. **- tievine** **IPCO8**
Similar to *Ipomoea purpurea* except **leaves** with 2 basal lobes or entire, 4-9 cm long and 2-5 cm wide; **flowers** (Sep-Oct) clusters of 1-3, pink to purple with darker inner throat, 2.8-4.5 cm long and wide, sepals similar. **Range:** Native. TX to FL and north to NC, mainly in Cp. **Synonyms:** *I. trichocarpa* Ell., sharppod morning-glory.

Ipomoea coccinea L. **- red morning-glory or redstar** **IPCO3**
Similar to *Ipomoea purpurea* except **plant and stem** hairless; **leaves** more angular heart-shaped, 4-9 cm long and 2-5.5 cm wide, lower margins often weakly toothed; **flowers** (Aug-Nov) crimson or red and trumpet-shaped, 2.2-3 cm long and 1-2 cm wide, stamens and pistil exserted, sepals 4-7 mm long; **fruit and seeds** ovoid capsule, 5-6 mm wide and recurved. **Range:** Native. TX to FL and north to NY and west WI and NE.

Ipomoea quamoclit L. **- cypressvine** **IPQU**
Similar to *Ipomoea coccinea* except **leaves** divided into many linear segments, 1 mm or less wide; **flowers** (Sep-Nov) petals also crimson to red, 2.7-4 cm long. **Range:** Introduced from tropical America. TX to FL and north to MD and west to MO.

Wildlife: The morning-glories are little used by White-tailed Deer. The large seeds are taken infrequently by Northern Bobwhite and seed-eating songbirds. Flowers are used by some of the larger butterflies such as swallowtails and fritillaries and the Ruby-throated Hummingbird.

B. Craft (Sep)

Ipomoea cordatotriloba

B. Craft (Sep)

Ipomoea quamoclit

T. Bodner (Sep)

Ipomoea coccinea

Jacquemontia **Jacquemontia** **Convolvulaceae**

Annual or perennial vines (herbs or upright shrubs in other regions). Leaves with entire margins, rarely toothed or lobed. Flowers blue or white in heads, cymes, or corymbs and all with 2-lobed stigma. Fruit a capsule with 4 or 8 sections. 4 tropical species in south FL, in addition to *Jacquemontia tamnifolia*.

Common Species

Jacquemontia tamnifolia (L.) Griseb. **- smallflower morning-glory** **JATA**

Plant: Annual trailing vine, up to 2 m long, with young plants upright to 30 cm tall, often branched near base, from a shallow taproot with lateral roots along the entire length.

Stem: Stout, trailing or twining (or upright when young), hairy and greenish-gray.

Leaves: Alternate, heart shaped (cordate), 3-12 cm long and 2-9 cm wide, prominent indented center and lateral veins, margins entire and hairy, soft hairy above and beneath, petioles 1-6 cm long, hairy.

Flowers: Jun-Nov. Dense clusters on long flower stalks in leaf axils, rounded hairy heads, 1-5 cm wide, corolla funnel-shaped, blue, 12-16 mm long (often closed in mid-day), sepals very hairy with numerous small to large leafy bracts beneath. Flowers in the same head do not bloom at the same time.

Fruit and seeds: Jul-Dec. Spherical capsule, 4-6 mm wide, light brown, splitting at maturity to release 4 seeds, 3 mm long and 2 mm wide, light brown and dull.

Range: Migrating north from tropical America. TX to FL and north to VA and west to KY and OK.

Ecology: Primary invader of disturbed soils, such as new forest plantations, log decks, and croplands, decreasing in abundance after first year. Troublesome mid-summer invader of intensive tree farms. Spreads by seeds that persist in soil.

Synonyms: tievine, false morning-glory, hairy clustervine.

Other species in the SE: only those in tropical FL.

Wildlife: Jacquemontia seeds are taken infrequently by Northern Bobwhite and likely seed-eating songbirds.

Jacquemontia Jacquemontia **Convolvulaceae**

T. Bodner (Aug)

Jacquemontia tamnifolia

T. Bodner (Aug)

Jacquemontia tamnifolia

T. Bodner (Aug)

Jacquemontia tamnifolia

Kummerowia **Kummerowia** **Fabaceae**

Low growing and spreading annuals with 3-leaflet leaves having distinct parallel-striate veins and persistent stipules. Tiny flowers, solitary in upper leaf axils, pea-type, bicolor pink-purple. Legume (pods) within a persistent calyx, sessile, not splitting, 1 seeded. Differs from *Lespedeza*, where previously listed, by annual habit, conspicuous stipules, striate leaflets, and solitary flowers. 2 species in the SE.

Common Species

Kummerowia striata (Thunb.) Shindl.- **Japanese clover** **KUST2**

Plant: Ascending to reclining to prostrate annual, 10-40 cm long or tall, often forming mats, from a semi-woody taproot with diffuse laterals.

Stem: Reclining stems, many branched, brownish, white downwardly appressed or recurved hairy.

Leaves: Alternate, 3 leaflet, leaflets obovate to elliptic or oblong, 0.5-1.4 cm long and 2-6 mm wide, hairless, margins entire, petioles 1-2 mm long, persistent stipules, 4-6 mm long, light brown with lighter nerves.

Flowers: Jul-Sep. Solitary along stem in upper leaf axils, not aggregated, pink to pinkish purple, 4-6 mm long, calyx 2.5-3 mm long.

Fruit and seeds: Aug-Nov. Flat legume (pod), 3-4 mm long, tip pointed, partially covered by calyx containing 1 seed, seed released within pod.

Range: Native to Asia and widely planted in US. TX to FL and north to NJ and west to KS.

Ecology: Nitrogen fixer. Occurs in new forest plantations, open forests, pastures, old fields, and right-of-ways, and also in lawns and urban waste places. Forms mats and grows in small patches. Spreads by animal-dispersed seeds.

Synonyms: *Lespedeza striata* (Thunb.) Hook. & Arn., annual lespedeza, common lespedeza.

Kummerowia stipulacea (Maxim.) Makino **- Korean clover** **KUST**

Similar to *Kummerowia striata* except **plant** reclining or ascending, basally branched, 10-20 cm long or tall; **stem** upwardly appressed hairy; **leaves** spatulate to obovate, tip notched, margins and lower midvein hairy, petioles 4-10 mm long; **fruit and seed** legume (pod) rounded tipped. **Range:** Native to Asia and widely planted in US. TX to FL and north to PA and west to IA.

Synonyms: *Lespedeza stipulacea* Maxim., kobe lespedeza, Korean lespedeza.

Wildlife: Along with *Lespedeza*, kummerowias are likely the most important seed producers for Northern Bobwhite in the Southeast. The seeds and foliage also are consumed by Ruffed Grouse, Wild Turkey and Eastern Cottontail. Interestingly, the seeds apparently are only little used by seed-eating songbirds. Kummerowias are a fairly common summer White-tailed Deer forage of moderate to high preference.

Kummerowia **Kummerowia** **Fabaceae**

T. Bodner (Sep)

Kummerowia striata

T. Bodner (Aug)

Kummerowia striata

Lactuca **Lettuce** **Asteraceae**

Annual, biennial, or perennial herbs with erect leafy stems, having white or brown latex sap sometimes at injuries. Leaves spirally alternate, unlobed to pinnately toothed and sometimes spiny-margined. Flower heads in erect panicles, heads cylindric or tapering towards the apex, petals yellow, blue, or rarely white. Seeds in pappus-plumed nutlets (achenes) being largest at pappus attachment. 10 species in the SE.

Common Species

Lactuca canadensis L. **- tall lettuce** **LACA**

Plant: Upright leafy-stemmed annual or usually biennial, 1-3 m tall, with an earlier leafy rosette, from a taproot.

Stem: Stout, leafy, scattered long hairy or hairless, slightly white waxy, green becoming reddish.

Leaves: Alternate to appearing whorled, well distributed along the stem, oblanceolate to lanceolate, 10-35 cm long and 1.5-12 cm wide and smallest upward, margins entire to toothed and pinnately lobed or cleft, green with white midvein, bases tapering on lower leaves and clasping on upper leaves.

Flowers: Jun-Nov. Cylindric heads in spike-like or very broad panicles, 30-60 cm long and 8-50 cm wide, orange-yellow flowers slightly extending from an urn-shaped involucral cylinder, 10-14 mm long and 4-6 mm wide, green or tinted with purple.

Fruit and seeds: Jul-Dec. Flattened nutlet (achene), 3-4 mm long and 4-6 mm wide, brown-black, 1-ribbed, topped with white bristles, 4-7 mm long, nutlet containing 1 seed. Bracts reflex to release bristle-plumed seeds.

Range: Native. TX to FL and north to Canada and west to MI and MO.

Ecology: Early invader of disturbed soils and occurs as scattered plants in young forest plantations and as small groups along forest roads and right-of-ways. Spreads by wind- and animal-dispersed seeds.

Synonyms: wild lettuce, Canada lettuce.

Other species in the SE: *L. biennis* (Moench) Fern., *L. floridana* (L.) Gaertn., *L. graminifolia* Michx., *L. hirsuta* Muhl. *ex* Nutt., *L. ludoviciana* (Nutt.) Riddell, *L. saligna* L., *L. sativa* L., *L. serriola* L., *L. tatarica* (L.) C.A. Mey.

Wildlife: The seeds of the various species of wild lettuce are occasionally eaten by songbirds and small mammals. In some areas of the Southeast, wild lettuce is a common late-spring and early-summer White-tailed Deer forage.

Lactuca **Lettuce** **Asteraceae**

T. Bodner (Apr)

Lactuca canadensis

T. Bodner (May)

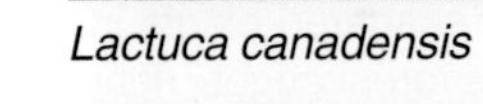

Lactuca canadensis

T. Bodner (Jun)

Lactuca canadensis

J. Miller (Jun)

Lactuca canadensis

Lepidium **Lepidium or Pepperweed** **Brassicaceae**

Winter and summer annual herbs, occasionally biennials, from basal rosettes that wither at maturity. Stems erect to ascending with linear, elliptic, or deeply dissected leaves. Flowers small, petals white or yellow, less than 1 mm long, often lacking, occurring clustered at the end, or spiraling along, an elongated flower-seed stalk. Seed pods small, many, flattened, nearly circular, notched at apex, and containing 2 tiny, obovoid seeds. 6 species in the SE.

Common Species

Lepidium virginicum L. **- Virginia pepperweed** **LEVI3**

Plant: Erect to ascending, winter or summer annual (occasionally a biennial), 2-50 cm tall, much branched at maturity from a leafy rosette having a slender taproot. Young leaves and capsules with peppery taste (thus the common name).

Stem: Slender, single to much branched, smooth to slightly hairy, green.

Leaves: Basal rosette (absent on mature plants) and alternate on stem, 2-10 cm long and 0.5-2 cm wide and smaller upward, basal leaves doubly serrate to lobed or deeply dissected often to the midvein and stem leaves linear to lanceolate and coarsely toothed or irregularly lobed, tapering base.

Flowers: Apr-Jun. Spiraling along an elongate bottlebrush-like raceme, 4-10 cm long, petals 4 (or absent), 3-3.5 mm long, white to green, sepals 4. Often flowers above and seed pods below.

Fruit and seeds: Jun-Nov. Pod nearly-circular, flat, notched at tip, 3-4 mm long, light-green to white, containing 2 seeds, oval, 1.5 mm long, minutely winged, peppery taste.

Range: Native. Throughout US (except AZ, NM, WY, and w MT) and into Mexico and Canada.

Ecology: Occurs as scattered plants in new forest plantations and persists under pine canopies, also inhabits forest edges, right-of-ways, fields, and disturbed sites. Spreads by seeds dispersed by animals, wind, and water.

Synonyms: poor-man's pepper, pepper-grass, Virginian peppercress, bird's pepper, tongue-grass.

Other species in the SE: *L. campestre* (L.) R. Brown, *L. densiflorum* Schrad., *L. lasiocarpum* Nutt., *L. perfoliatum* L., *L. ruderale* L.

Wildlife: Although common throughout the Southeast, both the seeds and foliage of the pepperweeds are used infrequently by wildlife.

Lepidium Lepidium or Pepperweed Brassicaceae

J. Miller (Jun)

Lepidium virginicum

T. Murphy (Jun)

Lepidium virginicum

J. Miller (Jun)

Lepidium virginicum

Lespedeza **Lespedeza** **Fabaceae**

Perennial leguminous herbs (and shrubs), with 3-leaflet leaves, each leaflet broadly ovate to narrowly oblong having entire margins. Leaves with pair of stipule bracts at the petiole base. Few to many small pea-type flowers, often in terminal or axillary stalked clusters, with basal bracts. Flowering in late summer and early fall (except for *L. repens*). Flowers violet, purplish, roseate, yellowish, or whitish. Fruit a small 1-seeded legume (pod) often in clusters, flattened, rounded to elliptic, and not splitting to release the seed. 13 native species and 3 Asiatic introduced species in the SE, with numerous hybrids. Japanese and Korean lespedezas now classified in the genus *Kummerowia*.

Common Species

Lespedeza virginica (L.) Britt. **- slender lespedeza** **LEVI7**

Plant: Ascending to upright perennial, 30-80 cm tall, often numerous stems, from a persistent rootcrown and often in colonies.

Stem: Single to several, wand-like, often branched above mid-plant, leaning and spreading, moderately hairy in lines.

Leaves: Alternate, crowded and numerous, 3 leaflet, leaflets oblong to linear, 1.5-3 cm long and 5-10 mm wide, densely hairy below, tip pointed, petioles hairy, 5-15 mm, stipules narrowly linear.

Flowers: Jul-Oct. Crowded clusters of 4-10 flowers in upper leaf axils, shorter than leaves, pea-type, 4-7 mm long, pinkish to purple with white marks, calyx hairy, 5-lobed, lobes 1.2-2.4 mm long, shorter than petals.

Fruit and seeds: Aug-Feb. Flatten legume (pod), clustered in terminal axils and scattered along the stem, ovate to rounded, 3.5-7 mm long (longer than persistent calyx), hairy on margins, containing 1 seed.

Range: Native. TX to FL and north to MA and west to Ont, WI, and KS.

Ecology: Nitrogen fixer. Persists in new and older forest openings, dry upland forests to moist savannas, old fields, right-of-ways, and urban disturbed sites. Persists, colonizes, and spreads by animal-dispersed seeds.

Lespedeza cuneata (Dum.-Cours.) G. Don **- Chinese lespedeza** **LECU**

Similar to *Lespedeza virginica* except **plant** 0.8-2 m tall; **stem** often gray-green with lines of hairs; **leaves** 3 leaflet, often gray-green, leaflets oblanceolate, 0.8-1.5 cm long and smaller upward, blunt tipped, lower petioles 1-2.5 cm long, upper leaves without petioles; **flowers** 2-3 in clusters, whitish with purple marks; **fruit and seeds** (Oct-Nov) pods 3-4 mm long. **Range:** Introduced from Asia and planted extensively for erosion control, along roadsides, and as a forage crop. Throughout Eastern US, except DL and NJ.

Ecology: Nitrogen fixer. Plantings can form dense colonies in open fields, roadsides, and right-of-ways. Invades nearby forests after disturbance.

Synonyms: sericea lespedeza.

Lespedeza hirta (L.) Hornem. **- hairy lespedeza** **LEHI2**

Similar to *Lespedeza virginica* except **plant** 0.8-1.5 m tall; **stem** long hairy (dense or scattered); **leaves** leaflets vary greatly in shape, round to ovate-elliptic or obovate, 1-4 cm long, with petioles of middle leaves 1-1.5 cm long and those of upper leaves, short to none; **flowers** (Aug-Oct) 10-20 in clusters, longer than leaves, whitish to cream yellow, with a hairy calyx having lobes much longer than the petals; **fruit and seeds** (Sep-Mar) legume (pod), 5-8 mm long, shorter than the lobes of the long calyx. **Range:** Native, with 2 varieties in the SE. Throughout the Eastern US.

Ecology: Nitrogen fixer. Persists in new and older forest openings, old fields and right-of-ways. Spreads by animal-dispersed seeds.

Lespedeza Lespedeza Fabaceae

T. Bodner (Aug)

Lespedeza virginica

J. Miller (Oct)

Lespedeza virginica

J. Miller (Jul)

Lespedeza cuneata

T. Bodner (Sep)

Lespedeza virginica

T. Bodner (Oct)

Lespedeza hirta

J. Miller (Jul)

Lespedeza cuneata

Lespedeza **Lespedeza** **Fabaceae**

Lespedeza procumbens Michx. **- trailing lespedeza** **LEPR**

Plant: Trailing or slightly ascending perennial, 30-80 cm long, with numerous stems, from a slender woody root and forming small patches or colonies.
Stem: Slender, frequently branched, downy short hairy, green becoming brown basally.
Leaves: Alternate, 3 leaflet, leaflets ovate to oblong to elliptic, 0.8-2.5 cm long and 0.5- 1.6 cm wide, tip and base rounded, downy short hairy both surfaces and petioles.
Flowers: Jun-Sep. Ascending slightly-branched racemes, terminal or axillary, 4-10 cm long, pea-type, 5-8 mm long, purplish to rosy-purple.
Fruit and seeds: Aug-Dec. Flat legume (pod), round to elliptic, 4-6 mm long, short hairy.
Range: Native. Throughout SE.
Ecology: Nitrogen fixer. Occurs on moist to dry sites and open to shady habitat. Forms colonies in new forest plantations, forest openings, and forest roads and larger right-of-ways. Scattered plants occur in the pine needle litter under developing stands. At times planted for wildlife food and cover. Persists by rootcrowns and spreads by animal-dispersed seeds.

Lespedeza repens (L.) W. Bart. **- creeping lespedeza** **LERE2**

Similar to *Lespedeza procumbens* except **stem** sparsely to densely upward appressed hairy or not; **leaves** smaller, leaflets elliptic to obovate or oblong, 1-2 cm long and 0.5-1 cm wide, fine appressed hairy below and sometimes above, stipules 2-4 mm long, petioles slightly hairy; **flowers** (May-Oct) racemes with 2-6 loosely spaced pink-lavender to violet flowers, 5-8 mm long; **fruit and seeds** (Jun-Dec) flat legume (pod), broadly oval to round, 4-6 mm long, appressed hairy. **Range and ecology:** same as above.

Lespedeza bicolor Turcz. **- shrubby lespedeza or bicolor** **LEBI2**

Plant: Perennial, ascending shrub, 1-3 m tall, much branched, from woody rootcrown.
Stem: Upright stems, 0.5-2 cm in diam, appressed hairy to hairless, often gray-green.
Leaves: Alternate, 3 leaflet, leaflets elliptic to ovate with a distinct tip, 2-6 cm long and 1- 3 cm wide, petioles 2-4 cm, lower surface lighter green, stipules narrowly linear, 1-8 mm long.
Flowers: Jun-Sep. Raceme of 5-15 well-spaced flowers in upper axils and exceeding the upper leaves, raceme 10-40 cm long, subtended by an ovate bract, 0.5-1.5 mm long, pea-type, rosy-purple to white (bicolor), 8-11 mm long, calyx sparsely to very hairy, lobes 2.5-4.5 mm long with lower lobe 1 mm longer.
Fruit and seeds: Aug-Nov. Flat legume (pod), broadly elliptic, 6-8 mm long, densely appressed hairy.
Range: Introduced from Japan. Widely planted throughout SE and escaping.
Ecology: Nitrogen fixer. Initially planted and becoming invasive in adjacent forests. Forming dense colonies even under shade. Promoted by burning.

Other species in the SE: *L. angustifolia* (Pursh) Ell., *L. capitata* Michx., *L. cyrtobotrya* Miq. [VA only], *L. frutescens* (L.) Hornem., *L. stuevei* Nutt., *L. thunbergii* (DC.) Nakai, *L. violacea* (L.) Pers.

Wildlife: The lespedezas likely are the most important Northern Bobwhite seed producers in the Southeast. Seeds apparently are only little used by other seed-eating songbirds. *Lespedeza bicolor* and *L. thunbergii* have been planted extensively as a Northern Bobwhite forage and cover plant. *Lespedeza cuneata*, another widely-planted introduced plant, is somewhat less palatable but provides extensive nesting and foraging cover. Several of the lespedezas are preferred White-tailed Deer forages while others, such as *L. cuneata*, *L. violacea* and *L. virginica*, are consumed less frequently.

Lespedeza **Lespedeza** **Fabaceae**

J. Miller (Sep)

Lespedeza procumbens

T. Bodner (Jun)

Lespedeza procumbens

J. Miller (Oct)

Lespedeza repens

J. Miller (Oct)

Lespedeza repens

J. Miller (Jul)

Lespedeza bicolor

T. Bodner (Aug)

Lespedeza bicolor

Liatris **Blazingstar or Gayfeather** **Asteraceae**

Slender perennial herbs, from corm-like rootstocks, with one to several arching stems in late summer. Leaves alternately spiraling, simple, narrow and sessile, covered with small sunken glands. Flowers mostly in terminal racemes and spikes, heads small to fairly large, pink or pink-purple to whitish, tubular with frilly petals. Fruit a plumed, hairy nutlet (achene), blackish and tapered from base to apex with many ridges. 22 species in the SE.

Common Species

Liatris pilosa (Ait.) Willd. - **shaggy blazing star** **LIPI7**

Plant: Upright, arching, or leaning, slender perennial, 0.8-1.5 m tall, rarely branched, from a spherical corm with fibrous roots at base.

Stem: Stout below and becoming slender above, green to brownish, faint whitish lengthwise stripes, rough hairy to hairless.

Leaves: Alternately spiraling, numerous, linear with tapering tip, 6-20 cm long and 2-7 mm wide and much smaller upward, sessile, covered with sunken glands (requires hand lens), slightly rough.

Flowers: Sep-Nov. Numerous stalked heads (only 2-4 flowers per head) spaced and spiraling along an elongated spike, 10-60 cm long, violet to purple, frilly thin petals, exserted branched styles, violet to purple, bristles tan to purplish, involucre 7-12 mm long, rows of overlapping rounded green bracts, some with purplish margins.

Fruit and seeds: Nov-Jan. Nutlet (achene) 2.5-3 mm long, hairy with ribs, topped with a feather-like plume, 4-5.5 mm long, purplish to tan, nutlet containing 1 seed.

Range: Native. AL to nw FL and north to NJ, mainly in Cp and lower Pd.

Ecology: Occurs as scattered plants in open forests, forest openings, plantation forests, and right-of-ways. Persists by corms and spreads by wind- and animal-dispersed seeds. Several varieties.

Synonyms: *L. graminifolia* Willd., *Lacinaria graminifolia* (Willd.) Kuntze, purple featherflower.

Liatris elegans (Walt.) Michx. - **pinkscale blazing star** **LIEL**

Similar to *Liatris pilosa* except **stem** green, brown, or purple and hairy; **leaves** narrowly oblanceolate, often drooping; **flowers** whitish (1-3 per head), enclosed within showy involucral bracts, white-hairy with upper margins pinkish-white, 15-25 mm long, elongated and twisting, tips tapering, lower bracts green; **fruit** nutlet (achene) 4-6 mm long, pappus-plume, 6-8 mm long, purplish to tan. **Range:** TX to FL and north to VA and west to MO.

Ecology: Scattered plants in open forests, young forest plantations, and along forest right-of-ways. Persists by corms and spreads by seeds. Several varieties.

Synonyms: *Lacinaria elegans* (Walt.) Kuntze, pinkscale gayfeather, white featherflower.

Wildlife: *Liatris* are low preference White-tailed Deer forages. No other wildlife use is reported, although seeds likely are taken occasionally by songbirds.

Liatris **Blazingstar or Gayfeather** **Asteraceae**

J. Miller (Oct)

Liatris pilosa

J. Miller (Oct)

Liatris elegans

Lobelia **Lobelia** **Campanulaceae**

Erect perennial or annual herbs, with leafy stems, and flowers crowded along and projecting out from a terminal stalk (raceme), each flower from a small leafy bract. Corolla tubular-like and split along the sides to the base, upper lip 2-lobed and lower lip 3-lobed; red, blue, purple, lavender, or white. Stamens 5, with anthers usually colored, ringing the style, and lower 2 having tufts of hairs. Fruit a capsule, 2 chambered, opening at the top, with yellowish-brown seeds, oblong, 0.6-1 mm long. 12 species in the SE.

Common Species

Lobelia cardinalis L. **- cardinalflower** **LOCA2**

Plant: Upright perennial, 0.5-2.5 m tall, 1-several stems, from a rootcrown. Found growing by water.

Stem: Stout, green, smooth or short hairy, often not branching or branching just below the flower, pith white.

Leaves: Alternate, lanceolate to elliptic, 5-20 cm long and 2-5.5 cm wide and smaller upward, acute tipped and tapering base, margins serrate to dentate with large and small teeth.

Flowers: Jul-Oct. Deep red flowers crowded along and around a terminal stalk (raceme), 10-30 cm long, with small bracts below each flower, tubular petals, 1.7-2.2 cm long in halves, upper 2-lobed and lower 3-lobed, white hairy tufts on joined stamen-column very visible, sepals 5, spike-shaped, 6-22 mm long. Flowering progressing from base to apex.

Fruit and seeds: Nov-Feb. Capsule, spherical to ovoid, 0.8-1.2 cm wide, opening at the top to release many seeds, 0.7-1 mm long.

Range: Native. TX to FL and north to New Brunswick and west to s Ont and MN.

Ecology: Occurs along forested streams, rivers, ponds, and springs. Persists by rootcrowns and spreads by water-dispersed seeds.

Lobelia puberula Michx. **- downy or blue lobelia** **LOPU**

Similar to *Lobelia cardinalis* except **plant** grows in uplands and lowlands, 0.4-1.2 m tall; **stem** slender, short hairy; **leaves** elliptic to oblanceolate, 3-12 cm long and 1-4 cm wide and smaller upward, slightly hairy, margins coarsely to finely dentate to serrate; **flowers** (Jul-Nov) spaced along a terminal stalk (a raceme) and to one side, bright blue to violet, 7-15 mm long, often hairy except for inside lower petal, sepals 5-10 mm long, entire or sharp-tooth margins; **fruit and seeds** (Nov-Jan) capsule 6-8 mm wide. **Range:** TX to FL and north to NJ and west to OH.

Ecology: Occurs as scattered plants in forest openings and plantations, along forest edges and right-of-ways, both upland and lowland sites. Present on moist to dry sites and open to shady habitats. Persists by rootcrowns and spreads by seeds.

Lobelia brevifolia Nutt. *ex* A. DC. **- shortleaf or wavy-leaf lobelia** **LOBR3**

Similar to *Lobelia puberula* except **plant** only in Cp, 30-80 cm tall, with short hairs; **stem** slender, sometimes branching, hairy; **leaves** fleshy, linear to spatulate, 0.5-2.5 cm long and 3-8 mm wide, margins dentate with callous-tips (often black); **flowers** (Jul-Nov) few, 3-20, widely spaced, blue to lavender, 7-15 mm long, hairy on all surfaces, sepals entire or with callous-tipped teeth; **fruit and seeds** (Nov-Jan) capsule 5-7 mm broad. **Range:** LA to n FL, only Cp.

Ecology: Occurs in new open forests, forest opening, and along forest roadsides.

Wildlife: Cardinalflower and several other species of lobelia are toxic. Nevertheless, several species of lobelia are listed as moderate preference White-tailed Deer browses. Lobelia are common nectar sources for the Spicebush Swallowtail, Pipevine Swallowtail, and Dogface butterflies and the Ruby-throated Hummingbird.

Lobelia Lobelia Campanulaceae

J. Miller (Sep)

Lobelia puberula

T. Bodner (Aug)

Lobelia cardinalis

J. Miller (Sep)

Lobelia puberula

J. Miller (Oct)

Lobelia brevifolia

Ludwigia Primrose-willow or Ludwigia Onagraceae

Erect or reclining short-lived perennial (rarely annual) herbs, some shrubby in appearance, and usually found on wet soils. Leaves alternate or opposite and entire or shallowly toothed, not divided. Flowers, usually yellow or lacking petals, solitary in upper leaf axils, sessile or short-stalked, with 4-7 petals and sepals. Fruit a 4-chambered capsule, often 4-sided or cylindrical or spherical, with 4 persistent sepals around the top. Capsules split along sides or top releasing many small yellowish seeds, rarely reddish. 23 species in the SE.

Common Species

Ludwigia alternifolia L. - **seedbox** LUAL2

Plant: Upright, branched perennial to 1.2 m tall, shrub-like, from corm-like roots, usually found in wet to moist soils.
Stem: Stout, smooth to minutely hairy and often slightly wing-angled, green to brown to tan.
Leaves: Alternate, elliptic or linear-lanceolate, 3-12 cm long and 0.8-2 cm wide, gradually tapering to the base, veins parallel the margins, often light green with whitish midvein above and finely scattered hairy especially on veins beneath.
Flowers: May-Oct. Solitary in upper leaf axils, on stalks 3-5 mm long having 2 small leaf-like bracts to 3 mm long, petals 4, yellow, 3-6 mm long, sepals 4 of equal length.
Fruit and seeds: Nov-Mar. Capsule 4-angled or winged but rounded at base, 4-5 mm long and 4-6 mm wide, with the 4 flared sepals on the top edges, opening by terminal pore releasing numerous seeds, ovoid, about 0.5 mm long, light brown.
Range: Native. TX to FL and north ME and west to MI, IA, and KS.
Ecology: Inhabits wet soils in new forests, right-of-ways, and older forest openings. Seeds spread when plant is shaken by wind and possibly carried by water.

Ludwigia decurrens Walt. - **winged primrose-willow** LUDE4

Similar to *Ludwigia alternifolia* except **plant** only to 1 m tall; **stem** 4-angled and usually winged; **leaves** attached to stem, lanceolate, to 14 cm long and 2.5 cm wide and smaller upward; **flowers** petals 4-7, yellow, 6-12 mm long and 4-6 mm wide, sepals 4-7, ovate, 5-10 mm long and 2-3 mm wide, no bracts below; **fruit** capsule, 4 sided, tubular to 2 cm long and 3-4.5 mm wide, tipped with 4 flared bracts (sepals), to 1 cm long. **Range:** Native. Common throughout the SE.
Synonyms: bushy waterprimrose, winged waterprimrose.

Wildlife: Species of ludwigia range from very poor (*L. peploides*) to moderate or high (*L. alternifolia*, *L. glandulosa*, and *L. hirtella*) preference White-tailed Deer forages. *Ludwigia leptocarpa* has been reported to be a minor fall food of Wood Ducks in South Carolina. Common Muskrat reportedly consumes the stems of *L. alternifolia* and perhaps other species of ludwigia.

Ludwigia Primrose-willow or Ludwigia Onagraceae

T. Bodner (Aug)

Ludwigia decurrens

J. Miller (Jul)

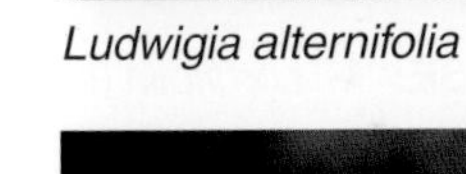

Ludwigia alternifolia

T. Bodner (Aug)

Ludwigia decurrens

T. Bodner (Sep)

Ludwigia alternifolia

Mikania **Hempvine or Mikania** **Asteraceae**

Perennial twining vines, or rarely erect shrubs or herbs. Leaves opposite, simple, heart-shaped or arrowhead-shaped. Flowers in 4's in small heads, all tubular and perfect, petals white to pink to yellowish, within a narrow involucre of 4 equal bracts. Nutlets (achenes) 5-angled, topped by pappus bristles in a single series. 2 species in the SE.

Common Species

Mikania scandens (L.) Willd. **- climbing hempvine** **MISC**

Plant: Perennial, twining or sprawling, herbaceous vine, to 5 m or longer, often covering vegetation, from a persistent semi-woody rootcrown with diffuse roots below.

Stem: Slender above, round to slightly 4-angled, soft hairy or hairless.

Leaves: Opposite, deeply heart-shaped (cordate) with tapering tips, 3-14 cm long and 2-9 cm wide, margins entire to undulate or few toothed, palmately veined, sparsely hairy, petioles 2-10 cm.

Flowers: Jul-Oct. Flower heads numerous in clusters, 2-8 cm wide, from each leaf axil, 4 flowers per head, petals white or pinkish, about 4 mm long, involucral bracts linear, 4-6 mm long and 1 mm wide or less, margins whitish and whitish hairy.

Fruit and seeds: Nov-Feb. Nutlets (achenes) in tight clusters, green maturing to black, 1.5-2.5 mm long, tapered from base to apex, 4-5 ribbed, resinous dotted, topped with curling series of whitish barbed bristles, 2.5-3 cm long, very fuzzy, conspicuous, nutlet containing 1 seed.

Range: Native. TX to FL and north to ME and west to MI and MO.

Ecology: Inhabits wet soils in new forests, right-of-ways, and older forest openings. Seeds spread by water, animals, and wind.

Synonyms: *M. angulosa* Raf., mikania vine, climbing hempweed.

Other species in the SE: *M. cordifolia* (L. f.) Willd. [found mainly in s FL and but also in s LA].

Wildlife: No wildlife value reported.

Mikania **Hempvine or Mikania** **Asteraceae**

T. Bodner (Sep)

Mikania scandens

J. Miller (Oct)

Mikania scandens

T. Bodner (Sep)

Mikania scandens

Mimosa — Mimosa or Sensitive plant — Fabaceae

Reclining, perennial, leguminous herbs or shrubs, with alternate bipinnate-compound leaves (twice divided and small leaflets in rows), leaflets often fold together when touched. Flattened and curved prickles (or none) on stems and fruit pods. Flowers rounded to spherical heads in leaf axils, white or pink-purple. Fruit a legume (pod), flattened or tubular. 7 species in the SE.

Common Species

Mimosa microphylla Dry. **- littleleaf sensitive-briar** **MIMI22**

Plant: Reclining perennial legume with fine prickles, vine-like, 1-3 m long, from a woody rootcrown with lateral and fibrous roots.
Stem: Trailing or arching, numerous recurved-flattened prickles, short hairy, light- and dark-green lengthwise stripes.
Leaves: Alternate (leaflets opposite), bipinnately compound (twice divided with leaflets in rows), 4-8 pairs, leaflets oblong-linear, 2-8 mm long, prickles on leafstalks, with stipules and stipels. Leaflets folding slowly when touched or at darkness.
Flowers: Jun-Nov. Spherical flower heads in leaf axils, 1.5-2 cm wide, pinkish-purple petals and bristly stamens, with yellow anthers, flowers on slender stalks, 3-6 cm long.
Fruit and seeds: May-Dec. Prickly pod, 6-12 cm long and 3-5 mm wide, tapering tipped, splitting along 4 sides to release single or clustered seeds.
Range: Native. TX to FL and north to VA and west to KY, rare in Lower Cp.
Ecology: Nitrogen fixer. Occurs in new forest plantations, open forests, and forest openings, and along right-of-ways. Most common and abundant on dry sites with sandy or gravelly surface soils. Persists by rootcrowns and spreads by animal-dispersed seeds.
Synonyms: *Schrankia microphylla* (Dry.) J.F. Macbr., *Leptoglottis microphylla* (Dry.) Britt. & Rose., sensitivebrier, catclaw sensitivebrier, bashful briar. 10 varieties.

Other species in the SE: *M. latidens* (Small) B.L. Turner, *M. nuttallii* (DC.) B.L. Turner, *M. pigra* L. [aggressive exotic invasive], *M. pudica* L., *M. quadrivalvis* L., *M. strigillosa* Torr. & Gray.

Wildlife: Seeds of littleleaf sensitive-briar are used sparingly by Northern Bobwhite and likely seed-foraging songbirds. It is a poor White-tailed Deer forage plant. The foliage is reportedly a common item in the diet of the Gopher Tortoise.

Mimosa **Mimosa or Sensitive plant** **Fabaceae**

T. Bodner (Jun)

Mimosa microphylla

Monarda **Monarda or Horsemint** **Lamiaceae**

Perennials with stolons, or annual herbs, all with square stems and simple, opposite, petioled leaves, having a mint fragrance. Large conspicuous flowers in one to several stacked terminal whorls. Corolla tubular and 2-lipped, an elongated upper lip and a 3-lobed spreading lower lip, subtended by showy bracts, mingled with other linear bracts. Fruit a compound nutlet, cylindric and smooth, splitting into 4 parts at maturity. 7 species in the SE.

Common Species

Monarda punctata L. **- spotted beebalm** **MOPU**

Plant: Upright perennial herb, 0.4-1m tall, simple to branched stems, from a persistent rootcrown.

Stem: Square in cross-section, purplish to brown hairy, with or without branches from upper leaf axils.

Leaves: Opposite, lanceolate to narrowly ovate, 3-9 cm long and 5-17 mm wide, margins finely serrate, tapering tipped, petioles 0.5-2.5 cm long.

Flowers: Jul-Oct. Terminal whorls or heads, 1-3, spaced along and encircling a stalk, showy yellow and purple dotted petals, upper lip hairy and arching, lower lip spreading, sepals tubular with 15 stripes, all subtended by whorls of bristles and leafy bracts, green to pink to lavender to white.

Fruit and seeds: Oct-Nov. Compound nutlet, 1.3-1.5 mm long, ellipsoid, brown to blackish, splitting into 4 parts, each a 1-seeded nutlet.

Range: Native. TX to FL and north to NY and west to IL and KS.

Ecology: Occurs as scattered plants or small groups along forest roads and openings and spreads into disturbed forests. Seeds shaken from plants by wind and probably spread by birds.

Synonyms: dotted monarda, horsemint. 4 subspecies and one with 9 varieties.

Other species in the SE: *M. citriodora* Cerv. *ex* Lag., *M. clinopodia* L., *M. didyma* L., *M. fistulosa* L., *M. media* Willd., *M. russeliana* Nutt. *ex* Sims.

Wildlife: Horsemints are a moderate preference White-tailed Deer forage that is used intensively in heavily populated areas. They are a nectar source for the Pipevine Swallowtail butterfly, particularly *M. fistulosa*.

Monarda **Monarda or Horsemint** **Lamiaceae**

T. Bodner (Aug)

Monarda punctata

Nuttallanthus **Toadflax** **Scrophulariaceae**
(formerly *Linaria*)

Erect annual or biennial herbs, with tall slender stems from basal shoots. Stem leaves alternate and basal leaves opposite, with blades numerous and narrow, sessile, and with entire margins. Flowers along an erect stalk (raceme), corolla violet and white, irregular, tubular with 2-lips, each lobed. The tube with a downward small pointed spur at the base, behind the flower. Fruit a capsule, spherical or cylindrical, with 2 cells that split at the apex. Seeds numerous, angled and wingless. 3 species in the SE, until recently classified under *Linaria*.

Common Species

Nuttallanthus canadensis (L.) D.A. Sutton **- oldfield toadflax** **NUCA**

Plant: Smooth biennial (or winter annual), with leafy basal rosettes of small linear leaves, and erect to ascending flowering stems, 15-70 cm tall.

Stem: Slender and pale green, single upright and several reclining, widely spreading branches from the base.

Leaves: Alternate and scattered on the erect stem, linear, 5-20 mm long and 2-3 mm wide, opposite or whorled on the spreading stems, linear, 3-12 mm long, all hairless.

Flowers: Mar-May (-Aug). Crowded at the end and around a long stalk (raceme), violet to blue, irregular with two lips, the upper lip 2-lobed, the lower lip 3-lobed with 2 white raised spots, a curved elongated spur projecting downward through the calyx, 5-15 mm long including the spur, sepals 5, lanceolate, 2-3 mm long.

Fruit and seeds: May-Jun (-Sep). Capsule nearly spherical, 3-4 mm wide, with 2 cells, splitting at top to release numerous seeds, angular with flat sides, about 0.4 mm long, brown.

Range: Native to Eurasia and naturalized. Throughout US (interestingly not in Canada as sci. name implies).

Ecology: Scattered plants, mainly noticed in the spring, in new plantations, fields, and pastures, and along forest edges and right-of-ways, especially on sandy soils. Spreads by seeds.

Synonyms: *Linaria canadensis* (L.) Chaz, Canada toadflax.

Other species in the SE: *N. floridanus* (Chapman) D.A. Sutton [formerly *Linaria floridana* Chapman], *N. texanus* (Scheele) D.A. Sutton [formerly *Linaria canadensis* var. *texana* (Scheele) Pennell and *L. texana* Scheele].

Wildlife: Toadflaxs are moderate preference White-tailed Deer forages and a larval food plant of the Buckeye butterfly.

Nuttallanthus **Toadflax** **Scrophulariaceae**

J. Miller (Apr)

Nuttallanthus canadensis

J. Miller (Apr)

Nuttallanthus canadensis

T. Murphy (Jun)

Nuttallanthus canadensis

Oenothera **Evening-primrose** **Onagraceae**

Annual, biennial, or perennial herbs with alternate, mostly narrow leaves becoming smaller upward on ascending to spreading stems, all hairy. Small to large flowers from upper leaf axils, commonly yellow (or pink or white), with 4 pointed sepals and 4 broad petals, usually blooming at dusk to early morning and dropping (thus the common name). Fruit a capsule, short to elongated, round to 4-angled in cross-section, 4-chambered, splitting from top to base and releasing numerous seeds. A long calyx tube (deciduous) connects the flower with the developing capsule. 9 species in the SE.

Common Species

Oenothera biennis L. **- common evening-primrose** **OEBI**

Plant: Erect biennial from a first-year's leafy rosette, to 1.5 m tall, branched basally, from a fleshy taproot. Erect leafless stems with erect capsules, persist in early winter.
Stem: Stout, densely to scattered hairy, tannish-green with reddish streaks or dots.
Leaves: Alternately spiraling, ascending or spreading around the stem at close spacings, lanceolate to oblong, up to 15 cm long and 4 cm wide and smaller upward, hairy, acute tip and tapering base, margins with small widely-spaced teeth, petioled below and sessile above.
Flowers: Jun-Oct. Several to many in a terminal cluster (raceme) each from small leafy bracts (later deciduous), 4 broad petals with notched tips, 1-2.5 cm long, bright yellow becoming reddish with age (opening in evening to morning), with yellow anthers extended and visible, sepals 1.5-3 cm long, flower stems 3-6 cm long with the base becoming the capsule.
Fruit and seeds: Oct-Apr. Cylindric capsule, 1.5-4 cm long and 4-5 mm wide, straight or curved, tapering to 4-toothed tip with appressed hairs, gradually splitting from tip downward to base to progressively release numerous seeds, angled, 1.4-2 mm long, reddish.
Range: Native. TX to FL and north to s Canada and west to ND and OK.
Ecology: Resides on right-of-ways and early invader of new forest plantations. Occurs commonly along forest roads, as single plants or small colonies. Spreads by bird-dispersed seeds that can remain viable in the soil for decades.
Synonyms: sundrops, weedy eveningprimrose.

Oenothera laciniata Hill **- cutleaf evening-primrose** **OELA**

Similar to *Oenothera biennis* except **plant** weakly ascending winter annual or biennial with spreading outer branches up to 75 cm long, from a fibrous taproot; **stem** single or many branched from base, prostrate or weakly ascending; **leaves** oblanceolate to elliptic, up to 8 cm long and smaller upward, coarsely toothed to irregularly lobed or deeply incised; **flowers** (Mar-Aug) single in upper leaf axils, 4 broad petals with notched tips, 0.8-2.5 cm long, yellow becoming purplish with age, with bright yellow anthers, sepal lobes 0.6-1.2 cm long, flower stalks 2-2.5 cm long with the base becoming the capsule; **fruit and seeds** cylindric capsule, 2-4 cm long and 3-4 mm wide, straight or curved, hairy becoming smooth, with seeds ellipsoid, angled, variable shapes, 1.2-1.4 mm long and 0.8 mm wide, pale brown and pitted. **Range:** Native. CA to FL and north to ME and west to IA and KS. **Ecology:** Most common on forests roadsides and margins.

Oenothera **Evening-primrose** **Onagraceae**

T. Bodner (Aug)

Oenothera biennis

T. Bodner (Sep)

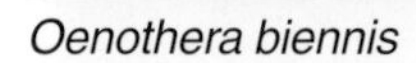

Oenothera biennis

J. Miller (May)

Oenothera laciniata

T. Bodner (Aug)

Oenothera biennis

Oenothera **Evening-primrose** **Onagraceae**

Oenothera speciosa Nutt. **- pinkladies or showy evening-primrose OESP2**
Similar to *Oenothera laciniata* except **plant** perennial, branched basally and spreading, 20-40 cm long, from a creeping rootstock; **stem** slender, greenish; **leaves** oblanceolate to linear, to 8 cm long and 3 cm wide, with lobes usually near the tapered base; **flowers** (Apr-Aug) petals white to pink with yellow centers, without notched tips, 2-4 cm long, sepal lobes 1.5-3 cm long, flower stalks 1-2.5 cm long; **fruit** (Oct-Jan) capsule, obovate, angled and winged, 6-14 mm long, tapering to both ends. **Range:** Native. TX to FL and north to PA and west to IL and KS.
Synonyms: *Hartmannia speciosa* (Nutt.) Small, buttercups.

Oenothera fruticosa L. **- sundrop OEFR**
Similar to *Oenothera speciosa* except **stem** slender and brown; **leaves** to 11 cm long and 2.5 cm wide, margins entire or serrate; **flowers** (Apr-Aug) petals yellow, 1-2 cm long, sepal lobes 8-18 mm long; **fruit** capsule, obovate to oblong-obovate, 5-9 mm long. **Range:** Native. MS to FL and north to Nova Scotia and southwest to WV and TN.
Synonyms: narrowleaf evening-primrose.

Other species in the SE: *O. drummondii* Hook., *O. humifusa* Nutt., *O. parviflora* L., *O. perennis* L., *O. rhombipetala* Nutt. *ex* Torr. & Gray.

Wildlife: Evening-primrose seeds are used by the Northern Bobwhite and Mourning Dove, along with seed-eating songbirds such as the American Goldfinch and Dark-eyed Junco. They are poor to moderate preference White-tailed Deer forages. Flowers are nectar sources for various moths and butterflies.

Oenothera Evening-primrose Onagraceae

T. Bodner (Apr)

Oenothera speciosa

T. Bodner (Jun)

Oenothera fruticosa

Oxalis **Woodsorrel or Sheepsorrel** **Oxalidaceae**

Low-growing perennial herbs with rhizomes or bulbs, or annuals with fibrous roots. Leaves alternate, 3-leaflet leaves, leaflets heart-shaped and partly folded, having a sour taste. Flowers 5-petaled, joined at base or not; yellow, white, pink or lavender and withering soon after expansion, and 5 persistent sepals. Capsules splitting along sides to expel tiny red seeds. 6 species in the SE.

Common Species

Oxalis stricta **L. - yellow woodsorrel** **OXST**

Plant: Upright to reclining to trailing perennial, 6-20 cm tall (up to 50 cm tall) or long, from slender rhizomes. Sour tasting foliage.

Stem: Slender, much branched and often mat-forming, green to yellow green, fine white hairy.

Leaves: Alternate, 3 leaflet, leaflets heart-shaped and joined at tips, 1-2 cm wide, often folded in mid-day, white hairy on margins and under surface, long petioles with oblong stipules at base.

Flowers: Mar-Oct (some year-round). Clusters of 4 or less from upper leaf axils (umbels), petals 5, yellow often with red dots at base, slightly funnelform to tubular at times, 5-10 mm long, 5 hairy sepals

Fruit and seeds: May-Oct. Erect "okra-like" capsule, 5-angled, slender with point, 1- 1.5 cm long, hairy or not, green, splitting at maturity and forcibly expelling numerous seeds, flat and many ribbed, 1-1.3 mm long, brown.

Range: Native. Throughout US.

Ecology: Common early invader of new forest plantations, disturbed sites, pastures, and right-of-ways, on dry soils. Occurs as scattered plants and plant mats. Persists and colonizes by rhizomes and spreads by ejected seeds.

Synonyms: lady's sorrel, sheepsorrel, sheep sour, common yellow oxalis.

Oxalis violacea **L. Jacq. - violet woodsorrel** **OXVI**

Similar to *Oxalis stricta* except **plant** perennial, 6-12 cm tall, in small tufted clumps, from scaly bulbs having stolons; **stem** stemless, long leaf petioles and flower stalks radiating from the base, hairless; **leaves** 3 leaflet, leaflets broadly heart-shaped, 1.5-3.5 cm wide, blue-green often with purple markings, smooth and hairless, petioles 6-12 cm long; **flowers** (Apr-May and sometimes Aug-Nov) umbel clusters on erect stalks, stalks 10-20 cm tall, extending above leaves, petals 5, lavender to white with white and green centers, 10-18 mm long, sepals 5, with orange glands on tips; **fruit and seeds** (Apr-May and Aug-Nov) capsule, spherical, 4-6 mm long, expels seeds, about 1 mm long. **Range:** Native. NM to FL and north to NY and west to ND and CO.

Ecology: Scattered plants and small groupings in new forest plantations and forest openings, on moist soils (clay and loam soils). Colonizes by stolons and spreads by animal-dispersed seeds.

Other species in the SE: *O. corniculata* L., *O. grandis* Small, *O. montana* Raf., *O. rubra* St.-Hil.

Wildlife: Woodsorrel seeds are consumed by Northern Bobwhite and songbirds, including the Dark-eyed Junco and various sparrows. Seeds and leaves are used sparingly by Ruffed Grouse and Wild Turkey in the spring. *Oxalis* are low preference White-tailed Deer forages. Flowers are a nectar source for the Falcate Orangetip butterfly.

Oxalis **Woodsorrel or Sheepsorrel** **Oxalidaceae**

T. Bodner (May)

Oxalis stricta

T. Bodner (Jul)

Oxalis stricta

T. Bodner (May)

Oxalis violacea

Passiflora Passionflower Passifloraceae

Perennial herbaceous (or woody) vines, with alternate leaves, and unbranched tendrils. Leaves simple, entire, or lobed. Flowers complex with 5 sepals and 5 petals below a double or single ring of thread-like fringe that encircles below 5 elevated anthers, each having a broad stigma, that encircle 3 styles. Fruit a 1-chambered, pod-like berry with numerous seeds. 3 species in the SE.

Common Species

Passiflora incarnata L. - **purple or maypop passionflower** PAIN6

Plant: Perennial trailing or rarely climbing herbaceous vine, often erect when young, 20 to 120 cm long, sparsely branched, from an irregular-shaped woody rootcrown.
Stem: Moderately stout, fine hairy (or not), twining tendrils at leaf axils.
Leaves: Alternate, deeply 3-lobed (rarely 5-lobed), 6-15 cm long and 6-15 cm wide, green glands at base, margins serrate, fine hairy or not, petioles 1-4 cm long with deciduous stipules at base, 2-3 mm long.
Flowers: May-Oct. Showy, single complex flowers in leaf axils on stout stalks, stalks 5-10 cm long with 3 glandular bracts, 3-6 mm long; sepals 5, scooped, sharp-tipped, 2.5-3.5 cm long, green on back and white to lavender on front; petals 5, scoop-shaped, 3-4 cm long, lavender to white; a whorl of fringe, filaments lavender with purple and white bands, to 3 cm long; anthers 5, attached at middle, 0.8-1.1 cm wide, yellowish white; and 3 extended styles. Flowers open and close daily.
Fruit and seeds: Aug-Oct. Large pod-like berry, ovoid, 4-6 cm long, green ripening to yellow, containing numerous dark-brown seeds, ovoid, 4-6 mm long, enclosed in a membranous envelope.
Range: Native. TX to FL and north to PA and west to IL, MO, and OK.
Ecology: Occurs on moist to dry sites, open to semi-shady habitats, as single plants or small groupings. Inhabits young and old forests, roadsides, right-of-ways, and old fields. Persists by rootcrowns and spreads by animal-dispersed seeds.
Synonyms: wild apricot vine.

Passiflora lutea L. - **yellow passionflower or broadleaf maypop** PALU2

Similar to *Passiflora incarnata* except **stems** softly hairy becoming smooth with age; **leaves** shallowly 3-lobed, 2-10 cm wide, wider than long, often mottled with silvery gray, no glands at base, petioles 1-6 cm long; **flowers** (Jun-Oct) sepals green, 7-12 mm long, petals yellowish green, 9-15 mm long, fringe green, rarely purple at base, 7-12 mm long; **fruit and seeds** (Aug-Nov) berry black at maturity, 0.8-1.5 cm long, seeds 3-4.5 mm long.
Range and ecology: Occurs usually on wet to moist sites, often in shady habitat.

Other species in the SE: *P. morifolia* Masters [rare introduction].

Wildlife: Passionflower seeds are consumed in small amounts by Northern Bobwhite and various songbirds. The plants are a larval food source of the Mexican Silverspot, Variegated Fritillary, Gulf Fritillary, Zebra Longwing, and Isabella Tiger butterflies.

Passiflora **Passionflower** **Passifloraceae**

T. Bodner (Jul)

Passiflora incarnata

J. Miller (May)

Passiflora incarnata

J. Miller (Jul)

Passiflora incarnata

Phytolacca | **Pokeweed or Pokeberry** | **Phytolaccaceae**

Herbs, shrubs, and trees, generally tropical, with alternate leaves having entire margins. Flowers in racemes and lack petals. Fruit a juicy, 5- to 15-celled berry. 1 species in the SE, with 2 varieties.

Common Species

***Phytolacca americana* L. - American or common pokeweed** **PHAM4**

Plant: Shrub-like herb, 1-3 m tall, often a single branched stem or multiple stems when browsed, from a perennial fleshy taproot (up to 15 cm wide). Emits a distinct odor. Roots especially, but also mature leaves, stems, and seeds are poisonous to humans and livestock.

Stem: Stout, 2-6 cm wide at base, branched upward, purplish to reddish (stains clothing), smooth, hollow pith.

Leaves: Alternate, smooth, lanceolate to elliptic-lanceolate, 8-30 cm long and 3-12 cm wide and smaller upward, margins entire to slightly wavy, petioles 1-5 cm long, very green.

Flowers: May-Nov. Erect to drooping elongated racemes, stalks 5-20 cm long, opposite to upper leaves, green turning reddish, petals lacking but sepals 5, green to white, persistent on fruit.

Fruit and seeds: Aug-Dec. Berry, spherical and squat, 4-6 mm long and 7-10 mm wide, green maturing to purplish black, producing red juice (stains clothing), containing 10 seeds, rounded, about 3 mm wide, glossy and black.

Range: Native. TX to FL and north to Que and west to Ont, MN, and OK.

Ecology: Common in new forest plantations, forest openings, forest margins, and disturbed sites. Rootcrowns can remain viable for years after canopy development. Seeds are spread mostly by birds, and seeds can remain viable for at least 40 years. Seedlings emerge midspring through early summer.

Synonyms: *P. americana* var. *americana* [formerly *P. decandra* L.], *P. americana* var. *rigida* (Small) Caulkings & Wyatt [formerly *P. rigida* Small], pokeberry, poke, inkberry, pigeonberry, scoke, garget.

Wildlife: Pokeweed fruit is an important soft mast species in the late summer and early fall. Fruits are consumed by many songbirds, particularly the Northern Mockingbird, Gray Catbird, and Cedar Waxwing, and several mammalian species such as the Black Bear, Raccoon, Virginia Opossum, and Gray Fox. Both the fruit and foliage are readily consumed by White-tailed Deer. The seeds are an important component of the Mourning Dove diet in fall and winter.

Phytolacca Pokeweed or Pokeberry Phytolaccaceae

T. Bodner (Aug)

Phytolacca americana

T. Bodner (May)

Phytolacca americana

T. Bodner (Jun)

Phytolacca americana

Pityopsis **Silkgrass** **Asteraceae**
(formerly *Chrysopsis* or *Heterotheca*)

Perennial, biennial, or annual herbs, with densely hairy leaves and late-summer yellow flowers in panicles or corymbs. Leaves alternate, blades elongated, entire or toothed. 5 species in the SE, genus formerly classified under *Chrysopsis* and previously *Heterotheca*.

Common Species

Pityopsis graminifolia (Michx.) Nutt. **- narrowleaf silkgrass** **PIGR4**

Plant: Perennial, resembling clumps of grass in early to mid-summer having blade-like leaves, silky dense hairy beneath, in late-summer to fall growing a stiffly erect flower stalk, 30-90 cm tall, sparsely to densely branched, from slender rhizomes having fibrous roots at clumps, often forming small to extensive colonies.

Stem: Absent in early summer with only clumps of grass-like leaves, in late-summer flowering stalks, 1 to several from base, stiffly erect, dense silver hairy (appressed to stem).

Leaves: Grass-like blades early, parallel veined, 20-35 cm long and 1-2 cm wide, silver-silky hairy, in late-summer stem leaves alternately spiraling, linear with a tapering tip, 4-10 cm long, similar size upward, appressed or curved outward, often with fascicled small leaves in axils by fall.

Flowers: Jul-Oct. Small heads, 1 to several, terminating few to many branches from leaf axils (corymb or panicle), bright yellow ray petals and yellow disks (centers), 1-2 cm wide, involucre urn-shaped, bracts linear, overlapping in several series, glandular hairy (sticky).

Fruit and seeds: Nov-Feb. Linear nutlet (achene), 2-3 mm long, reddish-brown to black, topped with whitish-tan bristles, 4.5-6 mm long, nutlet containing 1 seed.

Range: TX to FL and north to DE and west to OH and OK.

Ecology: Inhabits young tree plantations, open forests, forest openings, and right-of-ways. Occurs in dense colonies on dry sites, especially degraded soils, or as scattered plants on better soils. Persists and colonizes by rhizomes and spreads by wind-dispersed seeds.

Synonyms: grassleaf goldaster, golden aster.

Other species in the SE: *P. flexuosa* (Nash) Small, *P. oligantha* (Chapman *ex* Torr. & Gray) Small, *P. pinifolia* (Ell.) Nutt., *P. ruthii* (Small) Small.

Wildlife: Important food for Gopher Tortoise.

Pityopsis Silkgrass Asteraceae

T. Bodner (Sep)

Pityopsis graminifolia

T. Bodner (May)

Pityopsis graminifolia

T. Bodner (Sep)

Pityopsis graminifolia

T. Bodner (Sep)

Pityopsis graminifolia

Plantago **Plantain** **Plantaginaceae**

Perennial or annual herbs, low-growing or stemmed. Most species having only basal leaves (or opposite on freely branched stem with *P. psillium*). Flowers and fruit crowded in a many bracted slender spike. Flowers irregular, sepals and petals 4 lobed, petals papery and persistent on the fruit. Fruit a capsule, 2-celled, splitting at the middle to release 2 to many seeds. 14 species in the SE.

Common Species

Plantago aristata Michx. - **largebracted plantain** **PLAR3**

Plant: Annual with long slender upright leaves from a short base, "bottle-brush" like bracted spikes of flowers and fruit, 15-40 cm tall, from a taproot.
Stem: Very short, almost lacking.
Leaves: Basal, appearing grass-like, linear to very narrowly oblanceolate with long tapering tips, 5-20 cm long and 2-8 mm wide, dark green, hairy or hairless.
Flowers: Apr-Jul. Spikes of crowded (nonshowy) flowers with many linear bracts on a round leafless stem, 10-30 cm tall, jutting bracts to 2 cm long and smaller upward, green, scurfy hairy, corolla 4-lobed and papery, sepals green, lobes about 2 mm long and rounded.
Fruit and seeds: Jun-Nov. Capsule, 3-3.5 mm long, crowded along the spike, splitting at the middle to release 2 seeds, about 2.5 mm long, brown, angled and grooved.
Range: Native. Throughout US and n Mexico.
Ecology: Early appearing plant in new forest plantations after soil disturbance, especially sandy soils, soon disappearing. Spreads by seeds that persist in the soil.

Plantago virginica L. - **Virginia or paleseed plantain** **PLVI**

Similar to *Plantago aristata* except **leaves** hairy, 2-15 cm long and 0.5-4 cm wide, margins irregularly toothed to entire, petioles purple at base; **flowers** (Mar-Jun) spikes and stalk 18-30 cm tall, bracts and sepal lobes hairy, less than 2.3 mm long, petals to 3 mm long; **fruit and seeds** capsules 2-3 mm long, seeds light brown, 1.5-1.8 mm long. **Range:** Native. AR to FL and north to NY and west to MI and KS.

Plantago major L. - **common or broadleaf plantain** **PLMA2**

Similar to *Plantago aristata* except **plant** perennial herb with spikes to 24 cm tall, from a rootcrown with fibrous roots; **leaves** basal, broadly elliptic to oval, 4-15 cm long and 3-10 cm wide, with tapering petioles 2-15 cm long, hairless except on petiole, margins shallow-toothed to entire, parallel veined; **flowers** (Jun-Oct) slender spikes of tiny crowded greenish flowers on a leafless stem, 4-24 cm tall, bracts 1-1.5 mm long and blunt tipped, brownish flower tube, 2-3 mm long and lobes 1-1.3 mm long, sepals 2.2-2.8 mm long; **fruit and seeds** (Jul-Nov) capsules, ellipsoid, 2-3 mm long (crowded along the spike), splitting at middle to release 12-18 seeds, 1-1.3 mm long, reddish brown, various shaped. **Range:** Naturalized from Eurasia. Throughout US and s Canada.
Ecology: Inhabits new forest plantations, forest openings, margins, and right-of-ways, and occurs in lawns, gardens, and pastures, especially damp soils. Persists by rootcrowns and spreads by abundant seeds.

Wildlife: Plantain seed heads are infrequently consumed by White-tailed Deer and the Northern Bobwhite. There apparently is little to no other wildlife value.

Plantago **Plantain** **Plantaginaceae**

J. Miller (Jul)

Plantago major

J. Miller (May)

Plantago aristata

T. Murphy (Aug)

Plantago virginica

Polypremum **Polypremum or Rustweed** **Buddlejaceae**

Annual or perennial herbs, low growing and much branched. Sepals and petals 4-lobed, with petals barely extending past the sepals and hairy in the throat. Fruit a capsule, somewhat flattened, multi-celled, and separating to release numerous, minute seeds. 1 species in the SE.

Common Species

Polypremum procumbens L. **- juniper leaf or rustweed** **POPR4**

Plant: Perennial herb, low growing in clumps, branches 10-30 cm long, with many small green leaves turning golden in late-summer (thus the common names), from a slender elongating rootcrown.

Stem: Ascending or prostrate, branched from the base and radiating out from the rootcrown, angled to finely winged, green turning golden in late summer to fall, hairless.

Leaves: Opposite, linear, 1-2.5 cm long and 0.5-2 mm wide, tip pointed, paired bases surround the stem, additional tufts of leaves appearing in the axils.

Flowers: May-Oct. Tufts of flowers and tiny leaves from upper leaf axils, corolla 4-lobed, white, lobes 1-3 mm long.

Fruit and seeds: Jul-Nov. Capsule oval, 2-3 mm long, with 2 halves separating to release numerous seeds, more or less square and pitted, 0.3-0.4 mm wide, yellow.

Range: Native. TX to FL and north to NJ and southwest to KY, MO, and CO, also in n Mexico.

Ecology: Common early inhabitant in new forest plantations, declining with canopy development, but persisting in openings. Occurs in low vegetation along forest margins and roads. Found in old fields and along right-of-ways. Mainly occurs in sandy soils of the Cp, and rare in the Mt. Persists by rootcrowns and spreads by seeds.

Synonyms: goldweed.

Wildlife: Rustweed is a low preference White-tailed Deer forage. No other wildlife value is reported.

Polypremum **Polypremum or Rustweed** **Buddlejaceae**

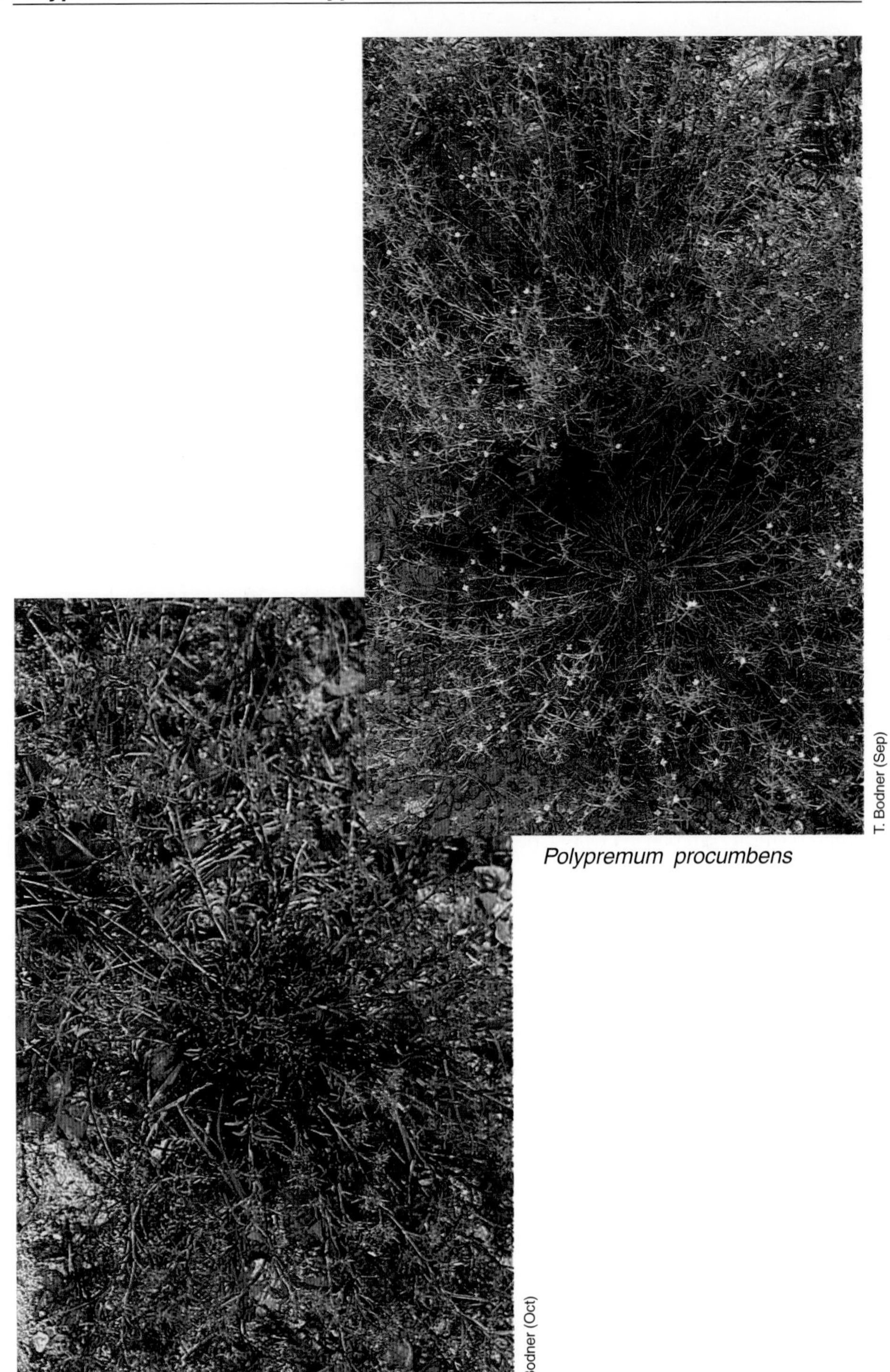

T. Bodner (Sep)

Polypremum procumbens

T. Bodner (Oct)

Polypremum procumbens

Potentilla **Cinquefoil or Five-fingers** **Rosaceae**

Perennial or biennial herbs or small shrubs, with compound leaves (usually 3, 5, 7, or 9 palmately-arranged leaflets) having stipules. Rose-type flowers, solitary or in cymes, with 5 notched petals (yellow, white, or purple) and 5 sepals with 5 similar bracts appearing like 10 sepals. Fruit a head of ovoid nutlets (achenes). 7 species in the SE.

Common Species

Potentilla simplex Michx. **- common cinquefoil** **POSI2**

Plant: Low-growing perennial herb, vine-like, 10-20 cm tall, with slender wiry stolons from a short vertical rhizome (to 8 cm long) having fibrous lateral roots. Dies back to ground in winter and young shoots and small leaves appear in Feb-Mar.

Stem: Ascending becoming prostrate, wiry stolons that root at nodes, 40-100 cm long, appressed hairy, often reddish.

Leaves: Alternate, 5 leaflets (palmately compound) on long petioles, leaflets lanceolate to elliptic, 1-4 cm long and 1-2 cm wide, glossy and dark green above and whitish-green beneath, margins coarsely toothed except near base, stipules linear-lanceolate and rolled.

Flowers: Apr-Aug. Solitary on long wiry stalks, 1.5-6 cm long, petals 5 with notched tips, golden-yellow, 10-15 mm wide, sepals 5, green.

Fruit and seeds: Jun-Sep. Head of nutlets (achenes), each nutlet yellowish brown, smooth, 1.2 mm long, with a tuft of long hairs.

Range: Native. TX to GA and north to MA and Que and west to MN, MO, and OK.

Ecology: Common ground cover in young forest plantations, occurring in small colonies. Inhabits forest openings and margins, right-of-ways, fields, and lawns. Persists by rhizomes, colonizes by stolons, and spreads by seeds.

Synonyms: creeping cinquefoil, five-fingers, oldfield cinquefoil.

Potentilla canadensis L. **- dwarf cinquefoil** **POCA17**

Similar to *Potentilla simplex* except **plant** smaller overall, rhizome is shorter (0.5-2 cm long) and erect; **stem** spreading hairy (not appressed hairy); **leaves** oblong-oblanceolate or oval, 2-5 cm long, with stipules oblong-lanceolate and flat; **flowers** Mar-May. **Range:** same but less common in Lower Cp.

Other species in the SE: *P. argentea* L., *P. inclinata* Vill., *P. intermedia* L., *P. norvegica* L., *P. recta* L.

Wildlife: The cinquefoils have relatively low wildlife value. The foliage is consumed commonly by Ruffed Grouse and infrequently by the Eastern Cottontail and White-tailed Deer, particularly during late winter and early spring.

Potentilla **Cinquefoil or Five-fingers** **Rosaceae**

T. Bodner (May)

Potentilla simplex

T. Bodner (May)

Potentilla simplex

Pycnanthemum **Mountain mint or Wild-basil** **Lamiaceae**

Upright perennial herbs from rhizomes, mints with 4-angled stems and opposite branched. Leaves opposite, often having a mint fragrance. Flowers purple or white and commonly spotted with purple, 2-lipped, upper lip entire or notched, and lower lip 3-lobed. Flowers in crowded or head-like cymes terminating stems and branches, often with whorls of whitish floral leaves. Fruit a small nut, smooth or hair-tufted, splitting at maturity into 4 nutlets. 8 species in the SE.

Common Species
Pycnanthemum incanum (L.) Michx. **- hoary mountain mint** **PYIN**
Plant: Erect perennial, 1-2 m tall, with many opposite branches above mid-plant, from a rhizome with fibrous roots. Mint fragrance or not.
Stem: Square in cross-section, softly hairy, light green turning dark brown to black, remaining standing during early to mid-winter.
Leaves: Opposite, ovate to elliptic to lanceolate, 3-10 cm long and 1.5-4 cm wide, green above and white beneath, increasingly dense whitish hairy toward flowers, acute tipped, margins shallow serrate, petioles 2-15 mm long.
Flowers: Jun-Oct. Clustered and compacted at terminals with whorled whitish floral leaves and whitish bracts, both long hairy, corolla 2-lipped, white to lavender and spotted with purple, 7-9 mm long, stamens extended. A very showy flower.
Fruit and seeds: Aug-Dec. Nutlet, broadly ovoid, 1-1.7 mm long, brown with hairy apex.
Range: Native. MS to FL and north to ME and west to Ont, IL, and MO.
Ecology: Persists after disturbance by rhizomes and occurs as scattered plants in new forest plantations. Inhabits forest openings and edges, right-of-ways, and fields. Spreads by seeds.
Synonyms: *Koellia incana* (L.) Kuntze.

Other species in the SE: *P. flexuosum* (Walt.) B.S.P., *P. montanum* Michx., *P. muticum* (Michx.) Pers., *P. setosum* Nutt., *P. tenuifolium* Schrad., *P. verticillatum* (Michx.) Pers., *P. virginianum* (L.) T. Dur. & B.D. Jackson *ex* B.L. Robins & Fern.

Wildlife: Mountain mints have little wildlife value. They occasionally are browsed by White-tailed Deer.

Pycnanthemum **Mountain mint or Wild-basil** **Lamiaceae**

T. Bodner (Jul)

Pycnanthemum incanum

T. Bodner (Sep)

Pycnanthemum incanum

T. Bodner (Jun)

Pycnanthemum incanum

Pyrrhopappus **Falsedandelion** **Asteraceae**

Annual or perennial herbs, with leaves in basal rosettes or sparsely alternate on slender flowering stems having milky sap. Similar in appearance to common dandelion but with more petals and often longer stalked. Leaves entire to minutely toothed to lobed or with projections along margins of up to 3 cm long. Flowers in heads with clustered petals, yellow, solitary or spaced along slender stalks. Nutlets (achenes) oblong to linear with bristle plumes for wind dispersal. 2 species in the SE.

Common Species

Pyrrhopappus carolinianus (Walt.) DC. **- Carolina falsedandelion** **PYCA2**

Plant: Winter annual or biennial, with rosette of leaves, later having erect branched flowering stems, 20-100 cm tall, from a well-developed taproot.

Stem: Slender, short hairy above and smooth hairless below, pale green, sap milky.

Leaves: Basal rosette and alternate on stem, oblanceolate to lanceolate, 8-25 cm long and 1-6 cm wide and smaller upward, margins entire to deeply dissected, whitish midvein. Basal leaves often absent later.

Flowers: Mar-Jul (sporadically year-round). Many-petaled head (dandelion-like but larger), 2-5 cm wide, bright yellow to greenish yellow, involucral bracts 2 types, uppermost bracts longer in 1 series and lower bracts shorter in several series becoming constricted into a cylinder, 1-2 cm high. Bracts reflexing when seeds released.

Fruit and seeds: Mar-Jun (Nov-Oct). Nutlet (achene), brown, cylindric and tapered at both ends, 4-6 mm long, 5 grooved, with beak 9-11 mm long, topped with whitish-tan bristles, 8-10 mm long, nutlet containing 1 seed.

Range: Native. TX to FL and north to PA and west to IA, NE, and OK.

Ecology: Early invader of new forest plantations for 1-4 years, persisting in forest openings. Occurs as scattered plants along forest edges and right-of-ways. Spreads by wind-dispersed seeds.

Synonyms: *Sitilias caroliniana* (Walt.) Raf., *Pyrrhopappus georgianus* Shinners, Carolina desert-chicory.

Other species in the SE: *P. multicaulis* DC. [TX and OK to AL].

Wildlife: The falsedandelions are used sparingly by White-tailed Deer in the early spring. No other wildlife value is reported.

Pyrrhopappus **Falsedandelion** **Asteraceae**

J. Miller (Nov)

Pyrrhopappus carolinianus

J. Miller (Jul)

Pyrrhopappus carolinianus

J. Miller (Jul)

Pyrrhopappus carolinianus

J. Millery (May)

Pyrrhopappus carolinianus

J. Miller (Mar)

Pyrrhopappus carolinianus

Rhexia Meadowbeauty Melastomataceae

Upright perennial herbs, branched or single stemmed, with opposite leaves having 3-5 veins paralleling margins, and no petioles or only to 2 mm long. Flowers from the terminal or upper leaf axils in cymes, petals 4, purplish or yellow or white, sepals 4, and stamens 8. Fruit a rounded urn-shaped capsule (4-celled) with persistent sepals around the tip, opening at the top to release numerous crescent-shaped seeds, brownish to yellowish. 8 species in the SE.

Common Species

Rhexia virginica L. - meadowbeauty RHVI

Plant: Upright perennial, 40-90 cm tall, unbranched or branched at base and upper plant, from a tuberous root.

Stem: Slender, 4-angled and 4-winged (to 1 mm wide), appears light- and dark-green striped (often with reddish tinges), hair tufts at nodes, otherwise scattered hairy or hairless, branched or unbranched in the upper plant.

Leaves: Opposite, ovate to lanceolate, 2-10 cm long and 1-4 cm wide and larger upward, 3 main veins with outer 2 paralleling margins, margins with small sharp teeth except near the base, tip acute, surfaces scattered hairy, dark green above and light green beneath.

Flowers: May-Oct. Single or multiple at upper leaf axils (cyme), purple to lavender, petals 4, 10-25 mm long, anthers 8, yellow, 5-7 mm long, sepals 4, 1.5-3 mm long.

Fruit and seeds: Aug-Dec. Capsule, round urn-shaped topped with 4 persistent sepals, capsule 4-8 mm long and 5.5-6.5 mm wide, 4-celled, top opening to release numerous seeds, 0.6-0.8 mm long, light-brown.

Range: Native. LA to FL and north to ME and west to Ont and southwest to MO, mainly in Cp and rare in Pd in the SE.

Ecology: Occurs in bottomland and moist forests in open to shady habitats, along ditches, and low fields. Persists by tuberous roots and spreads by gravity- and water-dispersed seeds.

Synonyms: deer-grass, handsome Harry, wing-stem meadowbeauty.

Rhexia mariana L. - Maryland or pale meadowbeauty RHMA

Similar to *Rhexia virginica* except **plant** very hairy, 20-80 cm tall, from a horizontal root (no tuber); **stem** winged; **leaves** linear to elliptic to obovate, 2-6 cm long and 1-2 cm wide, margins serrate and hairy; **flowers** purplish to pinkish (to white). **Range:** Native. TX to FL and north to PA and west to IL. 3 varieties.

Rhexia alifanus Walt. - savannah or tall meadowbeauty RHAL4

Similar to *Rhexia virginica* except **plant** unbranched or only from the base, 30-100 cm tall; **stem** slender, hairless, round and not winged; **leaves** lanceolate to narrowly elliptic, 3-8 cm long and 0.5-1 cm wide, 3-veined, margins entire, tip acute or tapering; **flowers** petals purple, 16-22 mm long, glandular hairy beneath, anthers curved, 5-11 mm long, flowers only last a few hours; **fruit and seeds** capsule, glandular hairy. **Range:** Native. LA to c FL and north to NC, only in Cp.

Ecology: Occurs in seasonally wet forests and plantations, also in savannas and ditches.

Other species in the SE: *R. aristosa* Britt., *R. cubensis* Griseb., *R. lutea* Walt., *R. nashii* Small, *R. parviflora* Chapman.

Wildlife: The *Rhexia* are moderately preferred White-tailed Deer browse in the early spring. No other wildlife value reported.

Rhexia Meadowbeauty Melastomataceae

J. Miller (Jul)

Rhexia mariana

J. Miller (Sep)

Rhexia virginica

J. Miller (Sep)

Rhexia alifanus

Rhynchosia **Rhynchosia or Snoutbean** **Fabaceae**

Erect, trailing, or climbing perennial herbs, legumes, from tuberous roots. Leaves simple or 3-leaflet, with stipules at the base of long petioles. Leaves and stem hairy. Small pea-type flowers, single or clustered in upper leaf axils, yellow, often partially hidden by a tubular 2-lipped calyx. Fruit a legume (pods), short and broad, somewhat flattened, tapering to a short beak, hairy with inconspicuous yellow resin glands, containing 1-2 seeds. 10 species in the SE.

Common Species

Rhynchosia reniformis DC. - **dollarleaf** **RHRE**

Plant: Short and erect perennial legume, 7-20 cm tall, from a slender woody rhizome and forming small colonies.
Stem: Moderately stout, strongly angled, densely hairy.
Leaves: Alternate, single (or rarely 3) leaflet, rounded or kidney shaped, 2-5 cm long and 2-8 cm wide, sparsely or densely brown hairy (tawny), wrinkled surface, petioles 1-5 cm long, with hair-like stipels at leaf bases, 1.5-2.5 mm, persistent linear-lanceolate stipules at petiole bases, 5-10 mm long.
Flowers: Jun-Oct. Terminal and axillary raceme clusters, pea-type, yellow, petals 6-8 mm long, calyx lobed, 6-9 mm long, hairy with amber glands (requires hand lens), each flower having a linear bract, 3-5 mm long.
Fruit and seeds: Jul-Jan. Legume (pod), 1-1.8 cm long and 6-7 mm wide, flat, densely hairy especially along suture, splitting to release 1-2 seeds, brownish-black, 3 mm wide.
Range: Native. TX to FL and north to NC and TN, only in Cp.
Ecology: Occurs as small colonies in dry forests, savannas, prairies, and right-of-ways. Inhabits forest edges and openings. Persists by rhizomes and spreads by animal-dispersed seeds.

Rhynchosia tomentosa (L.) Hook. & Arn. - **velvetleaf rhynchosia**

Plant: Erect perennial legume, 30-90 cm tall, single stem or branched from base and upper plant, from a woody rhizome.
Stem: Moderately stout, many-branched or not, densely (to sparsely) hairy, round or square in cross-section.
Leaves: Alternate, 3 leaflet, leaflets ovate to elliptic, 3-7 cm long and 1.5-3 cm wide, moderately to densely hairy beneath and velvety to touch, textured surface, stipules 4-10 mm long, linear to lanceolate, stipels absent or soon deciduous.
Flowers: Jun-Oct. Terminal and axillary raceme clusters on stalks 0.5-2 cm long, pea-type, yellow, petals 5-7 mm long, calyx hairy and lobed, 6-9 mm long, each flower having linear bracts, 2-8 mm long.
Fruit and seeds: Jul-Nov. Legume (pod), 1.2-2 cm long and 5-8 mm wide, flat, short hairy, splitting at maturity to release 1-2 seeds.
Range: Native. TX to FL and north to MD and southwest to TN.
Ecology: Same as *Rhynchosia reniformis* but occurs more as individual plants.
Synonyms: *R. erecta* (Walt.) DC., twining snoutbean.

Other species in the SE: *R. cinerea* Nash, *R. cytisoides* (Bertol.) Wilbur, *R. difformis* (Ell.) DC., *R. latifolia* Nutt. *ex* Torr. & Gray, *R. michauxii* Vail, *R. minima* (L.) DC., *R. parvifolia* DC., *R. swartzii* (Vail) Urban.

Wildlife: Rhynchosias are Northern Bobwhite food plants of moderate importance. The seeds likely are consumed by many songbirds and small mammals. They are a moderate- to high-preference White-tailed Deer forage, although rarely abundant.

Rhynchosia **Rhynchosia or Snoutbean** **Fabaceae**

T. Bodner (Jul)

Rhynchosia reniformis

T. Bodner (Jul)

Rhynchosia reniformis

T. Bodner (Aug)

Rhynchosia tomentosa

T. Bodner (Aug)

Rhynchosia tomentosa

Richardia Mexican clover or Pusley Rubiaceae

Annual or perennial hairy herbs, mostly many branched and reclining, with opposite leaves connected by fringed stipules. Flowers crowded in terminal bracted clusters, whitish to lavender, funnel-shaped with 4-8 short lobes, and calyx 4-8 lobed. Fruit separates longitudinally at maturity into 2-4 nutlets. 3 species in the SE, naturalized from tropics.

Common Species

Richardia scabra L. **- Florida pusley** **RISC**

Plant: Sprawling branched annual, branches to 8 cm long, short-hairy, from a taproot with fibrous laterals near the soil surface.
Stem: Reclining, densely hairy, green to reddish to brown, opposite branched.
Leaves: Opposite, ovate to elliptic, 2-6 cm long and 6-20 mm wide, sparsely to densely rough, hairy, especially margins, short petioles connected around the stem by a cup-collared stipule, having hairy projections.
Flowers: May-Nov. Terminal, compacted flat clusters, subtended by 4 leaf-sized paired bracts, corolla white, funnel-shaped, 4-6 mm long, 6 lobed, lobes less than 1/3 the length of tube, calyx 4-lobed, 1-1.5 mm long-stiff hairy.
Fruit and seeds: Jul-Jan. Nutlets leathery, separating into 3-4 parts with seeds 3-3.5 mm long, 2-grooved.
Range: Native to tropical America and naturalized. TX to FL and north to VA and west to IN and s AR, mainly in Cp and Pd in the SE.
Ecology: Spreads by animal-dispersed seeds. Seeds persist for long periods on abandoned crop lands and germinate after timber harvesting, site preparation, or burning, yielding dense colonies lasting only a year. Occurs also on mowed right-of-ways, Christmas tree and horticultural plantations, and cultivated fields.
Synonyms: Florida purslane, rough Mexican clover.

Richardia brasiliensis Gomes **- Brazil pusley** **RIBR2**

Similar to *Richardia scabra* except **plant** perennial from a thick woody rootcrown; **stem** much branched, becoming matted; **leaves** 1.5-4 cm long; **flower** corolla white, 3-4 mm long, lobes 1/3 as long as tube; **fruit and seeds** nutlets rounded and densely hairy.
Range and ecology: Similar to *Richardia scabra* but more common in Cp.
Synonyms: tropical Mexican clover.

Other species in the SE: *R. humistrata* (Cham.& Schlecht.) J.A. & J.H. Steud.

Wildlife: The seeds of Florida pusley are consumed occasionally by Northern Bobwhite and songbirds. It is a high preference summer and fall forage for White-tailed Deer and Eastern Cottontail on Coastal Plain sites that have been disturbed between April and August.

Richardia **Mexican clover or Pusley** **Rubiaceae**

T. Bodner (Sep)

Richardia scabra

Rudbeckia **Coneflower** **Asteraceae**

Annual or perennial herbs, with alternate leaves and branches. Flowers in sunflower-type heads, with disk flowers (centers) often purple or brown on domed or conical receptacle (thus the common name) and ray petals cream, yellow, orange, or sometimes purple, red and maroon. Involucral bracts unequal, spreading or reflexed, in 2-3 series. Nutlets (achenes) oblong to linear, 4-angled, with short pappus-crown or lacking. 14 species in the SE.

Common Species

Rudbeckia hirta L. **- hairy coneflower or blackeyed Susan** **RUHI2**

Plant: Upright biennial or short-lived perennial (rarely annual), 30-100 cm tall, sparsely branched and scurfy hairy, from a hard rootcrown with taproot or spreading roots. Prior year's stem often present.

Stem: Slender, branched from rootcrown or upper leaf axils or not, green turning brown.

Leaves: Basal rosette and alternate on stem, very variable (distinguishes 3 varieties), basal leaves often lanceolate to oblanceolate, 4-14 cm long and 1-7 cm wide and smallest upward, stem leaves linear to lanceolate to oblanceolate, bases long-tapering becoming sessile and almost clasping upward, margins entire to serrate to coarsely toothed, scurfy hairy, lighter green beneath.

Flowers: May-Oct. Terminal heads on long slender stalks, yellow or orange-yellow ray petals, 8-20, 2-4 cm long, disks (centers) dark purple to brown, hemispheric or ovoid, 1.2-2 cm wide, no pappus bristles, involucral bracts unequal in length, hairy and green, and overlapping in several rows.

Fruit and seeds: Aug-Dec. Oblong nutlet (achene), 4-angled, without plume, containing 1 seed.

Range: Native. TX to FL and north to Nova Scotia and west to CO.

Ecology: Occurs as scattered to frequent plants in new forest plantations, open forests, forest openings, and along forest edges, roads, and other right-of-ways. Persists by rootcrowns and spreads by animal-dispersed seeds.

Wildlife: Songbirds occasionally consume the seeds of *Rudbeckia*. The basal leaves remain green during winter and provide limited winter White-tailed Deer forage, with infrequent use during spring and early summer.

Rudbeckia **Coneflower** **Asteraceae**

T. Bodner (Jun)

Rudbeckia hirta

Rumex Dock or Sorrel or Sourgrass Polygonaceae

Perennial, biennial, or winter annual herbs with upright basal leaves and taller flowering stalks, all without hairs. Basal and alternate stem leaves, lanceolate to elliptic, with sheathing stipules. Flowers small, less than 3 mm broad, greenish or reddish, in small clusters along upright stems. Flowers unisexual or perfect, petals absent, and sepals 6 in 2 sizes, with the inner-3 enlarged and adhering to the nutlets. Nutlets shiny brown, rounded with 3 curved sides. 10 species in the SE.

Common Species

Rumex hastatulus Baldw. **- heartwing sorrel** **RUHA2**

Plant: Clumped perennial or winter annual, with erect 3-lobed basal leaves and small-leaved erect stems, 40-80 cm tall, from a taproot.
Stem: Slender, ridged, 1 to several from the basal rosette, branched above middle.
Leaves: Basal and alternate on stem, 3-lobed (arrowhead- or hastate-shaped) with the terminal lobe linear-oblong, with or without unequal lateral spurs, lower leaves 2.5-15 cm long and smaller upward, margins entire, petioles slender and tapering to stipules that sheath the stem.
Flowers: Apr-Jul. Tiny flowers on upper half of stems, 3-10 per tuft, greenish to yellow to red to purple, male and female flowers on different plants (dioecious), both 0.5-1.5 mm long, no petals, sepals 6 (2 sizes), inner 3 sepals enlarged to encase the nutlet on female plants, pedicels 1-3 mm long.
Fruit and seeds: Jun-Aug. Seed in a nutlet, rounded with 3 curved sides, 1.5 mm long, golden-brown, shiny, enlarged heart-shaped sepals encase the nutlet extending outward like wings (thus common name), 2.5-4 mm long and wide.
Range: Native. TX to FL and north to NJ and west to IL and KS.
Ecology: Common roadside and right-of-ways species that invades new forest plantations, forest edges as well as Christmas tree plantations and nurseries. Persists somewhat by roots and spreads by seeds.

Rumex acetosella L. **- common sheep or red sorrel** **RUAC3**

Similar to *Rumex hastatulus* except **plant** perennial, 10-50 cm tall, from creeping rhizome; **stem** 4 sided and branched above mid-stem; **leaves** 3-lobed, but terminal lobe elliptic to oblong and equal lateral spurs, acid to taste; **fruit and seeds** with an outer hull (sepals), reddish-brown, rough, often adhering to the seed (but not heart-shaped). **Range:** Native. Throughout US.
Ecology: Persists by rhizomes and spreads by seeds.
Synonyms: *Acetosella acetosella* (L.) Small, sourgrass, Indian-cane.

Other species in the SE: *R. altissimus* Wood, *R. chrysocarpus* Moris, *R. conglomeratus* Murr., *R. crispus* L., *R. obtusifolius* L., *R. patientia* L., *R. pulcher* L., *R. verticillatus* L.

Wildlife: Red sorrel is an important early summer producer of seeds that are used by a variety of sparrows and other songbirds. The seeds also are used heavily by Ruffed Grouse and Wild Turkey in June and July. Eastern Cottontail occasionally consume the foliage. Other *Rumex* species also are important seed producers but generally are less abundant.

T. Bodner (Apr)

Rumex hastatulus

Scutellaria Skullcap Lamiaceae

Upright, square-stemmed perennial herbs, mints but not aromatic, opposite leaves and branches, mostly from slender rhizomes. Flowers upward curving tubes, blue to pink to white and mostly white in the throat, corolla 2 lipped, the lower lip 3-lobed. Flowers single in terminal racemes subtended by leafy bracts. Calyx 2-lipped, persisting and covering the fruit, resembling a military hat on top, after seed release the upper half falls leaving a persistent "skull cap" (thus the common name). Nutlets 4 per flower, dark brown to black with wart-like or tubular protuberances. 12 species in the SE.

Common Species

Scutellaria integrifolia L. **- helmet flower or larger skullcap** **SCIN2**

Plant: Upright perennial, 20-80 cm tall, in clumps of 1 to several stems, from a rootcrown.
Stem: Slender, square in cross-section, branched from base often with small axillary branches, fine incurved hairy.
Leaves: Opposite, 3-8 pairs below flowers, lower leaves triangular-ovate, 0.7-3.5 cm long and 0.2-2 cm wide, margins crenate, petioles slender, early deciduous; middle leaves lanceolate to narrowly elliptic, 2-6 cm long and 0.4-15 mm wide, margins entire (to slightly crenate), base tapering, upper leafy-bracts narrowly elliptic and progressively smaller, sessile, all leaves green above and lighter green beneath.
Flowers: Apr-Aug. Terminal stalks in upper leaf axils (racemes), 25-60 cm long, many ascending floral tubes, blue to violet (rarely white), 1.3-2.5 cm long, 2 unequal lips with the lower spreading and notched with white marks, calyx 2-lipped, 2.5-3 mm long.
Fruit and seeds: Jun-Sep. Seed in a nutlet, 4 per flower, dark brown and rounded and warty, each 1-1.5 mm wide, beak attachment, enclosed within 2 enlarged calyx halves, a persistent lower dish (6-8 mm wide) and an upper deciduous half in the shape of a "military-hat."
Range: Native. TX to FL and north to MA and NY and west to OH and MO.
Ecology: Occurs as scattered plants in new forest plantations, forest openings, and along right-of-ways, on moist and dry sites. Persists by rootcrowns and spreads by seeds.
Synonyms: hyssop skullcap, rough skullcap.

Scutellaria lateriflora L. **- blue or mad-dog skullcap** **SCLA2**

Similar to *Scutellaria integrifolia* except **stem** much branched, hairless except on angles; **leaves** ovate to ovate lanceolate, 3-12 cm long and 1-5 cm wide, margins crenate to serrate, bases rounded, petioles 0.5-4 cm long; **flowers** (Jul-Sep) chiefly in lateral racemes from leaf axils, floral tube only, 5-7 mm long, along one-side of raceme.
Ecology: same, except only found on wet sites.

Other species in the SE: *S. elliptica* Muhl. *ex* Spreng., *S. floridana* Chapman, *S. glabriuscula* Fern., *S. incana* Biehler, *S. nervosa* Pursh, *S. ovata* Hill, *S. parvula* Michx., *S. racemosa* Pers., *S. saxatilis* Riddell, *S. serrata* Andr.

Wildlife: Skullcaps are low to moderate preference White-tailed Deer browse plants. No other wildlife value reported.

Scutellaria Skullcap Lamiaceae

J. Miller (Aug)

Scutellaria integrifolia

J. Miller (Aug)

Scutellaria integrifolia

Senna Senna or Sicklepod Fabaceae

Leguminous trees, shrubs, or herbs, both annuals and perennials, with pinnately compound leaves having few leaflets. Flowers in axillary racemes, yellow to orange-yellow, pea-type, and having 10 stamens of varying form. Fruit a legume (pod), oblong to long slender and arching, breaking apart irregularly or splitting open, with numerous smooth seeds. 12 species in the SE.

Common Species

Senna obtusifolia (L.) Irwin & Barneby **- sicklepod or Java-bean** **SEOB4**

Plant: Weedy annual or short-lived perennial legume, 5-100 cm tall, becoming bushy branched, having long-slender curved legume (pods), from a shallow woody root with many laterals.

Stem: Ascending, stout, yellow-green, smooth and only short scattered hairy, with tiny black glands (requires hand lens).

Leaves: Alternate, pinnately compound, 7 leaflet (terminal leaflet largest), leaves 4-15 cm long, leaflets obovate with a rounded tip, middle leaflets 1.5-5 cm long and 1-2 cm wide, stipules 5-10 mm long if not fallen.

Flowers: Jun-Sep. Drooping 1-2 racemes in upper leaf axils, pea-type but only slightly irregular, yellow-orange, 0.8-2 cm long and 1.5-2.5 cm wide, subtended by small bracts.

Fruit and seeds: Jul-Nov. Long legume (pod), slender and arching, 10-20 cm long and 3-5 mm wide, stalk 1-3 cm long, 1-2 per node, angular in cross-section, splitting to release many seeds, angular, about 5 mm long, shiny brownish.

Range: Exotic from tropical America. TX to FL and north to NJ and west to IL, MO, and NE.

Ecology: Nitrogen fixer. Weedy exotic annual that persists by long-viable seeds. Occurs in abandoned croplands to occupy and spread in new forest plantations, along forest edges and roads, and other right-of-ways. Spreads by water- and gravity-dispersed seeds.

Synonyms: *Cassia obtusifolia* L., coffeeweed.

Other species in the SE: *S. alta* (L.) Roxb. [formerly *Cassia alta* L.], *S. corymbosa* (Lam.) Irwin & Barneby [formerly *Cassia corymbosa* Lam.], *S. hebecarpa* (Fern.) Irwin & Barneby [formerly *Cassia hebecarpa* Fern.], *S. ligustrina* (L.) Irwin & Barneby [formerly *Cassia ligustrina* L.], *S. marilandica* (L.) Link [formerly *Cassia marilandica* L.], *S. mexicana* (Jacq.) Irwin & Barneby [*Cassia chapmanii* Isely], *S. occidentalis* (L.) Link [formerly *Cassia occidentalis* L.], *S. pendula* (Humb. & Bonpl. *ex* Willd.) Irwin & Barneby [formerly *Cassia pendula* Humb. & Bonpl. *ex* Willd.], *S septentrionalis* (Viviani) Irwin & Barneby [formerly *Cassia laevigata* Willd.], *S. spectabilis* (DC.) Irwin & Barneby [formerly *Cassia spectabilis* DC.], *S. surattensis* (Burm. f.) Irwin & Barneby [formerly *Cassia suffruticosa* Koenig *ex* Roth.

Wildlife: Both the foliage and seeds of sicklepod are toxic. It has no wildlife value.

Senna **Senna or Sicklepod** **Fabaceae**

T. Bodner (Aug)

Senna obtusifolia

T. Bodner (Aug)

Senna obtusifolia

T. Bodner (Aug)

Senna obtusifolia

Seymeria **Blacksenna or Seymeria** **Scrophulariaceae**

Upright, often much-branched, annual herbs (and perennials in other regions), probably parasitic on other plants. Aromatic and sticky from tiny glands on stems. Leaves opposite, compoundly divided into thin-linear leaflets. Flowers single in leaf axils, yellow, corolla 5-lobed. Capsules ovoid, splitting to release many winged seeds. 2 species in the SE.

Common Species

Seymeria cassioides (J.F. Gmel.) Blake **- smooth seymeria** **SECA4**

Plant: Upright annual, 50-100 cm tall, shrub-like and much branched, from a stout somewhat woody rootstock. Plant fragrant when rubbed

Stem: Slender branches and stout base, densely branched, opposite, ascending, squarish to angled in cross-section, sparse short hairy with sticky glands, yellow-green to purple to dark brown, turning black and persisting in late-season.

Leaves: Opposite, thread-like, pinnately to bipinnately divided to the midvein, 1-1.5 cm long and less than 0.5 mm wide, pale to yellowish green.

Flowers: Jul-Nov. Solitary and paired in upper leaf axils, yellow (purple mark inside), corolla with 5 rounded lobes and 2 partially united, flowers 7-9 mm long and 1 cm wide, hairless outside, anthers extended, calyx 5 lobed, each lobe linear, 2-3 mm long.

Fruit and seeds: Oct-Dec. Ovoid capsule, 4-7 mm long, hairless and smooth, splitting to release many furrowed seeds.

Range: Native. LA to FL and north to VA and TN, mainly in Cp.

Ecology: Occurs in small to extensive colonies in open forests, especially sandhills and loamhills, pocosin margins as well as right-of-ways. Parasitic on plant roots, including pines. Colonizes and spreads by seeds.

Synonyms: *Afzelia cassioides* J.F. Gmel., yaupon blacksenna.

Seymeria pectinata Pursh **- sticky seymeria** **SEPE2**

Similar to *Seymeria cassioides* except **plant** moderately branched with spreading branches, 20-60 cm tall; **stem** very hairy; **leaves** 1.5-3 cm long, leaflet segments 1-2 mm wide; **flowers** (Jun-Oct) hairy outside, calyx lobes 4-5 mm long; **fruit and seeds** capsule, 4-5 mm long, hairy with winged seeds. **Range:** Native. MS to FL and north to NC, dry sites.

Synonyms: comb seymeria, piedmont blacksenna.

Wildlife: No wildlife value reported.

Seymeria — Blacksenna or Seymeria — Scrophulariaceae

J. Miller (Sep)

Seymeria cassioides

J. Miller (Sep)

Seymeria cassioides

Silphium **Rosinweed** **Asteraceae**

Upright perennial herbs, large- to medium-sized, with alternate, opposite, or basal leaves. Flowers aster-type, yellow ray flowers with 3-notched tips, in flat heads, having unequal involucral bracts, in rows of 2-several series. Nutlets (achenes) broad, flat, and winged. 12 species in the SE.

Common Species

Silphium asteriscus L. **- starry rosinweed or false sunflower** **SIAS2**

Plant: Upright, large-stemmed perennial, 50-120 cm tall, sparsely branched, from a short rhizome or rootcrown with fibrous roots.
Stem: Stout, round in cross-section, smooth, yellowish-tan.
Leaves: Alternate or sometimes opposite, rough-hairy, ovate to lanceolate, 6-15 cm long and 1.5-5 cm wide and smaller upward, margins coarsely toothed to entire, green above and pale green beneath, long petioled below and sessile above.
Flowers: Jun-Sep. Several heads in leafy-bracted terminal panicles, 8-13 yellow ray petals, 1.5-3 cm long, 3-notched at end, disks yellowish, 1-3 cm wide, involucre bracts in several overlapping series, to 25 mm long, green, broad.
Fruit and seeds: Aug-Nov. Oval nutlet (achene), 4-8 mm wide and long, flat or curved, winged edges 1 mm wide with indented tip, nutlet containing 1 seed.
Range: Native. TX to FL and north to VA and west to IN, MO, and OK.
Ecology: Commonly occurs as scattered plants in new forest plantations, open forests, and along forest margins and roads, and other right-of-ways, especially dry upland sites. Persists by rhizomes and rootcrowns and spreads by seeds.
Synonyms: *Silphium dentatum* Ell.

Silphium compositum Michx. **- kidneyleaf rosinweed** **SICO5**

Plant: Perennial with an early-summer basal rosette of large deeply-lobed purple-veined leaves, producing an erect purple flower stalk, 1-3 m tall, from a coarse woody rootcrown with taproot.
Stem: Very tall and stout, erect, smooth and purple, leaves few and bract-like, appears in mid- to late-summer.
Leaves: Basal rosette, heart-shaped in outline, entire becoming deeply cleft or lobed, margins serrate (some entire and serrate), blades 10-30 cm long and 10-40 cm wide, smooth and hairless, green often with purple veins and petioles, petioles 20-30 cm, stem leaves only scattered short bracts.
Flowers: Jun-Sep. Terminal broad panicle of heads on long stalk, 1-3 m tall, heads with yellow ray petals, 5-10, 1-2 cm long, centers (disks) dark yellow, 8-12 mm wide, involucral bracts in several overlapping series, 6-10 mm long.
Fruit and seeds: Aug-Jan. Oval nutlet (achene), 6-9 mm long, flat with winged edges 1 mm wide having an indented tip, nutlet containing 1 seed.
Range: Native. AL to FL and north to VA and west to TN.
Ecology: Occurs as scattered plants in new forest plantations, open forests, and along forest margins and roads, especially dry upland sites. Persists by rootcrowns and spreads by seeds.
Synonyms: longtail silphium.

Wildlife: The rosinweeds are poor White-tailed Deer forage. No other wildlife value reported.

Silphium **Rosinweed** **Asteraceae**

T. Bodner (Jul)

Silphium compositum

T. Bodner (Jul)

Silphium asteriscus

T. Bodner (Jul)

Silphium compositum

J. Miller (Jul)

Silphium asteriscus

Solanum Nightshade Solanaceae

Annual or perennial herbs and vines, often prickly, with alternate and opposite leaves, deeply cleft, toothed, or entire. Flowers funnel-shaped, with 5 pointed lobes, spreading or reflexed, and exserted anthers. Fruit a 2-celled berry containing many seeds. 13 species in the SE.

Common Species

Solanum carolinense L. - Carolina horsenettle SOCA3

Plant: Upright perennial, 20-80 cm tall, low branching, with prickly yellow spines on stems and lower leaf veins, from a rhizome with taproots. Plant toxic to livestock and humans, berries are most toxic part of plant.
Stem: Ascending, few branches, scattered prickles, hairy.
Leaves: Alternate, thick, elliptic to ovate with 2-5 triangular lobes or teeth, 7-12 cm long and 3-8 cm wide, branched hairy (requires hand lens), folded at midvein, yellow prickles on lower veins and petiole.
Flowers: May-Sep. Terminal branched racemes, white to lavender, 2.5-3 cm wide, corolla 5 lobed, lobes pointed and spreading or reflexed, 5 extended yellow stamens.
Fruit and seeds: Jul-Oct. Spherical berry, green mottled to orange-yellow, 1-1.5 cm wide, smooth turning wrinkled, pulpy with numerous seeds, round and flattened, 1.5 mm wide, yellowish.
Range: Native. TX to FL and north to MA and west to MI and NE, and some western states.
Ecology: Occurs as scattered plants and small groupings in forest plantations and open forests, along forest edges and right-of-ways, especially dry sites. Commonly grows in pastures. Persists and colonizes by rhizomes and spreads by seeds.
Synonyms: Carolina nettle, bullnettle, Carolina nightshade.

Solanum americanum P. Mill. - American black nightshade SOAM

Plant: Upright annual, 10-100 cm tall, without prickles, sparsely or densely branched, from a taproot. Leaves and berries dangerously poisonous.
Stem: Ascending, few to many branches, slender, spineless and hairless.
Leaves: Alternate (and opposite), thin, ovate to broadly lanceolate, 3-10 cm long and 2-5 cm wide, margins wavy to toothed, hairless to scattered hairy, green above and pale green beneath.
Flowers: Jun-Nov. Axillary stemmed clusters, 3-5 flowers, white, 5 pointed spreading or reflected lobes, 6-8 mm across, 5 extended yellow stamens.
Fruit and seeds: Aug-Dec. Drooping clusters of spherical berries, shiny green turning black, 5-9 mm wide, containing many seeds, 1.2-1.8 mm long, and many hard granules.
Range: Native. TX to FL and north to ME and west to ND and NE.
Ecology: Leaves and berries poisonous to livestock and humans, although spreads by animal-dispersed seeds. Inhabits moist to wet sites and shady and semi-shady habitats, such as along stream terraces. Occurs in new forest plantations, open forests, and along forest edges and roads, and in old fields.
Synonyms: common nightshade.

Wildlife: Nightshade fruits or seeds are consumed by game birds including the Wild Turkey, Ruffed Grouse, and Northern Bobwhite as well as numerous songbirds including the Northern Cardinal, Gray Catbird, Northern Mockingbird, and others. White-tailed Deer occasionally browse the nightshades, especially *S. americanum*, primarily after the fruits ripen.

Solanum Nightshade Solanaceae

T. Bodner (Aug)

Solanum carolinense

J. Miller (Sep)

Solanum americanum

T. Bodner (Aug)

Solanum carolinense

Solidago **Goldenrod** **Asteraceae**

Erect to arching perennial herbs, from rootcrowns or short rhizomes and forming tufts of stems, and others from long rhizomes forming colonies of stems. Leaves alternate, simple, mainly basal or mainly stem, being larger towards the base or middle. Flowers yellow to gold (or a few white), in relatively small heads within yellow to whitish involucral bracts. Flowers in dense terminal branched clusters (few axillary), being flat- to round-topped or pyramid-shaped, or arching or nodding spikes with flowers on one side. Seeds within oblong, angled nutlets (achenes) with white bristle plume. 50 species in the SE.

Common Species

Solidago odora Ait. **- anisescented or fragrant goldenrod** **SOOD**

Plant: Upright slender perennial, 50-100 cm tall, with 1-few stems from a short rootcrown. Crushed leaves emit a licorice fragrance (thus the sci. and common names).

Stem: Erect to leaning, branched from base and unbranched above, slender, round in cross-section, short hairy in lines above becoming hairless below.

Leaves: Basal and alternate on stem, linear-lanceolate to elliptic, 3-10 cm long and 5-20 mm wide and smaller upward, sessile, margin entire and rough hairy, licorice odor.

Flowers: Jul-Nov. A nodding-tipped terminal panicle, slender, with horizontal lateral flower stems having many small, erect, yellow tubular heads, within yellow involucres, 3-6 mm tall.

Fruit and seeds: Oct-Mar. Nutlet (achene) 1.5-1.8 mm long, brown, topped with a tuft of white pappus bristles, 2-3 mm long, nutlet containing 1 seed.

Range: Native. TX to FL and north to VT and west to OH and OK.

Ecology: Commonly occurring as frequent scattered plants in forest plantations and persists but declining by mid-rotation. Common goldenrod in open forests, along forest margins and roads, and other right-of-ways. Mostly present on dry sites. Persists by rootcrowns and spreads by wind- and animal-dispersed seeds.

Solidago nemoralis Ait. **- gray goldenrod** **SONE**

Similar to *Solidago odora* except **plant** 20-100 cm tall, dense short-gray hairy; **leaves** thin, most and largest near base, oblanceolate, 5-25 cm long and 8-40 mm wide and smaller upward, margins slightly toothed to entire, long petioles below and sessile upward; **flowers** (Jul-Nov) in panicles or long narrow clusters, nodding at the tip.

Solidago canadensis L. **- Canada goldenrod** **SOCA6**

Similar to *Solidago odora* except **plant** 0.8-2 m tall with finely hairy stem from a creeping rhizome forming colonies; **leaves** basal rosettes in early spring and crowded stem leaves later, triple-veined, 5-15 cm long and 10-22 mm wide and smaller upward, rough hairy, margins slightly toothed; **flowers** (Jul-Nov) erect broad pyramid-shaped panicle, 10-40 cm high. **Range:** Native. Throughout Eastern US, s Canada, and n Mexico.

Ecology: Very common scattered plant on right-of-ways and in new forests.

Synonyms: *Solidago altissima* L.

Wildlife: White-tailed Deer, Eastern Cottontail, Ruffed Grouse, and Wild Turkey occasionally consume the basal rosettes of goldenrods during winter. Seeds are consumed in minor amounts by some small mammals and various songbirds, particularly the American Goldfinch.

Solidago **Goldenrod** **Asteraceae**

T. Bodner (Sep)

Solidago nemoralis

T. Bodner (Sep)/Inset - J. Miller (Oct)

Solidago odora

T. Bodner (May)

Solidago canadensis

T. Bodner (Oct)

Solidago canadensis

Sonchus — Sowthistle — Asteraceae

Annual or perennial herbs, with milky sap. Leaves often prickly-margined, occurring in spring rosettes, and in early summer becoming alternate along an erect stem. Flowers one to many in terminal clusters, in constricted urn-shaped heads, with overlapping linear bracts and yellow petals. Seeds in nutlets (achenes) topped with white bristle plumes for wind dispersal. 3 species in the SE.

Common Species

Sonchus asper (L.) Hill - **spiny sowthistle** — **SOAS**

Plant: Upright annual, 0.2-2 m tall, often with ascending branches on the upper half of the main stem, from a short taproot.
Stem: Moderately stout, fleshy, green often becoming purple, hairless, hollow, with milky sap.
Leaves: Basal and alternate on stem, margins sharp spiny and undulating, lower leaves deeply pinnately lobed becoming less so and smaller up the stem, 6-30 cm long and 3-10 cm wide, rounded ear-like projections at leaf bases clasping the stem, pale green with wide white midveins often becoming purplish with age.
Flowers: Mar-Jul (-Oct). Terminal radiating branched-clusters of small heads, 1.2-2.5 cm wide, yellow petals often hidden by rows of numerous involucral bracts, almost as long as petals and enclosing flowers tightly.
Fruit and seeds: May-Nov. Cylindric nutlet (achene), 2.7-3 mm long, tan-reddish brown, 3-5 ribbed per face, topped with white bristles, 3-4 mm long, nutlet containing 1 seed.
Range: Native to Europe and naturalized to entire US.
Ecology: Early invader of disturbed soils, such as occurs on new forest plantations, right-of-ways, and cultivated fields. Often occurs singly as scattered plants. Spreads by wind-dispersed seeds.
Synonyms: prickly sowthistle, sowthistle.

Other species in the SE: *S. arvensis* L. [perennial sow thistle], *S. oleraceus* L. [common sowthistle].

Wildlife: Sowthistles are poor White-tailed Deer forage, infrequently browsed.

Sonchus **Sowthistle** **Asteraceae**

J. Miller (May)

Sonchus asper

T. Bodner (Apr)

Sonchus asper

T. Murphy (May)

Sonchus asper

T. Bodner (May)

Sonchus asper

Stillingia **Toothleaf or Stillingia** **Euphorbiaceae**

Upright hairless perennial herbs or shrubs, with alternate leaves, sessile or short petioled, having glandular-serrate margins (minute red tipped). Flowers and fruit in terminal spikes. Female flowers at the spike base and petal-less yellow male flowers above. Fruit a 3-seeded capsule, with a 3-lobed base remaining on the spike base, spike top deciduous. 3 species in the SE.

Common Species

Stillingia sylvatica Garden *ex* L. **- queen's-delight** **STSY**

Plant: Upright hairless perennial, small shrub-like, 40-80 cm tall, with multiple stems from a woody rootstock. Milky juice where injured.

Stem: Stout, hairless, round in cross-section, light-green with reddish tinge, branched from base and unbranched above except for whorls immediately below terminal floral spike.

Leaves: Alternately spiraling, elliptic to lanceolate to oblanceolate, 3-9 cm long and 1- 4.5 cm wide, hairless, shiny green to yellowish green with light-green midvein, margins serrate with tiny incurved reddish glandular tips (requires hand lens), petioles 0-4 mm long.

Flowers: May-Sep. Stout terminal spike, 5-12 cm long, with small petal-less male flowers spiraling around the upper portion, yellow sepals, with 2 tiny saucer-shaped glands for each, few female flowers around spike base, spike deciduous at the base of male-flower portion.

Fruit and seeds: Jul-Nov. Woody capsule, 8-10 mm long, containing 2-3 seeds, ovoid, 5-8 mm long, wrinkled, a 3-pronged capsule base remaining on plant.

Range: Native. NM to FL and north to VA and west to MO and OK.

Ecology: Occurs as scattered plants and small colonies on dry upland sites. Present in new forest plantations and open forests, and along forest edges and roads, and other right-of-ways. Persists by woody rootcrowns and spreads by seeds. Apparently pollinated by ants.

Synonyms: *S. angustifolia* (Torr.) Engelm. *ex* S. Wats., *S. spathulata* (Muell.- Arg.) Small, queensroot.

Other species in the SE: *S. aquatica* Chapman [small shrub], *S. texana* I.M. Johnston.

Wildlife: Stillingia seeds are occasionally consumed by Northern Bobwhite. No other wildlife value is reported.

T. Bodner (May)

Stillingia sylvatica

T. Bodner (May)

Stillingia sylvatica

Strophostyles **Fuzzybean or Wild bean** **Fabaceae**

Twining or reclining, viney annual or perennial legumes. Leaves with 3 leaflets having long petioles, persistent stipules and stipels at bases of leaves and leaflets, respectively. Flowers in small clusters at the end of very long leafless flower stalks. Flowers pink-purple to white, pea-type but with a rounded back-petal, each with a pair of striate basal bracts. Fruit a legume (pod), sessile, straight and linear, splitting at maturity to release few to several woolly to hairy seeds. 3 species in the SE.

Common Species

Strophostyles umbellata (Muhl. *ex* Willd.) Britt. **- trailing fuzzybean STUM2**

Plant: Trailing or twining perennial, legume vine, 0.5-1.5 m long, from a woody rootcrown.
Stem: Vine-like, slender, hairy when young, twining among low plants and litter.
Leaves: Alternate, 3 leaflet, leaflets ovate to lanceolate to oblong, 2-4 cm long and 0.3- 2 cm wide, slightly hairy, tip round with a tiny hair-like end tip, persistent stipules and stipels at the base of leaves and leaflets, respectively.
Flowers: Jun-Sep. Tight clusters (often only 1 flowering at a time) on long-slender and erect leafless stalks, 10-25 cm long, pea-type, petals pinkish-white with purple center (fading to yellowish), back petals round in outline and curved backwards, 1-1.4 cm long.
Fruit and seeds: Aug-Dec. Legume (pod), 3-6 cm long and 3-5 mm wide, round, pointed tipped, slightly hairy, splitting at maturity to release several seeds, oblong, 3-6 mm long.
Range: Native. TX to FL and north to NY and west to IL and OK.
Ecology: Nitrogen fixer. Infrequent, occurring as single plants or small colonies in open forests, fields, and right-of-ways, especially dry sandy soils. Persists by rootcrowns and residual seeds, and spreads by animal-dispersed seeds.
Synonyms: sandbean, trailingbean, wild bean.

Other species in the SE: *S. helvula* (L.) Ell., *S. leiosperma* (Torr. & Gray) Piper.

Wildlife: Seeds of wild beans are an important food source for Northern Bobwhite and Mourning Dove across the Southeast. They likely are consumed by numerous seed-eating songbirds. They are a moderate to preferred summer White-tailed Deer forage, but rarely are abundant enough to contribute much to the total diet.

Strophostyles — Fuzzybean or Wild bean — Fabaceae

T. Bodner (Aug)

Strophostyles umbellata

T. Bodner (Aug)

Strophostyles umbellata

Stylosanthes **Pencilflower** **Fabaceae**

Low-growing, perennial herbs, legumes, with 3-leaflet leaves having sheath-like stipules on the petioles. Flowers in terminal, tight clusters (opening one at a time), with many bristly leaf-like bracts subtending small pea-type flowers, petals yellow to yellow-orange. Fruit an inflated legume (pod), 2-segmented (only 1 segment may be filled), ovoid to rectangular. 3 species in the SE, with 2 only in FL.

Common Species

Stylosanthes biflora (L.) B.S.P. **- sidebeak pencilflower** **STBI2**

Plant: Upright or sprawling, perennial legume, 10-40 cm tall or long, from a hard rootcrown having linear rootlets.

Stem: Erect to sprawling, stiff and wiry, green, round, bristly hairy (or not), 1 to few branches from base or sparingly above.

Leaves: Alternate, 3 leaflet, sheathing bristle-tipped stipules on petioles (often purple tinged), leaflets narrowly lanceolate to elliptic, 1.5-4 cm long and 3-7 mm wide, bristle tipped and bristles on some margins, green with veins on lower surface lighter green and paralleling the margins.

Flowers: Jun-Aug. Clusters of 2-6 at branch terminals, opening singly, small pea-type flowers, distinct yellow-orange petals, upper 2 lobes 6-7 mm long, within bristly bracts and sheaths.

Fruit and seeds: Jul-Oct. Legume (pod), ovoid with an incurved beak, 3-5 mm long, thinly hairy, short stalked, containing 1 rounded seed (does not split open to release seed).

Range: Native. TX to n FL and north to NY and west to IL and KS.

Ecology: Nitrogen fixer. Occurs commonly as widely scattered plants in new forest plantations, open forests, and along forest edges and right-of-ways. Inhabits mainly dry sites, but also occurs in some moist pine savannas. Persists by rootcrowns and spreads by animal-dispersed seeds.

Synonyms: *S. riparia* Kearney, *S. floridana* Blake.

Other species in the SE: Occur only in FL: *S. calcicola* Small, *S. hamata* (L.) Taubert.

Wildlife: Pencilflower seeds are consumed by Northern Bobwhite and songbirds. The plant is a moderately preferred White-tailed Deer forage.

Stylosanthes **Pencilflower** **Fabaceae**

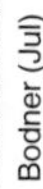

Stylosanthes biflora

T. Bodner (Sep)

Stylosanthes biflora

Tephrosia Hoarypea or Tephrosia Fabaceae

Upright perennial legumes, with odd-pinnate alternate leaves, having distinct lateral veins, pointed tips, and hairy under surfaces. Flowers in elongated racemes, pea-type, initially white, yellow, or pink, turning red-purple overnight. Fruit a legume (pod), oblong to linear and flat, splitting along 2 seams to release few to several rounded seeds. 9 species in the SE.

Common Species

Tephrosia virginiana (L.) Pers. **- goat's rue or Virginia tephrosia** **TEVI**

Plant: Erect perennial legume, 30-70 cm tall, often in dense colonies, from a branched woody rootcrown and long woody roots.

Stem: Upright, softly hairy, round in cross-section, green, 1-several from ground, often unbranched.

Leaves: Alternate, odd-pinnate with leafstalks 4-12 cm long, often curved, leaflets 11-23, gray hairy, elliptic to linear-oblong, 1-3 cm long and 4-8 mm wide, tip pointed (mucronate).

Flowers: May-Oct. Terminal and axillary long clusters (racemes), 4-8 cm long, bi-colored pea-type flowers, yellow-cream top petals (14-20 mm long) with pink-rose centers and outside of petals, bracts hairy.

Fruit and seeds: Jul-Dec. Legume (pods), oblong, 3-5 cm long and 3-5 mm wide, gray hairy, flattened, splitting along two seams to release 6-11 seeds, rounded, 3-4 mm wide.

Range: Native. TX to FL and north to MA and west to SD.

Ecology: Nitrogen fixer. Commonly occurs in young forests and open forests, fields, and right-of-ways, most often in small colonies. Scattered plants occur in shady pine plantations. Inhabits a wide range of sites but more common on dry sandy soils. Persists by rootstocks and residual seeds, and spreads by animal-dispersed seeds.

Synonyms: *Cracca virginiana* L., catgut, devil's-shoestring, hoarypea.

Tephrosia spicata (Walt.) Torr. & Gray **- spiked hoarypea** **TESP**

Similar to *Tephrosia virginiana* except **plant** ascending to sprawling, 30-80 cm long, few slender stems, rusty hairy, from a long cylindric root; **leaves** sparsely leafy, with hairy leafstalks 3-10 cm long, leaflets 9-17, oblong-obovate to obovate or elliptic, 1-3 cm long and 6-12 mm wide, rusty hairy on back, also with sharp-pointed tip; **flowers** (Jun-Oct) small clusters atop long slender stalks, 10-30 cm long, blooming singly, petals initially white, turning overnight to pink and maroon, 13-18 mm long; **fruit and seeds** legume (pods) brown, hairy becoming smooth. **Range:** Native. LA to all FL and north to s DE and se VA and southwest to KY and MO.

Ecology: Nitrogen fixer. Occurs commonly, but infrequently, as scattered plants in similar habitat as *T. virginiana*. Often occurs in small groups or scattered plants in openings.

Synonyms: sand pea, brownhair tephrosia.

Other species in the SE: *T. angustissima* Shuttlw. *ex* Chapman [only in FL], *T. chrysophylla* Pursh, *T. florida* (F.G. Dietr.) C.E. Wood, *T.* x *floridana* (Vail) Isely, *T. hispidula* (Michx.) Pers., *T. onobrychoides* Nutt., *T. rugelii* Shuttlw. *ex* B.L. Robins.

Wildlife: Tephrosia seeds are consumed infrequently by Northern Bobwhite and songbirds. The plants are low to moderate preference White-tailed Deer forages.

Tephrosia **Hoarypea or Tephrosia** **Fabaceae**

T. Bodner (Jun)

Tephrosia spicata

T. Bodner (May)

Tephrosia virginiana

J. Miller (Jun)

Tephrosia spicata

J. Miller (Jul)

Tephrosia spicata

Tradescantia **Spiderwort** **Commelinaceae**

Upright perennial herbs, somewhat succulent, with large alternate grass-like leaves having swollen basal sheaths. Flowers in terminal and axillary drooping clusters subtended by a leaf-like bract, petals 3, blue or rose, stamen 6, bearded. Tight clusters of spherical to obovoid capsules, with 2-3 cells, each splitting to release 1-2 seeds, gray, flat and ellipsoid. 9 species in the SE.

Common Species

Tradescantia ohiensis Raf. **- common spiderwort** **TROH**

Plant: Upright slender perennial, 20-60 cm tall, grass-like, with long arching leaves, from a rootcrown with only lateral roots.

Stem: Ascending, slender, green, round in cross-section, somewhat succulent, few alternate branches if any.

Leaves: Alternate, long linear and arching with long-tapering tips, flat and firm, 20-40 cm long and 6-12 mm wide, forming swollen sheaths at base, hairless and smooth.

Flowers: Apr-Jul (-Nov). Terminal or axillary tight drooping clusters on long stalks encased within a pair of arching leaf-like bracts, flowers opening singly, petals 3, blue or rose, broad-diamond shaped, 10-15 mm long and wide, sepals hairless or only hair tufts at tips. Flowers last at most 1 day before liquefying.

Fruit and seeds: Jun-Aug(-Dec). Ovoid capsule, 4-5.5 mm long and 3-4 mm wide, splitting to release gray seeds, 2-4 mm long.

Range: Native. TX to FL and north to MA and west to MN and NE.

Ecology: Occurs as scattered plants in new forest plantations, open forests, and along forest edges and right-of-ways. Inhabits fields and prairies, especially dry sites. Persists by rootcrowns and spreads by seeds.

Synonyms: *T. incarnata* Small, reflexed spiderwort, smooth spiderwort, dayflower, snotweed, bluejacket.

Other species in the SE: *T. ernestiana* E.S. Anderson & Woods, *T. fluminensis* Vell. [exotic - wandering jew], *T. hirsuticaulis* Small, *T. hirsutiflora* Bush, *T. ozarkana* E.S. Anderson & Woods, *T. roseolens* Small, *T. subaspera* Ker-Gawl., *T. virginiana* L.

Wildlife: Spiderworts have little wildlife value, although infrequent use of spiderwort seeds by songbirds and low to moderate use by White-tailed Deer have been reported.

Tradescantia **Spiderwort** **Commelinaceae**

T. Bodner (May)

Tradescantia ohiensis

Tragia **Noseburn** **Euphorbiaceae**

Erect, reclining, or twining perennial herbs, from rhizomes. Often armed with fine stinging hairs. Leaves alternate, with stipuled petioles. Male and female flowers separate on the same plant (monoecious). Flowers along delicate-slender stalks (spikes or racemes), arising opposite to leaves, petals absent, sepals greenish or purplish. Fruit a spherical capsule, 3 lobed, splitting to release 1 semi-spherical seed per lobe. 6 species in the SE.

Common Species

Tragia urticifolia Michx. **- nettleleaf noseburn** **TRUR2**

Plant: Upright or reclining perennial, 20-70 cm tall, slender stems with mildly stinging hairs (may only sting when touched to sensitive skin such as beneath the nose, thus common name), from a woody rhizome.

Stem: Branched from base or not, one to few branches, light green, bristled with white hairs that may sting.

Leaves: Alternate, triangular-lanceolate, with rounded or heart-shaped bases, 2-6 cm long and 0.7-4 cm wide, margins singly or doubly serrate (hair tipped), petioles 5-15 mm, white hairy, with tiny hairy stipules.

Flowers: May-Oct. Terminal or axillary spikes, 1-4 cm long, with 1-2 female flowers at the base, 11-40 male flowers above, no petals only 3 green sepals, persistent stalks and basal bracts (1.5-3 mm long).

Fruit and seeds: Jun-Nov. Spherical capsule, 4-10 mm wide, greenish, white bristly, clustered in 3's, each splitting to release 1 spherical seed, 3-4 mm wide, mottled.

Range: Native. AR to FL and north to VA and west to MO and CO.

Ecology: Commonly occurs as scattered plants in new forest plantations, forest openings, and along forest edges and right-of-ways. Inhabits old fields, especially dry upland sites. Persists and colonizes by rhizomes and spreads by seeds.

Synonyms: *T. urticaefolia* Michx.

Tragia urens L. **- wavyleaf noseburn** **TRUR**

Similar to *Tragia urticifolia* except **plant** smaller, 20-40 cm tall, usually branched, no stinging hairs, sparsely hairy; **leaves** narrowly elliptic to oblanceolate to linear, 2-10 cm long and 0.2-2 cm wide, narrowing base, margins wavy to irregularly serrate to entire, petioles 1-3 mm long. **Range:** Native. TX to FL and north to VA, mainly in Cp.

Synonyms: southeastern noseburn.

Other species in the SE: *T. betonicifolia* Nutt., *T. cordata* Michx., *T. smallii* Shinners, *T. saxicola* Small [only in FL].

Wildlife: *Tragia* produce seeds that are used by Northern Bobwhite and songbirds. Little other wildlife value has been reported.

Tragia **Noseburn** **Euphorbiaceae**

T. Bodner (Jul)

Tragia urticifolia

Verbascum **Mullein** **Scrophulariaceae**

Erect very hairy perennial or biennial herbs, forming leafy basal rosettes the first year and tall flowering and seeding stems later. Stem leaves alternately spiraling and sessile. Flowers yellow or white, 5-petaled, with bracts, along elongated stalks (spikes or racemes). Fruit a capsule, spherical to ovoid, splitting into halves to release numerous seeds. 4 species in the SE.

Common Species

Verbascum thapsus L. - common mullein **VETH**

Plant: Erect biennial, 1-2 m tall, usually unbranched and densely gray hairy, from an initial basal rosette yielding a stalk with crowded yellow fragrant flowers, from a fleshy taproot.

Stem: Stout, elongating during second year, densely and softly hairy.

Leaves: Basal rosette and alternately spiraling on stem, dense woolly hairy, leafy rosette 20-60 cm wide, leaves elliptic to oblanceolate, 10-30 cm long and 3-12 cm wide and much smaller upward and more pointed, yellow green, margins entire to irregularly toothed and extending down along the stem.

Flowers: Jun-Sep. Crowded around and along a tall leafy stalk, 20-50 cm high, rarely branched near top, corolla yellow and 5 lobed, 15-25 mm wide, within woolly 5-lobed sepals.

Fruit and seeds: Oct-Mar. Ovoid capsule, 6-10 mm long, 2 celled, splitting to release many seeds, 0.8-1 mm long, tan to brown, roughened and grooved resembling corn-cob parts. Erect stalk remains standing through winter.

Range: Naturalized from Europe. Throughout US and s Canada.

Ecology: Occurs commonly and frequently as scattered plants in open forests, new forest plantations, and disturbed sites, and along forest edges and right-of-ways. Spreads by seeds that germinate in winter.

Synonyms: woolly mullein.

Other species in the SE: *V. blattaria* L., *V. phlomoides* L., *V. virgatum* Stokes.

Wildlife: No wildlife value reported.

Verbascum **Mullein** **Scrophulariaceae**

T. Bodner (Jun)

Verbascum thapsus

T. Bodner (Aug)

Verbascum thapsus

Verbena **Vervain** **Verbenaceae**

Erect, ascending, or reclining annual or perennial herbs, with square stems and opposite leaves and branches. Flowers with bracts crowded into slender terminal spikes, either singly or many branched, 5 petals, blue, violet, pink, or white. Nutlets 4 per flower, held within the persistent calyx, splitting apart at maturity. 13 species in the SE.

Common Species

Verbena brasiliensis Vell. - **Brazilian vervain** **VEBR2**

Plant: Upright perennial, with rough green square stems, 1-2 m tall, spindly 1-few stems from the base and often branched at mid-plant, from a woody rootcrown or rhizome.
Stem: Slender, square in cross-section, green, rough hairy on edges, opposite branched in upper plant.
Leaves: Opposite in widely-spaced pairs, obovate to elliptic to lanceolate, 4-10 cm long and 0.8-2.5 cm wide, margins coarsely serrate, base tapering, dark green above with lighter green veins and whitish green beneath.
Flowers: Apr-Oct. Terminal fascicles of compact slender spikes, 0.5-4 cm long and 4-5 mm across, atop long slender leafless stalks, also square and rough, corolla lavender to purple, 2.5 mm wide, within hairy sepals and bracts.
Fruit and seeds: Jul-Mar. Tiny nutlets along spikes, each oblong, 1-2 mm long, 4 per flower, within persistent calyx and bracts.
Range: Naturalized exotic from South America. LA to FL and north to VA and west to AR.
Ecology: Occurs commonly as scattered plants and small colonies in new forest plantations, along forest edges and right-of-ways. Inhabits fields, especially on dry Cp sites. Persists by rootcrown and spreads by seeds.
Synonyms: purple vervain, blue vervain.

Verbena rigida Spreng. - **tuberous or rigid vervain** **VERI2**

Plant: Upright to spreading perennial, 10-50 cm tall, infrequently branched, rigid stems (thus the sci. name), from underground stolons.
Stem: Rigid, round in cross-section, rough hairy.
Leaves: Opposite, lanceolate to oblanceolate, 4-11 cm long and 0.7-3 cm wide, flaring coarse teeth, sessile and partially clasping stem, spine tipped, hairy, prominently veined.
Flowers: Mar-Sep. Stiffly-erect spike, 1-6 cm long and 1-3 cm wide, corolla purple to violet, funnelform, 2-5 times longer than hairy calyx and subtending hairy bract, corolla as long or longer than calyx, bristly appearing.
Fruit and seeds: Jun-Nov. Nutlets loosely clustered, 1.7-1.9 mm long, warty.
Range: Naturalized exotic from South America. TX to n FL and north to NC, mainly Cp.
Ecology: Occurs in patches along forest edges and right-of-ways. Persists and colonizes by stolons and spreads by seeds.

Wildlife: Vervains can be locally abundant but have limited wildlife value. Seeds are consumed in small amounts by Northern Cardinals and various sparrows. The foliage is a low to moderate preference White-tailed Deer forage. The flowers are attractive to numerous species of butterflies including the Gray Hairstreak, Common Checkered Skipper, Gulf Fritillary, Malachite, Monarch, and Pipevine Swallowtail.

Verbena **Vervain** **Verbenaceae**

J. Miller (Jun)

Verbena brasiliensis

T. Bodner (May)

Verbena brasiliensis

T. Bodner (Oct)

Verbena brasiliensis

J. Miller (Jun)

Verbena rigida

Vicia **Vetch** **Fabaceae**

Trailing or sprawling annual or perennial legumes, with angled stems. Pinnately compound leaves with numerous small leaflets, terminal branched tendrils, and small clasping cleft stipules. Flowers single to many on stalks in leaf axils, pea-type, reddish-purple to violet to yellow to white, with a cleft tubular calyx. Fruit a legume (pod), oblong and flattened to rounded, splitting to release 2 to many pea-type seeds. 20 species in the SE, about half being introduced.

Common Species

Vicia sativa* ssp. *nigra (L.) Ehrh. **- garden or narrowleaf vetch** **VISA**

Plant: Trailing to sprawling annual, 30-80 cm long, in spring often forming dense entanglements by twining tendrils, from a taproot.

Stem: Reclining or erect or tending to climb, slender and angled, green to purple, slightly hairy or not, branched near base.

Leaves: Alternate, pinnately compound (3-8 leaflet pairs) with a branched tendril at the end, leaflets oblong to elliptic to obovate, 3-5 cm long and 1.5-7 mm wide, tip pointed (mucronate), softly hairy, small stipules at petiole base cleft and clasping stem.

Flowers: Mar-Jul. Paired in upper leaf axils, pea-type, pink-purple (to whitish), 1-1.8 cm long, nearly sessile, calyx tubular, 7-12 mm long, with pointed tips.

Fruit and seeds: Apr-Aug. Legume (pod), 4-6 cm long and 5-7 mm wide, flattened, smooth, brown turning black, splitting to release 4-7 round seeds, 3 mm wide, velvety, brown-black or olive-brown with fine black spots.

Range: Naturalized from Europe. Throughout US.

Ecology: Nitrogen fixer. Early spring to early summer small colonies occur along forest margins and right-of-ways. Spreads by animal-dispersed seeds.

Synonyms: *V. angustifolia* L., *V. sativa* var. *angustifolia* (L.) Ser., *V. sativa* var. *nigra* L., *V. sativa* var. *segetalis* (Thuill.) Ser., spring vetch.

Vicia grandiflora Scop. **- large yellow vetch** **VIGR**

Similar to *Vicia sativa* except **flowers** yellow often purple-streaked, 2.3-3 cm long.

Vicia hirsuta (L.) S.F. Gray **- tiny or hairy vetch** **VIHI**

Similar to *Vicia sativa* except **leaves** 6-8 pairs of leaflets with branched tendril (or simple) at end, leaflets linear to elliptic, 0.5-2 cm long; **flowers** white, 3-6 in raceme, each 2-3 mm.

Vicia caroliniana Walt. **- Carolina or wood vetch** **VICA2**

Similar to *Vicia sativa* except **plant** perennial from rhizomes; **leaves** leaflets (10-18 pairs) 1-2.5 cm long, tendrils unbranched, stipules entire and non-clasping; **flowers** white and lavender-tipped, 0.8-1.2 cm long, 8-20 on one side of stalk (3-10 cm long), calyx tube 2-2.5 mm long. **Range:** Native. LA to GA and north to NY and west to MN and OK, mainly Mt and Pd.

Ecology: Nitrogen fixer. Occurs as single plants or small colonies in open sites, alluvial bottoms to dry uplands sites. Persists and colonizes by rhizomes and spreads by animal-dispersed seeds.

Wildlife: Vetch seeds are used occasionally by gamebirds and songbirds. The foliage is browsed by White-tailed Deer, Eastern Cottontail, Wild Turkey, and Northern Bobwhite, particularly in the spring and early summer. Vetches are larval food plants for the Dogface and Little Yellow butterflies.

Vicia **Vetch** **Fabaceae**

J. Miller (Dec)

Vicia sativa ssp. *nigra*

J. Miller (Apr)

Vicia sativa ssp. *nigra*

J. Miller (Apr)

Vicia hirsuta

J. Miller (Apr)

Vicia grandiflora

Viola Violet or Pansy Violaceae

Low growing annuals or perennials, from rhizomes and stolons. Leaves in basal rosettes, or alternate on stems in some species. Familiar violet flowers, irregular with 5 petals, white to yellow to violet or purple, on stalks having 2 small lateral bracts. Some species also with small inconspicuous flowers under leaves in the summer producing most of the viable seeds. Fruit a 3-celled capsule, splitting to release numerous seeds. 28 species in the SE, with many hybrids.

Common Species

Viola affinis Le Conte **- sand or common violet** **VIAF2**

Plant: Basal rosette of heart-shaped, dark-green leaves and stalked flower, 4-10 cm tall, from a much-rooted rhizome.
Stem: None.
Leaves: Basal rosette, flat against the ground, blades roundish to broad heart-shaped, thick, 3-8 cm long and 2-5 cm wide, slightly hairy, margins with minute teeth, stipules.
Flowers: Feb-May. Solitary on stalk, petals 5, violet (rarely white) with white throat and fringe, 1-2 cm wide, whitish petal-spur extending back between 5 sepals, 2 tiny reddish bracts mid-stalk.
Fruit and seeds: Mar-Jun. Ellipsoid capsule, purple-dotted or green, 5-8 mm long within persistent sepals to half-length, splitting into 3 segments releasing many minute seeds.
Range: Native. TX to FL and north to s Canada and west to WI and AR.
Ecology: Occurs as single plants or small colonies under open to semi-shady forests, and along margins. Persists by rhizomes and spreads by seeds. Insect and self pollinated.
Synonyms: *V. floridana* Brainerd.

Viola blanda Willd. **- sweet white violet** **VIBL**

Similar to *Viola affinis* except **plant** from slender trailing rhizome; **leaves** also heart-shaped, thin, ascending rosette, margins crenate; **flowers** (Mar-Jun) white with brown-purple veins on lower 3 petals, lateral petals turned back and often twisted, stalk usually reddish tinged, stalk bracts minute. **Range:** Native. AL to GA and north to Canada and west to MN.

Viola pedata L. **- birdfoot violet** **VIPE**

Similar to *Viola affinis* except **leaves** deeply lobed (5-11), usually in 3's (bird-foot shaped, thus the common name); **flowers** (Mar-Jun) reddish-purple with white centers, petals all in one plane, 3-4.5 cm wide, with orange stamens in center. **Range:** Native. TX to GA and north to ME and west to MI and MO.

Viola bicolor Pursh **- field pansy** **VIBI**

Plant: Erect annual, 5 -10 cm tall (up to 40 cm tall), often occurring in small colonies.
Stem: Slender, angled, green turning dark purple, often branched near base.
Leaves: Alternate, rounded with long petioles (1-3 cm long), stipules large, cleft and lobed.
Flowers: Feb-May. Axillary, lavender to white with purple veins and yellow center.
Fruit and seeds: Mar-Jun. Capsule, 5-7 mm long, seeds light brown, 1.1-1.3 mm long.
Range: Native. TX to n FL and north to NY and west to MI and KS.
Ecology: Occurs singly or in small colonies in open forests and margins. Spreads by seeds.
Synonyms: *V. rafinesqueii* Greene, Johnny-jump-up, wild pansy.

Wildlife: During summer, violet seeds are used commonly by Northern Bobwhite and sparingly by songbirds. Leaves are eaten during winter occasionally by White-tailed Deer and Ruffed Grouse, but rarely at other times of the year. Wild Turkey forage on the tuberous roots and several of the smaller species of butterfly visit violet flowers.

Viola Violet or Pansy Violaceae

T. Bodner (Apr)

Viola pedata

T. Bodner (Mar)

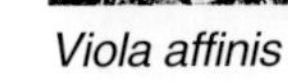

Viola affinis

T. Bodner (Mar)

Viola bicolor

T. Bodner (Mar)

Viola blanda

Grasses, Sedges, and Rushes

Andropogon Bluestem or Broomstraw or Beardgrass Poaceae

Erect tufted or bunched perennial grasses, from a hardened base or short rhizome. Leaves long-linear, basal becoming alternate, having rough margins and membranous ligules (with or without fringe). Flowers and seeds in stemmed spikes, branched and long hairy, within or extending from leaf-like sheaths (spathes). Seed grain purplish or yellowish, tapered linear, 2-3 mm long, with awns and stalked hairy bristles for wind dispersal. 12 species in the SE (some with several varieties), with 5 mainly in FL. Previously included *Schizachyrium* (such as *S. scoparium* - little bluestem) and *Bothriochloa* (such as *B. saccharoides* - silver bluestem), now separate genera.

Common Species

Andropogon virginicus L. **- broomsedge bluestem** **ANVI2**

Plant: Commonly-occurring bunch grass, 50-100 cm tall, upright flattened tufts of leaves and later stems, greenish-yellow or blue-gray, with long-slender often reddish-tinged internodes, from a flattened hard base with fibrous roots. Dried straw-colored plants remain upright until following spring, and dried curled basal leaves remain most of the summer.

Stem (culm): Erect, tufted in flattened leafy bases (2-5 mm thick), greenish white to yellow, rounded above, almost hairless, internodes numerous and often reddish to purplish tinged becoming shorter upward, branched in the upper 2/3 with branches close to the main stem.

Leaves: Basal and alternate, 15-35 cm long and 3-6 mm wide, flat or folded, light yellowish green, sheaths shorter than internodes, hairy on the upper surface near the sheath (smooth or hairy elsewhere), membrane a ligule with fringe, 0.5-0.8 mm long.

Flowers: Sep-Nov. White-hair tufted spikes, each 2-3 cm long, thin wavy stemmed, 1-4 spikes only partially extending from a 2-6 cm long small leaf (spathe) overlapping on a stalk and few to several branched, terminal and axillary, overall forming a raceme.

Seeds: Dec-Apr. Grain slender with tapered ends, 3-4.5 mm long, tipped with a straight thin hair-like awn, 1-2 cm long, and paired stemmed tufts of white to silvery bristles from the seed base.

Range: Native. TX to FL and north to ME and west to ND, also s Canada and n Mexico.

Ecology: Common grass species in new forest plantations and forest openings, and along forest margins and right-of-ways. Forms extensive colonies in old pastures and fields. Occurs in all physiographic provinces, but local in abundance. Often increases in abundance after burning and forest herbicide treatments. Persists by rootcrowns, increases abundance and size by rootcrown sprouting, and spreads by wind-dispersed seeds.

Synonyms: broomstraw.

Schizachyrium scoparium (Michx.) Nash **- little bluestem** **SCSC**

(formerly *Andropogon scoparius* Michx.)

Similar to *Andropogon virginicus* except **plant** 50-150 cm tall, true bunch grass with few to many roundish stems in clumps, bases not as flattened, loosely branching and leaning in mature plants; **stem** long and slender with mostly hairy sheaths and mostly reddish to purplish internodes in late summer and fall; **leaves** 25-60 cm long with ligule unfringed to 1 mm long; **flowers** a single spike extending well beyond a narrow spathe; **seeds** distinct due to less tufted bristles, alternate, tipped by hair-like awn 7-14 mm long. **Range:** Native. Throughout U.S., s Canada, and n Mexico.

Ecology: Most common grass in young forests and on right-of-ways in many locales.

Andropogon **Bluestem or Broomstraw or Beardgrass** **Poaceae**

J. Miller (Nov)

Andropogon virginicus

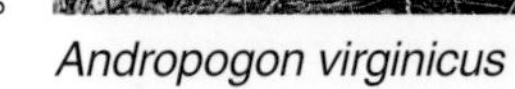

J. Miller (Jul)

Andropogon virginicus

right J. Miller (Oct) left (Nov)

Andropogon virginicus

J. Miller (Dec)

Schizachyrium scoparium

Andropogon Bluestem or Broomstraw or Beardgrass Poaceae

Andropogon ternarius Michx. - splitbeard bluestem ANTE2

Similar to *Andropogon virginicus* except **plant** 80-120 cm tall, only slightly or not flattened at the base, base 2-4 mm thick, white-haired stemmed tufts in pairs remaining on branch tips all winter (splitbeards, thus common name); **stem** often purplish to reddish green with green upper sheaths, all silver waxy; **leaves** and sheaths often densely hairy and silver waxy, ligules brown 1-3 mm long; **flowers** (Aug-Nov) white-hair tufted spikes, paired (rarely 3), each 3-6 cm long, flat stemmed, mostly fully extending from a longer leaf-like spathe, smooth and hairless; **seeds** grain 5-6 mm long, often greenish-purple, tipped with a twisted or spiraling awn, 1.5-2.5 cm long, base with paired stems of white bristles. **Range:** Native. TX to FL and north to DE and southwest to KY and KS.
Synonyms: *A. cabanisii* Hack., silver bluestem, paintbrush bluestem.

Andropogon glomeratus (Walt.) B.S.P. - bushy bluestem ANGL2

Similar to *Andropogon virginicus* except **plant** 50-150 cm tall, slightly flattened at base, distinctly bushy-branched above; **stem** stout, base 5-8 mm thick; **leaves** usually more than 35 cm long, membrane ligule with fringe, 1-2 mm long, no hairs near sheath; **flowers** in terminal dense fan-shaped clusters of densely silky spikes, each spike 2 branched, 1-3 cm long, extending from crowded and overlapping leafy bracts (spathes) about the same length or shorter; **seeds** grain 3-4 mm long, tipped with a straight awn, 1.0-1.5 cm long, base with paired stems of white bristles. **Range:** Native. CA to FL and north to MA and west to WI.
Ecology: Occurs on low wet sites, along ditchbanks, pond margins, and swamps where tree canopies are absent or sparse.
Synonyms: *A. virginicus* var. *glaucopsis* (Ell.) A.S. Hitchc., *A. virginicus* var. *abbreviatus* (Hack.) Fern. & Grisc, bushy beardgrass.

Andropogon gyrans Ashe - Elliott's bluestem ANGY2

Similar to *Andropogon virginicus* except **plant** 30-90 cm tall, base slightly flattened and scattered hairy, 1-few stems; **stem** base 2.5-4 mm thick; **leaves** to 50 cm long and 2.5-5 mm wide; **flowers** in terminal clusters of white-hair tufted spikes, paired, each 3-5 cm long, thin wavy stemmed, both extending from and enclosed within a large leaf-like spathe, 8-15 cm long and 4-15 mm wide, few to several overlapping on a stalk, remaining dried and obvious in winter; **seeds** grain 4-5 mm long, tipped by a loosely twisted hair-like awn, 1-2.5 mm long, base with paired stems of white bristles. **Range:** Native. TX to FL and north to NJ and west to IL and MO.
Ecology: Occurs with lower frequency on sites with *Andropogon virginicus.*
Synonyms: *A. elliottii* Chapman, *A. campyloracheus* Nash.

Other species in the SE: *A. arctatus* Chapman [only in FL], *A. brachystachyus* Chapman [only in FL], *A. capillipes* Nash., *A. floridanus* Scribn. [only in FL], *A. gerardii* Vitman, *A. longiberbis* Hack. [only in FL], *A. mohrii* (Hack.) Hack. *ex* Vasey, *A. tracyi* Nash [only in FL].

Wildlife: Broomsedge and bluestem seeds are consumed sparingly by Northern Bobwhite and seed-eating songbirds, particularly the Field Sparrow and Dark-eyed Junco. Clumps of broomsedge provide excellent nesting sites for Northern Bobwhite. *Andropogon* spp. are poor White-tailed Deer forage. Stems are a primary food of the Cotton Rat.

Andropogon Bluestem or Broomstraw or Beardgrass Poaceae

J. Miller (Nov)

Andropogon ternarius

J. Miller (Dec)

Andropogon ternarius

J. Miller (Apr)

Andropogon gyrans

T. Bodner (Sep)

Andropogon glomeratus

Aristida **Threeawn grass** **Poaceae**

Tufted annual or perennial grasses, with persistent long and very narrow leaves, flat or margins rolled inward, with minute fringed ligules. Leaves and branches mainly from lower nodes. Flowers and seeds in terminal, long and slender, spike-like panicles, each flower and grain tipped by 3 hair-like awns (thus the common name), awns straight or bent and twisted equally or not. Grain long-linear, round and pointed. Inhabit dry, nutrient-poor (often sandy) soils. 19 species in the SE.

Common Species

Aristida purpurascens Poir. **- arrowfeather threeawn** **ARPU8**

Plant: Common, tufted perennial grass, 50-80 cm tall, with dense tufts of long-thin leaves and slender stems, from a hardened knotty rootcrown with abundant fibrous roots. Dry leaves curl and persist.

Stem (culm): Erect to ascending, round in cross-section, hairless, green to light-green with base often purplish.

Leaves: Alternate, long and thin, 15-30 cm long and 1-2 mm wide, long tapering tip, curling with age forming ringlets, hairless except for long hairs tufted near base (throat), margins rough, ligule a fringed membrane, 0.1-0.5 mm long.

Flowers: Aug-Nov. Spike-like slender panicle, 15-30 cm long, 1/3 to 1/2 length of stem, branchlets short and close to stalk, with 2-3 spikelets each, spikelets with 3 straight to spreading awns, 12-25 mm long.

Seeds: Oct-Jan. Grain, 3-6 mm long, purplish to greenish, tipped with 3 awns.

Range: Native. TX to FL and north MA and Ont and west to MI and KS.

Ecology: Occurs as scattered plants, becoming most abundant on dry sandy sites, especially in the Cp. Found in open forests and along forest edges and right-of-ways. Persists by rootcrowns and spreads by wind-tumbled seeds.

Synonyms: *A. affinis* (J.A. Schultes) Kunth, *A. tenuispica* A.S. Hitchc., *A. virgata* Trin.

Aristida lanosa Muhl. *ex* Ell. **- woollysheath threeawn** **ARLA6**

Similar to *Aristida purpurascens* except **plant** 70-120 cm tall; **leaves** 20-50 cm long and 2-6 mm wide, sheaths woolly hairy (thus sci. and common names) and longer than internodes; **flowers** panicle 30-50 cm long, branchlets apparent, spikelets brown, central awn 17-30 mm long, twisted horizontal, lateral awns 10-20 mm long, erect or slightly deflected.

Aristida **Threeawn grass** **Poaceae**

J. Miller (Oct)

Aristida purpurascens

T. Bodner (Sep)

Aristida purpurascens

J. Miller (Sep)

Aristida lanosa

J. Miller (Oct)

Aristida purpurascens

J. Miller (Oct)

Aristida lanosa

J. Miller (Oct)

Aristida purpurascens

Aristida **Threeawn grass** **Poaceae**

Aristida beyrichiana Trin. & Rupr. **- wiregrass** **ARBE7**

Plant: Densely-tufted perennial grass, 50-100 cm tall, with dense clumps of mainly low or basal leaves, long and wire-like, from hardened rootcrowns with fibrous roots.

Stem (culm): Tufted, erect, wiry, short within leaves or long flower stalks.

Leaves: Densely tufted, margins rolled inward, wire- or needle-like, 10-40 cm long and 1-1.5 mm wide, long hairy when young, hairy along margins and at collars, sheaths longer than internodes, hairless, ligules 0.2 mm long

Flowers: Sep-Nov (or any season in FL). Spike-like panicle, brownish, 15-40 cm long, branchlets close to the stalk with 2-3 spikelets, spikelets (grains) 1-1.2 cm long, sharp tipped with 3 hair-like awns, 10-15 mm long (laterals slightly shorter), straight or bent horizontal.

Seeds: Fall after a summer fire. Spikes with yellow grain, grain 4 mm long, enclosed within husks with 3 hair-like awns, straight or bent and twisted.

Range: Native. MS to FL and north to s SC, only the Cp.

Ecology: Occurs as abundant understory cover in frequently burned and open longleaf pine stands, and persists after harvesting in new forest plantations. Damaged or eliminated by tillage. Summer burns are required for seed production, with July burns reportedly yielding the most viable seeds.

Synonyms: pineland threeawn, Beyrich threeawn. The other closely-related species of wiregrass, *A. stricta*, occurs in n SC and NC, more recently separated in classification.

Other species in the SE: *A. basiramea* Engelm. *ex* Vasey [forked threeawn], *A. condensata* Chapman [big threeawn], *A. desmantha* Trin. & Rupr., *A. dichotoma* Michx. [churchmouse threeawn], *A. gyrans* Chapman [corkscrew threeawn], *A. longispica* Poir. [slimspike threeawn], *A. mohrii* Nash [Mohr's threeawn], *A. oligantha* Michx. [prairie or oldfield threeawn], *A. palustris* (Chapman) Vasey, *A. patula* Chapman *ex* Nash [tall threeawn], *A. purpurea* Nutt. [purple threeawn], *A. ramosissima* Engelm. *ex* Gray, *A. simpliciflora* Ell. [Chapman's threeawn], *A. spiciformis* Ell. [bottlebrush threeawn], *A. stricta* Michx. [wiregrass], *A. tuberculosa* Nutt. [seaside threeawn].

Wildlife: Seeds of threeawn grasses are consumed infrequently by songbirds. Although not an important wildlife forage species, wiregrass is a keystone understory ground cover in the fire-dominated longleaf pine-wiregrass ecosystem of the Southeast. It also is used heavily by the Gopher Tortoise.

Aristida **Threeawn grass** **Poaceae**

T. Bodner (Sep)

Aristida beyrichiana

T. Bodner (Sep)

Aristida beyrichiana

Carex Sedge Cyperaceae

Grass-like perennial herbs, from rhizomes, having stems with a triangular cross-section and thus 3 edged ("sedges have edges"). Flat grass-like leaves, with the base and sheath enclosing around the stem, with no sheath seam. Stems without nodes or joints (unlike grasses). Flowers and seeds in spikes. Flowers without petals or sepals, male and female flowers in separate spikes or separate parts of the same spike. Each flower in the spike subtended by a minute bract (sheathing the flower or not), with a special shape by species. Each female flower and nutlet (achene) encased within a minute sac (perigynium), having a special shape and markings by species. Thus fruiting specimens are required for species identification. Nutlets round, lens-shaped, or 3-sided with no attachments for wind dispersal (unlike some grasses) and no long beak (unlike beakrushes). Often found on wet, open sites. 122 species in the SE, the most species of any genera of plants.

Common Species

Carex cephalophora Muhl. *ex* Willd. **- oval-leaf or wood sedge** **CACE**

Plant: Densely tufted clumps of mainly basal leaves, with long slender stems, 30-60 cm tall, from fibrous-rooted rhizomes.

Stem (culm): Erect, triangular in cross-section, slender, smooth.

Leaves: Basal or alternate on stem, mainly on lower 1/4 of the stem, flat and grass-like, 2-4.5 mm wide, long-tapering tip, hairless and soft, whitish-tan at plant base and green upward.

Flowers: Apr-Jul. Terminal dense spikes, ovoid, 1-2 cm long, with a thin leaf-like bract at the base, 1-2 cm long, with many female flowers protruding outward in all directions from the spike, each enclosed within a sac (perigynium), round-flat, each 2.5 mm long, light-green with a copper caste, having a short twin-pointed beak and often twin hair-like stigma extending from this, and a few scattered spherical male flowers of the same size interspersed.

Seeds: Jul-Dec. Terminal spikes of nutlets (achenes) each within a minute sac (husk).

Range: Native. TX to FL and north to Que and west to MI, all physiographic provinces, rare in Lower Cp.

Ecology: Occurs in small colonies in new forest plantations, open forests, and forest openings, and along forest edges and right-of-ways, wet to dry sites. Persists and colonizes by rhizomes and spreads by seeds.

Carex glaucescens Ell. **- southern waxy sedge** **CAGL5**

Similar to *Carex cepalophora* except **plant** common sedge in Cp, 60-100 cm tall, gray-green waxy appearing; **stem** also triangular in cross-section; **leaves** 3-8 mm wide, slightly gray-green waxy, sheaths purplish and smooth; **flowers** (Jul-Sep) spikes 2 types, terminal upright male spike, 3-5.5 cm long, green turning brown, and lateral dangling female spikes, 3-6, cylindric, 2.5-4.5 cm long and 7-10 mm wide, flower sac waxy gray, 2.8-3.5 mm long, minute bracts, purplish, rough margins; **seeds** (Aug-Mar) gray nutlets, ovoid, 3-sided with a short beak, on drooping spikes. **Range:** Native. LA to FL and north to VA, only in Cp. **Ecology:** Occurs in seasonally wet forests, swamps, and open savannas.

Wildlife: Because of their abundance and wide distribution, sedges are important wildlife plants. Seeds (achenes) are of value to Wild Turkey, Ruffed Grouse, and various songbirds, particularly sparrows. Sedges also provide valuable nesting and escape cover. Sedges are important items in the diets of Swamp Rabbits and Marsh Rabbits, but they are poor White-tailed Deer forage in the Southeast.

Carex **Sedge** **Cyperaceae**

T. Bodner (Apr)

Carex cepalophora

J. Miller (Sep)

Carex glaucescens

D. Lauer (Aug)

Carex glaucescens

Chasmanthium Oatgrass or Spikegrass or Spanglegrass Poaceae

Tufted and tall perennial grasses, from short rhizomes. Spikelets distinctly flattened, in V-shaped clusters, scattered along slender stalks or drooping in terminal panicles. Leaves mostly on short stems or seed stalks, blades narrowly lanceolate, flat or folded, with a fringed-membrane ligule. Leaves persist green or brown during winter. Grain ovoid and flat, husk-less, awnless, and plumeless. 4 species in the SE.

Common Species

Chasmanthium sessiliflorum (Poir.) Yates - **longleaf uniola or woodoats CHSE2**

Plant: Commonly-occurring, upright, tufted perennial, 0.5-1.2 m tall, clumps loose to dense with a curved topped, from knotty rhizomes. Leaves persistent and often growing during winter.

Stem (culm): Thin, round in cross-section, long-white hairy (or hairless) nodes, only 1-3 leaves per stem and mainly basal, stem much extended from sheath.

Leaves: Alternate, broadly long-attenuate with rounded base, 20-40 cm long and 5-10 mm wide, sheaths long hairy, collars hair-tufted, ligule a fringed membrane, 2-4 mm long.

Flowers: Jun-Oct. Thin stalks, 10-40 cm tall, V- or U-shaped spikelet clusters spaced 1-3 cm apart, spikelets 4-8 mm long and 4-7 mm wide, green to greenish-white.

Seeds: Nov-Feb. Grain, flat-ovoid, 2-3 mm long, blackish, without plumes or hair tufts.

Range: Native. TX to FL and north to VA and west to TN and OK.

Ecology: Common grass in the shaded understory of pine plantations and mixed forests. Occurs along forest edges and right-of-ways, on wet to dry sites. Dense clumps are common in cleared forested areas, singly or in small colonies. Increases in abundance after burning. Persists and colonizes by rhizomes and spreads by animal-dispersed seeds.

Synonyms: *C. laxum* var. *sessiliforum* (L.) Yates, *Uniola sessiliflora* Poir.

Chasmanthium laxum (L.) Yates - **slender woodoats** **CHLA6**

Similar to *Chasmanthium laxum* var. *sessiliflorum* except **leaves** sheaths hairless, blades 3-6 mm wide. **Range:** Native. TX to FL and north to NY and west to KY, MO, and OK. **Synonyms:** *Uniola laxa* (L.) B.S.P., spike uniola.

Chasmanthium latifolium (Michx.) Yates - **Indian woodoats** **CHLA5**

Plant: Tufted grass, 0.5-1.5 m tall, with loose clumps of broad leaves, from a rhizome.

Stem (culm): Ascending to arching, stout, round, slightly swollen nodes, green to purplish.

Leaves: Alternate, lanceolate, 10-20 cm long and 1-2.5 cm wide, hairless, margins rough.

Flowers: Jun-Oct. Drooping panicles of dangling flat spikelet clusters, 1.5-4 cm long and 8-15 mm wide, 5-10 spikelets in V-shaped pairs, on thin flexible stems.

Seeds: Nov-Feb. Grain, flat-ovoid, 4-5 mm long, dark red to brown.

Range: Native. TX to n FL and north to s Canada and west to IL and KS.

Ecology: Occurs in colonies along streams, wet forests, and bluffs, in shade and open.

Synonyms: *Uniola latifolia* Michx., wildoats, broadleaf uniola.

Other species in the SE: *C. nitidum* (Baldw.) Yates, *C. ornithorhynchum* (Steud.) Yates.

Wildlife: White-tailed Deer graze the spikegrasses infrequently. The seeds are consumed occasionally by Northern Bobwhite and many songbirds.

Chasmanthium Oatgrass or Spikegrass or Spanglegrass Poaceae

T. Bodner (Nov)

Chasmanthium laxum

T. Bodner (Jun)

Chasmanthium sessiliflorum

J. Miller (Oct)

Chasmanthium latifolium

T. Bodner (Jun)

T. Bodner (Aug)

Chasmanthium sessiliflorum

Cyperus **Umbrella sedge or Flatsedge** **Cyperaceae**

Grass-like annual herbs, as well as perennial herbs from rhizomes or thickened bases, having stems with a triangular cross-section, and thus 3 edged ("sedges have edges"). Leaves mostly basal or low on stems, channeled at the midvein (corrugated), and sheaths merely enclosing around the stem, with no sheath seam. Stems without nodes or joints (unlike grasses). Spikelets (flowers) round, quadrangular or flattened, in terminal clusters or heads, simple or branched, subtended by leaf-like bracts. Flowers perfect (unlike *Carex*), without petals or sepals, and each flower in a spike subtended by 2 minute bracts. Nutlets lens-shaped or 3-sided with no attachments for wind dispersal (unlike some grasses) and no long beak (unlike beakrushes). 32 species in the SE.

Common Species

Cyperus echinatus (L.) Wood - **roundhead sedge** **CYEC2**

Plant: Commonly-occurring, loosely-tufted, grass-like perennial, 15-80 cm tall, from a thick short rhizome.

Stem (culm): Ascending to erect, triangular in cross-section, slender, smooth, no joints or nodes.

Leaves: Basal with overlapping bases upward, linear grass-like, 8-30 cm long and 3-8 mm wide, channeled by a depressed and prominent midvein, green above and whitish to purplish at the sheath base.

Flowers: Jul-Sep. Terminal, spherical spike clusters, 8-15 mm wide, greenish, each on thin stems up to 10 cm long, radiating from a whorl of leaf-like bracts of unequal length, 3-15 cm long.

Seeds: Aug-Jan. Tubular pointed spikelets, each with 1-3 nutlets (achenes), 1.5-2 mm long.

Range: Native. TX to n FL and north to se NY and west to IL and KS, mainly in Pd and Cp in the SE, less frequent in the Lower Cp.

Ecology: Occurs singly or in small colonies in new forest plantations, open forests, prairies, and along forest edges and right-of-ways. Persists and colonizes by rhizomes and spreads by animal-dispersed seeds.

Synonyms: *C. ovularis* (Michx.) Torr., *C. ovularis* var. *sphaericus* Boeckl., globe flatsedge.

Cyperus retrorsus Chapman - **cylindric sedge** **CYRE5**

Similar to *Cyperus echinatus* except **plant** often densely tufted clumps; **leaves** 3-5 mm wide; **flowers** (May-Nov) spikes short cylinders, 6-25 mm long and 6-12 mm wide, green drab. **Range:** Native. TX to s FL and north to NY and west to IN and MO, all provinces.

Ecology: Common in open sandy pine forests, especially sandhills, and around fresh and brackish marshes, and in floodplains. Persists and colonizes by rhizomes and spreads by animal-dispersed seeds.

Synonyms: *C. cylindricus* (Ell.) Britt., *C. deeringianus* Britt. *ex* Small, *C. litoreus* (C.B. Clarke) Britt., *C. pollardii* Britt. *ex* Small, *C. winkleri* Britt. *ex* Small, *C. nashii* Britt., *C. torreyi* Britt. *C. globulosus* var. *robustus* (Boeckl.) Shinners, *Mariscus cylindricus* Ell., pine barren flatsedge.

Cyperus Umbrella sedge or Flatsedge Cyperaceae

D. Lauer (Aug)

Cyperus retrorsus

T. Bodner (Jul)

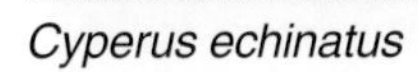

Cyperus echinatus

J. Miller (Sep)

Cyperus retrorsus

J. Miller (Aug)

Cyperus echinatus

Cyperus Umbrella sedge or Flatsedge Cyperaceae

Cyperus odoratus L. - fragrant flatsedge CYOD

Plant: Erect, loosely tufted, grass-like annual, 10-90 cm tall, pale green, from a thick base with fibrous roots.
Stem (culm): Moderately stout, triangular in cross-section, smooth, no joints or nodes, equal to or longer than basal leaves.
Leaves: Basal mostly, linear grass-like, 8-30 cm long and 3-8 mm wide, flat and channeled by a depressed and prominent midvein, green above and whitish at sheath base with purple spots.
Flowers: Jun-Oct. Terminal, bottle-brush shaped spike clusters, 3-10, each 2-3 cm long and 1.5-4 cm wide, gray brown or tan, each on thin stems up to 15 cm long, radiating from a whorl of leaf-like bracts, 5-30 cm long and 3-7 mm wide.
Seeds: Aug-Jan. Tubular long-linear spikelets, 0.5-1.5 cm long and 1 mm wide, sharp tipped, brown-red to golden with a green midvein, each containing 10-20 nutlets (achenes), nutlets 1-2 mm long, silvery brown, obovoid-oblong.
Range: Native. CA to FL and north to ME and west to MN.
Ecology: Occurs commonly near fresh and saltwater marshes, and in and near ditches and other moist and wet soils.
Synonyms: *C. acicularis* Schrad. *ex* Nees, *C. eggersii* Boeckl., *C. engelmannii* Steud., *C. ferax* L.C. Rich., *C. ferruginescens* Boeckl., *C. longispicatus* J.B.S. Norton, *C. macrocephalus* Liebm., *C. odoratus* var. *acicularis* (Schrad. *ex* Nees) O'Neill, *C. speciosus* Vahl.

Cyperus rotundus L - purple nutsedge or nutgrass CYRO

Similar to *Cyperus odoratus* except **plant** perennial, 10-50 cm tall, deep green, from a trailing rhizome with bulbs; **stem (culm)** slender, rarely branching, extending well above basal leaves; **leaves** basal mostly, 5-25 cm long and 2-6 mm wide, tips abruptly tapering; **flowers** (May-Nov) terminal radiating cluster of spikes, 3-10, each 1-2.5 cm long, reddish and yellow ribbed, each on thin stems up to 10 cm long, radiating from a whorl of leaf-like bracts, 2-4, 5-20 cm long and 2-6 mm wide; **seeds** (Jul-Jan) flattened and long-linear spikelets, 0.8-3 cm long and 1.5 mm wide, sharp tipped, reddish brown or reddish purple, each with 10-20 nutlets (achenes), nutlets 1-2 mm long, 3-angled, olive gray to brown to blackish with gray lines. **Range:** Exotic from Eurasia. TX to FL and north to VA and west to AR, also CA.
Ecology: Occurs in forest seedling nurseries and along right-of-ways as an invasive weed. Persists by rhizomes and spreads by trailing rhizomes and seeds.
Synonyms: purple nutgrass, coco-grass, coco weed.

Cyperus esculentus L. - yellow nutsedge or chufa flatsedge CYES

Similar to *Cyperus rotundus* except **plant** 30-80 cm tall, tubers spherical, sweet to taste; **stem** longer or shorter than basal leaves; **leaves** 5-6 mm wide, tips long tapering; **flowers** (Jun-Oct) spikelets golden and/or brown. **Range:** Naturalized from Europe. Throughout US.
Ecology: Troublesome weed in forest seedling nurseries and gardens. Planted for wildlife.

Wildlife: *Cyperus* are poor White-tailed Deer forages. In wet areas, the seeds are important waterfowl foods. The underground tubers of chufa (*C. esculentus*) are highly preferred by Wild Hogs, Raccoon, and Wild Turkey. Chufa often is planted in wildlife openings, particularly in alluvial soils.

Cyperus **Umbrella sedge or Flatsedge** **Cyperaceae**

J. Miller (Nov)

Cyperus rotundus

T. Bodner (Jul)

Cyperus odoratus

J. Everest (Oct)

Cyperus esculentus

Danthonia Oatgrass or Povertygrass Poaceae

Tufted and tall perennial grasses, from rootcrowns. Narrow terminal clusters of large V-shaped spikelets on long slender stalks, turning whitish and persisting after seed release. Leaves mostly basal or lower on stems, blades narrow and rolled inward on some, ligules with a fringe of hair. Grain reddish, ovoid, grooved, enclosed by a hairy husk with a long hair-like awn, curled at the base and straight above. 4 species in the SE.

Common Species

Danthonia sericea Nutt. **- downy oatgrass or danthonia** **DASE2**

Plant: Erect tufted perennial grass, 40-100 cm tall, from rootcrowns. Flowering in spring and early summer, and seed stalks and seed heads turning light tan to whitish soon afterward and persist standing for about a month. Hairy plant bases and new sprouts present from mid- to late summer.

Stem (culm): Upright, several in a clump, round in cross-section, nodes tan to brown bands, green basally and light tan to whitish upward, with 1-3 leaves towards the base.

Leaves: Alternate and mostly basal, flat and thin-linear, 5-20 cm long and 2-4 mm wide and reduced upward, margins rough and often rolled inward, lower surface and sheaths with long white hairs, ligules and throats white-hair tufted, 1 mm long.

Flowers: Apr-Jun. Terminal spike-like narrow panicles, 3-18 cm long and 1-6 cm wide, with ascending and appressed short branches (some spreading) having clusters of broad V-shaped spikelets, 2-6 per branch, each 13-20 mm long and 4-8 mm wide, green outer husks and white inner husks and hairy.

Seeds: May-Jul. Grain 2 mm long, within husks having basal white hairs, and tipped with a hair-like awn, 10-12 mm long, awn curled at the brown base when dry. V-shaped husks remain on the stalk after grain release.

Range: Native. LA to n FL and north to MA and southwest to KY, all provinces.

Ecology: Occurs in small to extensive colonies in new forest plantations, open forests, forest openings, right-of-ways, old fields and along forest margins. Persists by rootcrowns and dormant seeds, and spreads by animal- and wind-dispersed seeds.

Synonyms: *D. epilis* Scribn., *D. sericea* var. *epilis* (Scribn.) Blomquist, silky oatgrass, downy danthonia.

Danthonia spicata (L.) Beauv. *ex* Roemer & J.A. Schultes **- poverty oatgrass** **DASP2**

Similar to *Danthonia sericea* except **plant** 20-70 cm tall; **leaves** densely tufted at base and often curling, smooth and hairless to only sparsely hairy basally, hair tuft at throat; **flowers** (May-Jul) 1-2 spikelets per branch, each 10-15 mm long and 3-10 mm wide; **seeds** (Jun-Aug) awns 4-9 mm long. **Range:** Native. NM to n FL and north to ME and west to WA, rare in Cp.

Ecology: Most abundant on impoverished soils, overly farmed or grazed.

Synonyms: *D. thermalis* Scribn., Junegrass, white oatgrass, poverty grass.

Other species in the SE: *D. compressa* Austin *ex* Peck [only in Appalachian Mt], *D. epilis* Scribn.

Wildlife: Povertygrass is aptly named as it is a poor wildlife forage plant and an indicator of poor, infertile soils. Wild Turkey may consume very limited amounts of the foliage during early spring when other green foliage is scarce.

Danthonia | **Oatgrass or Povertygrass** | **Poaceae**

T. Bodner (May)

Danthonia sericea

J. Miller (May)

Danthonia sericea

J. Miller (May)

Danthonia sericea

T. Bodner (May)

Danthonia sericea

Dichanthelium Rosette grass or Low panicgrass Poaceae

Short to medium-tall, tufted, perennial grasses, from rootcrowns or short matted rhizomes. Broad short leaves in basal rosettes in fall and winter, yielding open panicles in spring to early summer. Most species changing in late spring and late summer to a form having many densely-leafed branches, with short panicles extending partly from leaf sheaths, and leaves becoming reduced or more slender or unchanged. Spikelets without awns or plumes. *Dichanthelium* until recently was classified as a subgenus of *Panicum*. In general the *Dichanthelium* species are shorter in height, with smaller and broader leaves, than those in *Panicum*. 26 species in the SE, with integradation between species common.

Common Species

Dichanthelium commutatum (J.A. Schultes) Gould **- variable panicgrass**
DICO2

Plant: Erect, ascending, or rarely sprawling perennial, 10-70 cm tall or long, much branched in late summer, with delicate small panicles, from a rhizome.
Stem (culm): Upright (to prostrate), round in cross-section, with frequent prominent nodes, often purple, hairy or not, and internodes hairless and green often tinged with purple, branching from the middle and upper nodes in late summer.
Leaves: Basal rosette and later alternate on stem, broadly lanceolate, 5-10 cm long and 1-2 cm wide and somewhat reduced later, green and jutting out from the stem, cordate-clasping base, hairless, sheath often purplish with hairy margins, membrane ligule to 0.5 mm long.
Flowers: Mar-Nov. Small terminal panicles early, and axillary panicles later, 5-15 cm long and 3-9 cm wide and reduced in late summer, branching ascending or spreading and alternate, wavy, with many scattered terminal spikelets, ellipsoid, 2.5-2.8 mm long, with outer greenish-purple husks, sparse short hairy.
Seeds: Aug-Jan. Grain 1.2-1.8 mm long, yellow or purple within outer husk.
Range: Native. TX to FL and north to MA and west to MI and MO, all physiographic provinces in the SE.
Ecology: Common grass, often in loose colonies in open habitat or scattered plants in shady habitat, occurring on moist to dry sites. Occurs in open forests and new forest plantations, and along forest margins and right-of-ways. Persists by rhizomes and spreads by animal- and ant-dispersed seeds.
Synonyms: *Panicum commutatum* J.A. Schultes, *P. ashei* Pearson *ex* Ashe, *P. equilaterale* Scribn., *P. joorii* Vasey, *Dichanthelium joorii* (Vasey) Mohlenbrock.

Dichanthelium aciculare (Desv. *ex* Poir.) Gould & C.A. Clark **- needleleaf rosette grass**
DIAC

Similar to *Dichanthelium commutatum* except **plant** usually stiffly erect in mid- to late-summer and leaning later, 10-45 cm tall, bushy-branched above midstem; **leaves** short ovate-lanceolate in winter rosettes, in summer 3-10 cm long and less than 12 mm wide, thin in fascicles, long-tapered to a point, margins often rolled inward (resembling short pine needles), stiff at mid-plant, internodes and nodes long hairy or hairless (but not short hairy), ligule hairs less than 1.5 mm long or hairless; **flowers** (Apr-Nov) spikelets 1.8-3 mm long, on open panicles earlier and later in constricted panicles within leafy fascicles. **Range:** Native. TX to FL and north to MA and west to IN and MO, uncommon in Mt.
Ecology: Common panicgrass on dry forest sites.
Synonyms: *D. angustifolium* (Ell.) Gould, *Panicum aciculare* Desv. *ex* Poir., *P. angustifolium* Ell., *P. fusiforme* A.S. Hitchc., *P. neuranthum* Griseb., *P. ovinum* Scribn. & J.G. Sm., needleleaf panicgrass.

Dichanthelium **Rosette grass or Low panicgrass** **Poaceae**

T. Bodner (Apr)

Dichanthelium commutatum

T. Bodner (Oct)

Dichanthelium aciculare

J. Miller (Jun)

Dichanthelium aciculare

Dichanthelium **Rosette grass or Low panicgrass** **Poaceae**

Dichanthelium acuminatum (Sw.) Gould & C.A. Clark **- tapered rosette grass DIAC2**
Similar to *Dichanthelium commutatum* except **plant** usually in large clumps, 15-60 cm tall, densely hairy (to hairless); **leaves** long hairy on back, 5-10 cm long and 5-12 mm wide, sheaths hairy and shorter than internodes, ligules mostly hairs, 2-4 mm long; **flowers** spikelets finely hairy (requires hand lens). **Range:** Native. Throughout North and South America.
Ecology: Common panicgrass on dry forest sites.
Synonyms: 7 varieties, *Panicum acuminatum* Sw., woolly panicum.

Dichanthelium laxiflorum (Lam.) Gould **- openflower rosette grass DILA9**
Similar to *Dichanthelium commutatum* except **plant** without basal rosette, more a soft cushion, branched only at base, 10-45 cm tall; **leaves** blades soft and not stiff (lax, thus the sci. and common names), upper 4-7 cm long and at least 3/4 as long as those near the base and 4-10 mm wide, narrowed at base, hairy or not, margins a fine-toothed membrane (requires hand lens) and often long hairy, nodes tufted hairy, sheaths long hairy; **flowers** (Mar-Nov) early panicles, 4-9 cm long and 3-6 cm wide, later panicles partly hidden in leaf sheaths, spikelets 1.6-2.3 mm long; **seeds** (Jun-Jan) grain whitish to yellowish, broadly ellipsoid, 1-1.5 mm long within greenish husks. **Range:** Native. TX to FL and north to MD and west to MO and OK, all physiographic provinces in the SE.
Ecology: Occurs as scattered plants or small colonies in new forest plantations, open forests, and forest openings, and along forest margins and mowed right-of-ways, especially moist sites.
Synonyms: *Panicum laxiflorum* Lam., *P. xalapense* Kunth., lax panicgrass.

Dichanthelium scoparium (Lam.) Gould **- velvet or hairy panicum DISC3**
Similar to *Dichanthelium commutatum* except **plant** softly hairy, erect, 0.5-2 m tall, with an early leafy rosette; **stems** tall later, stout, hairy with a hairless band below the nodes; **leaves** to 20 cm long and 5-22 mm wide, softly hairy both surfaces and sheath, hair tufted collars, ligules a fringe of white hairs, 0.4-1.5 mm long; **flowers** (May-Nov) panicles 6-16 cm long and 5-10 cm wide, spikelets 2.2-2.8 mm long; **seeds** (Jul-Jan) grain yellow, ellipsoid, 1-1.2 mm, within hairy greenish husks. **Range:** Native. TX to nw FL and north to NY and southwest to KY, MO, and OK, rare in the Mt.
Ecology: Occurs as scattered plants in new forest plantations, forest edges, and along ditches, floodplains, bogs, and marshes—wet to moist sites. Persists by rhizomes and spreads by seeds.
Synonyms: *Panicum scoparium* Lam.

Wildlife: Because of their abundance and wide distribution, *Dichanthelium*, along with *Panicum*, are one of the most important sources of food for ground-feeding songbirds, small mammals, and gamebirds. Basal rosettes remain green through winter and provide winter herbage for Wild Turkey and White-tailed Deer, although most are considered poor quality White-tailed Deer forages.

Dichanthelium **Rosette grass or Low panicgrass** **Poaceae**

T. Bodner (May)

Dichanthelium laxiflorum

T. Bodner (Oct)

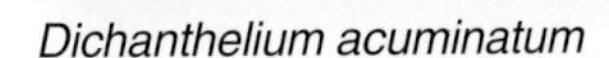

Dichanthelium acuminatum

J. Miller (Jul)

Dichanthelium scoparium

T. Bodner (Jun)

Dichanthelium scoparium

Digitaria Crabgrass Poaceae

Tufted and often stoloniferous annual grasses, branching from the base, having only flat and thin stem leaves (no basal rosettes), their margins a thin membrane. Tiny flowers and seeds in very narrow spikes, branching outward in many directions from the apex of a slender stalk. Seeds released within husks without awns, hairs, or plumes. 11 species in the SE.

Common Species

Digitaria ciliaris (Retz.) Koel. **- southern crabgrass** **DICI**

Plant: Sprawling, summer annual, 25-100 cm long or tall, rooting at nodes with dense fibrous roots.

Stem (culm): Ascending, reclining, or running, round in cross-section, green to whitish, jointed.

Leaves: Alternate, narrowly lanceolate, 5-18 cm long and 3-10 mm wide, margins a thin membrane (requires hand lens), blades hairless except near base and slightly hairy on upper surface, sheaths often purple with scattered long hairs especially near the throat, ligules a membrane with a frayed margin, 1-3 mm long.

Flowers: May-Nov. Terminal branched spikes, 2-9 fingers, thin finely-winged stalks, each 3-15 cm long, with paired spikelets, each long-ellipsoid and pointed, 2.5-3.5 mm long and 0.7-0.8 mm wide, husks whitish with green stripes, outer husks on curved side 2/3 length of spikelet.

Seeds: Jun-Jan. Grain black, 2-2.2 mm long, released within husks.

Range: Naturalized from Europe (or Asia). NM to FL and north to MA and west to NE and CO.

Ecology: Occurs as individual clumps to extensive colonies in forest plantations, abandoned fields, seed orchards, forest nurseries, and along forest margins and right-of-ways. Colonizes by stolons and spreads by animal- and ant-dispersed seeds.

Synonyms: *D. adscendens* (Kunth) Henr., *D. sanguinalis* (L.) Scop. var. *ciliaris* (Retz.) Parl., *D. sanguinalis* var. *marginata* (Link) Fern.

Digitaria sanguinalis (L.) Scop. **- hairy or large crabgrass** **DISA**

Similar to *Digitaria ciliaris* and difficult to distinguish differences; **plant** up to 70 cm long or tall; **leaves** hairy on sheath and blades, blades 3-20 cm long and 3-10 mm wide, ligule a membrane with uneven teeth or margin, 1-2 mm long; **flowers** (May-Nov) outer husks on curved side less than 1/2 length of spikelet (requires hand lens); **seeds** (Jun-Jan) grain yellow-brown, 2-3 mm long. **Range:** Naturalized from Europe. All US except FL and lower halves of TX, MS, and SC.

Other species in the SE: *D. bicornis* (Lam.) Roemer & J.A. Schultes *ex* Loud., *D. filiformis* (L.) Koel., *D. floridana* A.S. Hitchc., *D. horizontalis* Willd., *D. insularis* (L.) Mez *ex* Ekman, *D. ischaemum* (Schreb.) Schreb. *ex* Muhl., *D. serotina* (Walt.) Michx., *D. villosa* (Walt.) Pers., *D. violascens* Link.

Wildlife: Crabgrass seeds are used heavily by Mourning Dove, Northern Bobwhite, Wild Turkey, and numerous songbirds, particularly various sparrows and the Dark-eyed Junco. Crabgrass leaves also are used by Wild Turkey and to a lesser extent by White-tailed Deer.

Digitaria **Crabgrass** **Poaceae**

T. Bodner (Jun)

Digitaria ciliaris

T. Bodner (Sep)

Digitaria sanguinalis

T. Bodner (Aug)

Digitaria ciliaris

Eragrostis **Lovegrass** **Poaceae**

Short upright or mat-forming, tufted annual or perennial grasses, from rhizomes. Distinguished by densely- and finely-branched panicles, having numerous slightly to strongly flattened small spikelets in tight clusters, with alternating V-shaped husks. Leaf blades flat, or the margins rolled inward, sheaths commonly hairy at the summit. 16 species in the SE.

Common Species

Eragrostis spectabilis (Pursh) Steud. **- purple lovegrass** **ERSP**

Plant: Loosely to densely tufted, upright perennial grass, 30-70 cm tall, with finely- and densely-branched panicles that turn purple in late-summer, from stout horizontal rhizomes.
Stem (culm): Stout and short, rigid, overlapping leaf sheaths, densely hairy (hairs projecting outward) to hairless, light-green.
Leaves: Basal and alternate on lower stem, narrowly lanceolate, 20-30 cm long and 3-8 mm wide, stiff, flat and slightly corrugated, tapering tips, densely hairy (to hairless), margins rough, ligule a hairy fringe 3-6 mm long.
Flowers: Aug-Oct. Profusely- and finely-branched panicle, 15-45 cm long (about 2/3 height of plant) and 12-40 cm wide, rigid upright, turning purple in late summer, spreading and rough, spikelets flattened clusters of 2-10 flowers, alternating herring-bone arrangement, spikelets 1-6 mm long and 1-2 mm wide, turning purple.
Seeds: Oct-Feb. Grain reddish brown, 0.6-0.8 mm long, within purple husks. Panicles often break off and tumble to disperse seeds.
Range: Native. AR to FL and north to ME and west to MN, all physiographic provinces in the SE.
Ecology: Occurs as scattered plants in open forests, forest openings, forest plantations, old fields, and along forest margins and right-of-ways. Inhabits open to shady sites, moist to wet, being most abundant on open and moist sites. Persists and expands clumps by rhizomes and spreads by seeds dispersed by birds and wind-tumbled panicles.
Synonyms: tumblegrass.

Eragrostis hirsuta (Michx.) Nees **- bigtop lovegrass** **ERHI**

Similar to *Eragrostis spectabilis* except **plant** 60-120 cm tall, panicle not purple in late summer, from a vertical short rhizome or hardened base; **stem (culm)** densely long-hairy; **leaves** slender tapering, margins rolled inward, 20-60 cm long and 3-9 mm wide, lower sheath appressed hairy and upper sheath densely long-hairy, with projecting hairs near and around throat and along sheath margins, blades hairless except on margins near base, ligule a hairy fringe, 0.3-0.5 mm long; **flowers** (Jun-Nov) panicles about 2/3-3/4 height of plant and profusely branched, spikelets only 2-4 flowers, spikelets 3-4 mm long and 3-4 mm wide, greenish (not purplish); **seeds** (Aug-Feb) grain 0.7-1 mm long. **Range:** Native. TX to FL and north to MD and west to MO, mainly in Cp and Pd, rare in Mt.
Ecology: Mainly dry sandy sites as scattered plants or small colonies.

Wildlife: The seeds of the various species of lovegrass are consumed by gamebirds and some songbirds.

Eragrostis **Lovegrass** **Poaceae**

J. Miller (Oct)

Eragrostis hirsuta

J. Miller (Oct)

Eragrostis spectabilis

T. Bodner (Sep)

Eragrostis hirsuta

T. Bodner (Aug)

Eragrostis spectabilis

Festuca Fescue Poaceae

Upright perennial grasses, unbranched above, and smooth hairless (to slightly rough hairy), from rhizomes. Leaves flat or margins rolled inward, and margins rough. Flowers in spreading or contracted panicles. 6 species in the SE, most introduced for forage and now weedy. Many species that were formerly in *Festuca* are now classified as *Lolium.*

Common Species

Festuca arundinacea Schreb. **- tall fescue** **LOAR10**

Plant: Erect, tufted perennial, usually flowering in spring, 70-140 cm tall, stout unbranched stems, dark shiny green, from rootcrowns or short creeping rhizomes.
Stem (culm): Moderately stout, hairless, round in cross-section, with swollen light-green nodes.
Leaves: Alternate, whitish flared collars, flat and long-attenuate blades, 10-45 cm long and 3-8 mm wide, smooth to rough, white membrane ligule, 0.2-0.6 mm long, collar backs often at an angle.
Flowers: Mar-Oct. Terminal panicles, erect or nodding at tips, 10-30 cm long, narrow then spreading then narrow again, spikelets hairless, greenish-white and shiny, often with a purplish tinge, 4-7 flowered, spikelets 10-15 mm long, ellipsoid with a pointed tip, short stalked.
Seeds: May-Nov. Grain red, round, 3-5 mm long, grooved, within husks tipped with a short awn (or awnless).
Range: Naturalized from Eurasia. Throughout US, all physiographic provinces in the SE.
Ecology: Occurs as tufted clumps or small colonies along forest margins and right-of-ways, and invading new forest plantations and forest opens. Grows best on wet to moist sites. Established widely for forage, soil stabilization, and wildlife food plots, and now escaping. Persists by rootcrowns and spreads by plantings and animal-dispersed seeds.
Synonyms: *Lolium arundinaceum* (Schreb.) S.J. Darbyshire, *F. elatior* var. *arundinacea* (Schreb.) C.F.H. Wimmer, meadow fescue.

Other species in the SE: *F. filiformis* Pourret, *F. ovina* L., *F. paradoxa* Desv., *F. rubra* L., *F. subverticillata* (Pers.) Alexeev

Wildlife: Although a common livestock forage crop, especially in the Piedmont and Southern Appalachian Mountains, tall fescue has limited wildlife value. White-tailed Deer and Eastern Cottontail will consume tall fescue during winter when other forages are scarce. Most tall fescue is infected with a fungus that causes reproductive and nutritional problems for wildlife and livestock. This species can be troublesome on abandoned pastures and will prevent colonization of more valuable wildlife forage species.

Festuca **Fescue** **Poaceae**

T. Bodner (May)

Festuca arundinacea

T. Bodner (Apr)

Festuca arundinacea

J. Miller (May)

Festuca arundinacea

J. Miller (May)

Festuca arundinacea

Imperata — Imperatagrass or Satintail — Poaceae

Erect, tufted perennial grass, from rhizomes. Unbranched with mainly basal leaves having constricted bases and overlapping sheaths, and fringed ligule membrane. Narrow panicles of white woolly seed heads, spikelets all the same, awnless, in pairs (one pedicelled and one sessile) on a straight stalk surrounded by long silky hairs. 2 species in the SE, with cogongrass being one of the world's worst weeds.

Common Species

Imperata cylindrica (L.) Palisot **- cogongrass** **IMCY**

Plant: Aggressive exotic, colony-forming, perennial grass, 15-120 cm tall, clumps of long leaves, from sharp-tipped white-scaly rhizomes with fibrous roots.

Stem (culm): Upright to ascending, stout, hidden by overlapping leaf sheaths.

Leaves: Mainly from near the base with overlapping sheaths, long-lanceolate, 40-150 cm long and 6-18 mm wide and shorter upward, narrowing at the base and with a sharp tip often drooping, often yellowish-green, white midvein on upper surface with midvein slightly to mostly off-center, blades flat or cupped inward, margins translucent and minutely serrated (rough), outer sheaths often long hairy and hair tufts near the throat, ligule a fringed membrane, to 1.1 mm long.

Flowers: Mar-May (or year-round in southern part of range). Terminal, silky spike-like panicle, 3-20 cm long and 0.5-2.5 cm wide, cylindric and tightly branched, spikelets paired, each 3-6 mm long, obscured by silky to silvery white hairs to 1.8 mm long, 2 stamens or rarely 1 stamen (requires hand lens).

Seeds: May-Jun. Grain brown, oblong, 1-1.3 mm long, released within hairy husks for wind dispersal.

Range: Exotic, introduced from SE Asia and India, now distributed on all continents except Antarctica. Introduced to s AL (1912), s MS (1921), and FL (1940's). Currently scattered infestations in se LA, s and c MS, s and c AL, entire FL, s GA and s SC.

Ecology: Grows in full sunlight to partial shade, and thus can invade a range of sites. Aggressively invading right-of-ways, new forest plantations, open forests, old fields and pastures. Persists and expands colonies by rhizomes and spreads by wind-dispersed seeds. A single plant can produce as many as 3,000 seeds and the seed can travel over 80 kilometers. The sharp-tipped rhizome is able to grow through the roots of other plants. The branching, rapidly growing rhizomes form a dense mat able to exclude most other vegetation. The persistent rhizomes make cogongrass difficult to control and able to withstand burning.

Synonyms: *Lagurus cylindricus* L., *I. arundinacea* Cirillo.

Imperata brasiliensis Trin. **- Brazilian satintail** **IMBR**

Similar and difficult to distinguish from *Imperata cylindrica* except **plant** 20-70 cm tall; **leaves** hairy margined near apex; **flowers** panicle also, only 1 stamen (rarely 2). **Range:** Native to Central and South America. Currently rare in US, only s LA, s AL, and FL.

Synonyms: Brazilian bladygrass, silverplume.

Wildlife: No wildlife value reported.

Imperata Imperatagrass or Satintail Poaceae

C. Bryson (May)

Imperata cylindrica

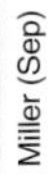
J. Miller (Sep)

Imperata cylindrica

J. Miller (Sep)

Imperata cylindrica

T. Bodner (Sep)

Imperata cylindrica

J. Miller (Sep)

Imperata cylindrica

T. Bodner (Sep)

Imperata cylindrica

Juncus **Rush** **Juncaceae**

Mostly perennial grass-like herbs, from rhizomes, with 2 to 3 annual species. Leaves grass-like and hairless, being either flat, rounded, or channeled, with open sheaths, or just sheaths around flower stalks. Plants upright or creeping in tufts or clumps or extensive patches in moist or wet habitat. Small flowers of similar 3 sepals and 3 petals, either green, whitish, yellowish, reddish and/or brownish. Flowers in dense heads or stemmed clusters, scattered in the axils of branched stalks, or a single lateral cluster on an unbranched stalk. Capsules spherical, ovoid, or oblong, splitting at the top to release many tiny seeds, rounded to oblong, ribbed. 28 species in the SE.

Common Species

Juncus effusus L. **- common or soft rush** **JUEF**

Plant: Densely-clumped, grass-like perennial, 0.5-1.5 m tall, with only tubular flower stalks, from stout rhizomes, forming extensive colonies in wet soils.
Stem (culm): Stout, tubular with a point, 3-7 mm wide, hairless and smooth, with a brown sheathed base.
Leaves: Absent or only a reddish-brown sheath enclosing the base of the flower stalk, up to 20 cm long, with a distinct hair-like tip.
Flowers: Apr-Sep. Lateral multi-branched cluster (cyme), 8-30 cm below stalk tip, 1-12 cm wide, with flowers terminal on branches, similar appearing 3 sepals and 3 petals, 2.5-3 mm long, pale green to brownish.
Fruit and seeds: May-Dec. Capsule within and same length as sepals and petals, ovoid with tip, splitting at top to release many tiny seeds, each 0.5 mm long.
Range: Native. Throughout US and Canada.
Ecology: Occurs as clumps or colonies on wet and very moist sites in open to semi-shady habitat. Persists and colonizes by rhizomes and spreads by seeds. Viable seeds can persist for at least 30 years.
Synonyms: needle rush, rice rush, pin rush, candle rush.

Juncus coriaceus Mackenzie **- leathery rush** **JUCO4**

Similar to *Juncus effusus* except **plant** spreading tufts, 30-50 cm tall; **stem (culm)** tubular and channeled above flower tuft; **leaves** some sheaths bladeless and others with slender blades of varying length; **fruit** capsule broadly ovoid within longer sepals and petals.
Range: Native. TX to FL and north to NJ and west to KY and OK, rare in Mt.
Synonyms: *J. setaceus* Rostk.

Juncus **Rush** **Juncaceae**

T. Bodner (Apr)

Juncus effusus

T. Bodner (Jul)

Juncus coriaceus

T. Bodner (Apr)

Juncus effusus

Juncus **Rush** **Juncaceae**

Juncus scirpodes Lam. **- needlepod or roundheaded rush** **JUSC**

Plant: Slender clumps, 30-90 cm tall, spreading tubular leaves and branched stalk, from a stout whitish rhizome.
Stem (culm): Round in cross-section (tubular), infrequent nodes whitish and slightly swollen.
Leaves: Basal and alternate on stem, channeled near throat and tubular above tapering to a point, with a distinct wide and open sheath from a swollen node having long filmy margins flaring at the throat, uppermost leaf much longer than its sheath.
Flowers: Jun-Nov. Terminal slightly-branched, with dense spherical clusters, 6-13 mm long and wide, 3 sharp-tipped sepals longer than petals, greenish turning dull brown at maturity.
Fruit and seeds: Jul-Jan. Capsule oblong and pointed, 2-3 mm long, longer than sepals and petals, with tips of valves remaining united after splitting, releasing many tiny seeds.
Range: Native. TX to FL and north to NY and west to MI and MO, rare in Mt.
Ecology: Occurs in moist and wet sites, such as flatwoods depressions, bog and marsh margins, and ditches. Also inhabits moist upland sites, such as new forest plantations, open forests, and along forest right-of-ways. Persists by rhizomes and spreads by seeds.
Synonyms: *J. scirpoides* var. *compositus* Harper, *J. scirpoides* var. *meridionalis* Buch.

Juncus marginatus Rostk. **- grassleaf rush** **JUMA4**

Plant: Slender, upright, grass-like perennial, 15-80 cm tall, 1-few stems with terminal compact branched clusters of loose-spherical flowers, from a short thick rhizome.
Stem (culm): Erect to leaning, tubular, slender, 1-2 mm wide at base.
Leaves: Basal and alternate, flat grass-like, basal leaves 4-20 cm long and 2-5 mm wide and very short below flowers, green with whitish vein(s), sheaths with thin rounded flares near throat, uppermost leaf shorter than sheath or sometimes bladeless.
Flowers: May-Sep. Terminal, densely-branched with loose spherical clusters, 2-20, each 4-6 mm wide, petals longer than sepals, green turning dark brown at maturity.
Fruit and seeds: Jun-Jan. Capsule broadly rounded including end, 2-3 mm long, same height of petals, reddish brown and lustrous, splitting to release tiny irregular seeds.
Range: Native. TX to FL and north to NY and west to MI, MO, and KS.
Ecology: Occurs as scattered plants and small colonies on open wet to moist sites, including new forest plantations, forest openings, and along forest margins and right-of-ways. Persists and colonizes by rhizomes and spreads by seeds.
Synonyms: *J. setosus* (Coville) Small, shore rush.

Wildlife: Rushes are a common forage plant of the Swamp Rabbit, Marsh Rabbit, Beaver, and Common Muskrat. They also provide cover for various wetland birds and small mammals.

Juncus **Rush** **Juncaceae**

T. Bodner (Jun)

Juncus scirpodes

T. Bodner (Jun)

Juncus scirpodes

D. Lauer (Aug)

Juncus marginatus

D. Lauer (Aug)

Juncus marginatus

Microstegium **Browntop or Microstegium grass** **Poaceae**

Reclining annual or perennial grasses, native to Asia, with linear to lanceolate leaves on culms branched at base (no basal rosette). Spikelets in pairs along spindly-stalked seedheads (racemes), with one spikelet of each pair stemmed (pedicelled) and the other sessile, otherwise alike. 1 exotic species in the SE.

Common Species

Microstegium vimineum (Trin.) A. Camus **- Nepalese browntop** **MIVI**

Plant: Sprawling annual grass, 30-100 cm tall, branched near the base and rooting at nodes to form dense colonies.

Stem (culm): Ascending to reclining, slender, wiry, up to 1.5 m long with hairless nodes and internodes, green to purple to brown.

Leaves: Alternate (none basal), lanceolate, flat blades to 10 cm long and 2-15 mm wide, white midveins, sparsely hairy on both surfaces and along margins, throat collar hairy and the ligule membranous with a hairy margin.

Flowers: Aug-Oct. Terminal, thin, spike-like racemes, 2-7 cm long, solitary or with 1-3 laterals, on an elongated wiry stem, spikelets paired, with one stemmed and the inner one sessile.

Seeds: Sep-Nov. Grain yellow to red, ellipsoid, 2.8-3 mm long, with seeds maturing over a period of about 2 weeks.

Range: Exotic from Asia. Introduced at Knoxville, TN, in about 1919. LA to FL and north to NY and west to OH and AR.

Ecology: Exotic. Very shade tolerant. Reproduces by seeds, and each plant may produce 100-1,000 seeds that can remain viable in the soil for 5 or more years. Seed dispersal is primarily by animals, flooding, and deposition of fill dirt. Ideal habitat is alluvial flood plains and streamsides, mostly colonizing flood-scoured banks. Other typical habitat includes forest edges, roadsides, and trailsides, as well as, damp fields, lawns, and along ditches. Now spreading to upland sites. Occurs up to 1,200 m elevation.

Synonyms: *Andropogon vimineum* Trin., *Eulalia viminea* (Trin.) Kuntze, *E. viminea* var. *variabilis* Kuntze, *Microstegium vimineum* var. *imberbe* (Nees) Honda, Japanese grass, Nepal microstegium, eualia, eulalia, Mary's grass, microstegium grass.

Wildlife: Microstegium grass has little wildlife value although seeds likely are consumed in minor quantities by some songbirds. This aggressive annual grass can outcompete more valuable wildlife plants and form extensive colonies, particularly in bottomland forests.

Microstegium Browntop or Microstegium grass Poaceae

T. Bodner (Jun)

Microstegium vimineum

T. Bodner (Sep)

Microstegium vimineum

T. Bodner (Sep)

Microstegium vimineum

Muhlenbergia **Muhly grass** **Poaceae**

Erect or reclining, tufted, perennial grasses from rhizomes or stolons, often many branched. Stem leaves only, narrow, flat or rolled, with rough surfaces and margins, having membranous ligules. Open panicles, or contracted and spike-like panicles, with small spikelets. 10 species in the SE.

Common Species

Muhlenbergia schreberi J.F. Gmel. **- nimblewill** **MUSC**

Plant: Delicate perennial, 10-70 cm long or tall, long stems often rooting at nodes, from a rootcrown.

Stem (culm): Weak, slender, hairless, ascending early, later reclining with upright terminals, becoming increasingly branched.

Leaves: Alternate (none basal), flat and narrow, 3-5 cm long and 1-4 mm wide, spreading or ascending, veins whitish, rough on surfaces and margins, sheaths hairless, throat spreading open, collars whitish with fine hair tufts, ligules 0.2-0.5 mm long.

Flowers: Aug-Nov. Panicles slender and spike-like, 6-18 cm long and 3-8 mm wide, terminal and axillary, erect or nodding, spikelet-bearing branches appressed-ascending, slightly rough, spikelets green or purplish.

Seeds: Sep-Dec. Grain reddish-brown, 1-1.4 mm long, released within husks having a slender hair-like awn, 2-5 mm long, and minute hairs encircling a broadened base.

Range: Native. TX to FL and north to MA and west to MI, IA, and CO.

Ecology: Occurs as small colonies within open forests, forest openings, new forest plantations, and along forest edges, right-of-ways, river and stream banks, especially wet to moist sites. Also, a lawn weed. Persists by rootcrowns, colonizes by rooting at nodes, and spreads by seeds.

Synonyms: *M. palustris* Scribn., *M. schreberi* var. *palustris* (Scribn.) Scribn.

Other species in the SE: *M. bushii* Pohl, *M. capillaris* (Lam.) Trin. [gulf muhly], *M. frondosa* (Poir.) Fern., *M. glabriflora* Scribn., *M. glomerata* (Willd.) Trin., *M. mexicana* (L.) Trin., *M. sobolifera* (Muhl. *ex* Willd.) Trin., *M. sylvatica* Torr. *ex* Gray, *M. tenuiflora* (Willd.) B.S.P.

Wildlife: Nimblewill has limited wildlife value although it is reported as a moderate to high preference White-tailed Deer forage in Mississippi. Some songbirds make light use of the seeds.

Muhlenbergia **Muhly grass** **Poaceae**

T. Murphy (Jul)

Muhlenbergia schreberi

T. Murphy (Jul)

Muhlenbergia schreberi

T. Murphy (Jun)

Muhlenbergia schreberi

Panicum **Panicgrass** **Poaceae**

Upright, usually large-tufted perennial grasses, from rhizomes, stolons, or rootcrowns, as well as some annual grasses. Flowers and seeds in open or narrow panicles (thus the common and sci. names). Leaves on stem, mostly near base, but not forming overwintering rosettes (in contrast to *Dichanthelium* spp.). Ligules either an unfringed to fringed membrane or a fringe of hair. 19 species in the SE.

Common Species

Panicum anceps Michx. **- beaked panicgrass** **PAAN**

Plant: Tall, upright perennial, 60-100 cm tall, often much branched from base forming clumps, from thick scaly rhizomes.
Stem (culm): Stout, several together, round to oval in cross-section and slightly angled at upper nodes, spacing between nodes increasing upward, green turning whitish green with purple tinge.
Leaves: Alternate and crowded near base (no rosette), flat, 20-40 cm long and 6-12 mm wide, erect to drooping tips, midvein (and other veins) whitish, both surfaces with scattered long hairs or smooth but usually hairy margins, sheaths hairy or not, collars constricted and often a purple-tinged band, ligule less than 0.5 mm long, an irregular fringe.
Flowers: Jun-Nov. Panicle pyramid-shaped, 10-50 cm long and 8-12 cm wide, branches ascending or spreading with branchlets short and appressed, spikelets 2.2-3.8 mm long, jutting from stalk, ovoid to lanceolate, outer husk minutely toothed on back rib and 1/3 to 1/2 length of spikelet (resembling a crab-claw), light-green often with faint purple tinge.
Seeds: Aug-Feb. Grain dark purple, ellipsoid, about 1.5 mm long, released within husks, whitish with light-green veins often with purplish tinge.
Range: Native. TX to FL and north to NJ and west to KS, frequent in the Cp.
Ecology: Occurs as scattered plants or small colonies in forest plantations, open forests, forest openings, and along forest margins and right-of-ways. Present on dry to moist sites, open or semi-shady habitat. Persists and colonizes by rhizomes and spreads by seeds.
Synonyms: *P. rhizomatum* A.S. Hitchc. & Chase, *P. anceps* var. *rhizomatum* (A.S. Hitchc. & Chase) Fern., beaked panicum.

Panicum virgatum L. **- switchgrass** **PANVI**

Similar to *Panicum anceps* except **plant** on wetter sites, 1-2 m tall, solitary or in clumps from long scaly rhizomes; **leaves** 20-60 cm long and 8-15 mm wide, sheaths often hairless (or long hairy) and gleaming white within, ligule 1-2 mm long, tan to brown membrane (later a hair fringe) with a patch of dense white hairs arising above it; **flowers** spreading panicle, 15-50 cm long and 6-20 cm wide, spikelets 3-5 mm long, often gaping, outer husk 2/3 length of spikelet and later purple. **Range:** Native. Throughout US except far west.

Other species in the SE: *P. amarum* Ell., *P. antidotale* Retz., *P. bergii* Arech., *P. bisculatum* Thunb., *P. brachyanthum* Steud., *P. capillare* L., *P. dichotomiflorum* Michx., *P. flexile* (Gattinger) Scribn., *P. gattingeri* Nash, *P. hemitomon* J.A. Schultes, *P. lacustre* A.S. Hitchc. & Elkman, *P. miliaceum* L., *P. philadelphicum* Bernh. *ex* Trin., *P. repens* L., *P. rigidulum* Bosc *ex* Nees, *P. tenerum* Bey. *ex* Trin, *P. verrucosum* Muhl.

Wildlife: Because of their abundance and wide distribution, panicgrasses (along with *Dichanthelium*) are one of the most important sources of food for ground-feeding songbirds, small mammals, and gamebirds. *Panicum dichotomiflorum* (fall panicgrass) and *P. capillare* (witch grass) are heavy producers of seed that are used by Mourning Dove, songbirds, and waterfowl. *Panicum miliaceum* (dove proso) is planted commonly in wildlife food plots, particularly for Mourning Dove and Northern Bobwhite, as is *Bracharia ramosa* (brown-top millet), until recently listed as *Panicum ramosum.*

Panicum Panicgrass Poaceae

J. Miller (Sep)

Panicum virgatum

T. Bodner (Oct)

Panicum anceps

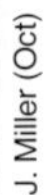

J. Miller (Oct)

Panicum virgatum

T. Bodner (Sep)

Panicum virgatum

Paspalum **Crowngrass or Paspalum grass** **Poaceae**

Erect, mostly mat-forming, perennial grasses, as well as annual tufted grasses. All with flat-rounded flowers and seeds, in rows on the under side of 1-many spike-like branches. Branches terminal, flattened or winged, ascending or spreading along a main slender stalk. Leaves mostly basal or low on stems, with membranous ligules having hair tufts behind it. 15 species in the SE, with several important imported forage grasses.

Common Species

Paspalum urvillei Steud. **- Vasey's grass or vaseygrass** **PAUR2**

Plant: Tall, erect, clump-forming perennial, 1-2 m tall, 1-few basal branches, from short and thick knotty rhizomes.
Stem (culm): Stout, 4-10 mm wide at base, round in cross-section, hairless above.
Leaves: Alternate and mostly near base, erect, flat, narrow lanceolate with a tapering tip, 15-55 cm long and 5-15 mm wide, whitish midvein (often whitish lateral veins), margins rough, hairless above, lowest sheaths often densely hairy and turning brown to purple later, ligule a long membrane, 4-8 mm long, with long white hair tufts behind, collar often a dark jointed band.
Flowers: Apr-Dec. Crowded terminal raceme of 12-25 ascending and overlapping branches (slightly winged), each 7-14 cm long, spikelets 4-rowed, each rounded with 1 flat side and 1 convex side, 2-3 mm long and 1-1.5 mm wide, margins long hairy, light-green often with dark flowering parts extending.
Seeds: Jul-Feb. Grain white, 1.2-1.7 mm long, released within white-hairy tan husks having a pointed tip.
Range: Exotic and naturalized, introduced from South America. TX to FL and north to MD and west to MO.
Ecology: Exotic introduced for forage. Occurs in seasonally wet to moist pastures, new forest plantations, open forests, and along stream and pond margins and right-of-ways.
Synonyms: *P. larranagai* Arech.

Paspalum dilatatum Poir. **- dallisgrass** **PADI3**

Similar to *Paspalum urvillei*, being an exotic, except **plant** 0.6-1.5 m tall; **leaves** 8- 28 cm long and 4-12 mm wide, sheaths sparsely hairy near base of plant and hairless above, ligule membrane 2-3 mm long; **flowers** terminal racemes with 3-7 alternate spreading branches, 8-12 cm long, with spikelets 4-rowed, each 3-4 mm long and 1-1.5 mm wide, margins long hairy; **seeds** grain flat-round, white, about 2 mm long and wide, released within silky-haired tan husks. **Range:** Exotic. Throughout SE.
Ecology: same as *P. urvillei.*

Paspalum floridanum Michx. **- Florida paspalum** **PAFL4**

Similar to *Paspalum urvillei* except **plant** 0.8-1.5 m tall, from long rhizomes; **leaves** 10-50 cm long and 4-15 mm wide, long hairy (especially lowest sheath) or not, uppermost sheath bladeless, membrane ligule 1.5-2.5 mm long; **flowers** (May-Nov) terminal racemes with 2-7 ascending branches, 6-15 cm long, wavy, long hairy at base, with large spikelets in 2 rows, each 3-4 mm long, flat-rounded, white waxy sometimes turning brown or peeling; **seeds** grain red, 2.5-2.8 mm long, released within grayish to brownish husks with distinct midveins. **Range:** Native. TX to FL and north to NJ and west to IL and KS.
Ecology: Occurs in moist to dry forest openings and margins, right-of-ways, prairies, marshes, and even in shallow water.
Synonyms: *P. difforme* Le Conte, *P. giganteum* Baldw. *ex* Vasey, *P. glabratum* (Engelm. *ex* Vasey) C. Mohr, tall paspalum.

Paspalum **Crowngrass or Paspalum grass** **Poaceae**

T. Bodner (Jun)

Paspalum urvillei

T. Bodner (Jun)

Paspalum urvillei

J. Miller (Oct)

Paspalum floridanum

T. Bodner (Jun)

Paspalum dilatatum

Paspalum **Crowngrass or Paspalum grass** **Poaceae**

Paspalum notatum Fluegge **- bahiagrass** **PANO3**

Plant: Mat- or sod-forming, erect or spreading, tufted perennial, 30-80 cm tall, from thick rhizomes covered with leaf sheath bases and having coarse fibrous roots.
Stem (culm): Upright along the rhizome, flat base of overlapping sheaths and round in cross-section above, hairless, light green.
Leaves: Basal erect and few reduced upward on stem, blades folded or flat, 20-35 cm long and 4-8 mm wide, long-tapering tips, much shorter than the flowering stem, basal sheaths flat and folded overlapping lengthwise, hairless or long hairy on or near the lower margins, ligules a membrane, 0.2-0.5 mm long, hairy behind ligule.
Flowers: May-Nov. Spike-like racemes in terminal pairs (rarely a third below), branches slightly winged, ascending to erect (often forming a V), each 8-14 cm long, spikelets 2-rowed, overlapping on one side of the stalk, each oval and pointed, 3-4 mm long and 2-3 mm wide, hairless, shiny light-green often with dark flower parts extending.
Seeds: Jun-Jan. Grains white, 2-3 mm long, released within tan shiny husks.
Range: Exotic and naturalized, native to South America. CA to FL and north to NC and west in NV, OR, and WA.
Ecology: Widely planted as a pasture and right-of-way grass, and can spread rapidly to exclude most native plants. Occurs singly or in small to large colonies in new forest plantations, forest openings, and open forests, and along forest margins, right-of-ways, and stream margins. Very competitive with planted pines, but can be grown as silvipastures.
Synonyms: *P. notatum* var. *saurae* Parodi [Pensacola bahiagrass], *P. notatum* var. *notatum*.

Paspalum laeve Michx. **- field paspalum** **PALA10**

Similar to *Paspalum notatum* except **plant** native and not planted, rhizomes short and forming close basal offshoots; **leaves** flat or folded, 10-20 cm long and 4-10 mm wide, long hairy on sheath margins and lower leaf margins (and often outer sheath), ligules 2 mm long; **flowers** (Jun-Nov) raceme branches alternate, 2-6, spreading, 3-7 cm long, spikelets 2-rowed, 2.5-3 mm long, hairless, dull light-green. **Range:** Native. TX to FL and north to MA and west to IL and KS.
Ecology: Occurs around depressions in open forests and along forest right-of-ways.
Synonyms: *P. circulare* Nash, *P. longipilum* Nash, *P. plenipilum* Nash.

Paspalum setaceum Michx. **- thin paspalum** **PASE5**

Similar to *Paspalum notatum* except **plant** native and not planted, tufted or spreading, 10-90 cm tall, rhizomes short or a knotted base, base brown or purplish and not flattened; **leaves** blades flat, ascending or spreading, 8-35 cm long and 7-17 mm wide, hairy or not but usually long hairs along margins, ligule membrane 1-3 mm long with long hairs behind; **flowers** raceme branches terminal, ascending, 1-5, 5-10 cm long, and/or axillary, being partially or completely hidden in leaf sheath, spikelets 2-rowed, 1.4-2.7 mm long and 1-2 mm wide, hairy or not. **Range:** Native. AR to FL and north to NY and west to MN.
Ecology: Occurs mainly on sandy sites in open to semi-shady situations as scattered plants.
Synonyms: 9 varieties, *P. bushii* Nash, *P. debile* Michx., bull paspalum

Wildlife: Paspalums are low preference White-tailed Deer forages. However, they are heavy producers of seed that are readily used by upland game birds, particularly Wild Turkey and Northern Bobwhite as well as various songbirds and small mammals. Bahiagrass is commonly planted in wildlife food plots for Wild Turkey.

Paspalum — Crowngrass or Paspalum grass — Poaceae

T. Bodner (Jun)

Paspalum laeve

T. Bodner (Jul)

Paspalum notatum

J. Miller (Oct)

Paspalum setaceum

T. Bodner (Jul)

Paspalum notatum

Piptochaetium **Speargrass or Needlegrass** **Poaceae**
(formerly *Stipa*)

Upright, tufted, perennial grasses, from hard rootcrowns. Mostly ascending basal leaves, having margins rolled inward and resembling a pine needle, with ligules of an unfringed membrane. Flowers and seeds in panicles with paired ascending branches, having spikelets with long-slender twisting awns, and persistent spikelet bases remaining on panicle. 2 species in the SE.

Common Species

Piptochaetium avenaceum (L.) Parodi - **blackseed speargrass** **PIAV**

Plant: Erect or ascending, densely-tufted perennial, 30-100 cm tall, with slender ascending basal leaves, from hard rootcrowns.

Stem (culm): Slender, round in cross-section, taller than twice the leaf height.

Leaves: Mostly erect, basal in dense tufts, and scattered alternate upward, 30-50 cm long and 1-1.5 mm wide, margins rough and often rolled inward (resembling pine needles), curling and persisting, ligules 2-4 mm long.

Flowers: Mar-Jun. Loose open panicles, 10-30 cm long and 3-6 cm wide, with wiry ascending branches in pairs, each having a single terminal spikelet, spindle-shaped, 11-13 mm long, tipped by a long needle-like awn projecting outward (thus the common name).

Seeds: Mar-Jun. Grain yellow, 5-6.5 mm long, released within distinct black husks having a long twisted thread-like terminal awn, 4-10 cm long, brown and finely rough, and a pointed hair-tufted base, 2 V-shaped transparent husks remaining on panicle branch.

Range: Native. TX to FL and north to MA and west to MI.

Ecology: Occurs as scattered plants or small groups, especially in forests with livestock use and along horse trails. Spreads by seeds attached to animals, mainly in fur. Seeds can become embedded in bird dog noses requiring surgery to remove. Mostly occurs on dry sites and open to shady habitats. Present in forest plantations and open forests, and along forest margins, horse trails, and right-of-ways.

Synonyms: *Stipa avenacea* L., blackseed needlegrass, black oatgrass.

Other species in the SE: *P. avenacioides* (Nash) Valencia & Costas.

Wildlife: The long, hard seeds of needlegrass are used occasionally by songbirds.

Piptochaetium **Speargrass or Needlegrass** **Poaceae**

T. Bodner (May)

Piptochaetium avenaceum

T. Bodner (Aug)

Piptochaetium avenaceum

T. Bodner (May)

Piptochaetium avenaceum

Rhynchospora Beaksedge or Beakrush Cyperaceae

Upright, mostly perennial, clump-forming, grass-like herbs, from rootcrowns or rhizomes. Leaves flat, hairless, and grass-like, narrow to wide, and basally erect. Stems triangular to nearly round in cross-section, with alternate small leaves on the stem. Pointed spikelets, beaked, usually brownish or reddish, in various clusters, mostly heads or some umbels and a few panicles, terminal and/or axillary. 54 species in the SE.

Common Species

Rhynchospora cephalantha Gray **- bunched beaksedge** **RHCE**

Plant: Erect grass-like perennial, 0.4-1 m tall, small tufted clumps, from a hard rootcrown or short rhizome.

Stem (culm): Slender, slightly angular, 1-several per tuft, nodes several, slightly swollen, with green color shades changing at nodes.

Leaves: Mostly erect basal and few reduced upward, basal 15-40 cm long and 1-5 mm wide.

Flowers: May-Nov. Spikelets in 1-6 round-topped dense clusters, 0.8-2 cm wide, terminal and axillary, spikelets lance-shaped and long pointed (projecting outward from cluster), 1.5-6 mm long, dark brown, fine bristles as long as beak, one fertile flower per cluster, small leaflets projecting outward at cluster base.

Seeds: Jun-Feb. Nutlets (achenes) lustrous brown with yellow centers, 1.4-2.5 mm long and 0.7-1.5 mm wide.

Range: Native. MS to FL and north to NJ, mainly in Cp.

Ecology: Occurs commonly on wet sandy soils in forest plantations, forest openings, open forests. Inhabits forest margins and right-of-ways, and around bogs and swamps.

Synonyms: *R. microcephala* (Britt.) Britt., *R. axillaris* (Lam.) Britt., roundheaded beakrush.

Wildlife: Little wildlife use reported.

Rhynchospora **Beaksedge or Beakrush** **Cyperaceae**

D. Lauer (Aug)

Rhynchospora cephalantha

D. Lauer (Aug)

Rhynchospora cephalantha

Saccharum — Plumegrass — Poaceae

Erect, stout and reed-like, perennial grasses, forming clumps, from harden bases or short rhizomes. Leaves mainly lower-stem, long-attenuate and flat, having long hairs near the base, and ligule a fringed membrane. Large terminal panicles, white to reddish hairy in most species, having spikelets in pairs, one stemmed and one sessile. Same genus as sugarcane. 6 species in the SE.

Common Species

Saccharum alopecuroidum (L.) Nutt. **- silver plumegrass** **SAAL9**

Plant: Erect, large grass, 1.5-3 m tall, with long flat leaves mainly from the lower stem, from harden bases or short rhizomes and having coarse shallow roots.
Stem (culm): Stout, round, green or purplish, densely hairy just below panicle, nodes usually densely hairy and purplish, internodes short-haired or hairless.
Leaves: Mostly on lower stem and scattered alternate upward, blades to 75 cm long and mostly 2-3 cm wide, white midvein above, white hair tufts at nodes and throat, long hairs on lower margins and upper surface of lowermost portion, ligule fringed membrane, 1-4 mm long.
Flowers: Sep-Nov. Panicle silvery-whitish or purplish woolly, 20-30 cm long, branches ascending or spreading, spikelets lanceolate, 5-6 mm long, cream to yellowish, awns thread-like, 1-2 cm long, flat and spirally twisted, bristle hairs from spikelet base longer than spikelet.
Seeds: Oct-Feb. Grain reddish, 2-2.5 mm long, released in husks with white or tawny hairs and twisted awn.
Range: Native. TX to FL and north to NJ and west to IL and MO, less frequent in Mt.
Ecology: Occurs commonly as individuals or colonies in new forest plantations, forest openings, and along forest margins and right-of-ways. Found on dry to moist sites and open to semi-shady habitats. Persists by rhizomes and spreads by wind-dispersed seeds.
Synonyms: *Erianthus alopecuroides* (L.) Ell., *E. divaricatus* (L.) A.S. Hitchc.

Saccharum giganteum (Walt.) Pers. **- sugarcane plumegrass** **SAGI**

Similar to *Saccharum alopecuroides* except **plant** 1-4 m tall; **leaves** blades flat to 50 cm long and to 2.5 cm wide, long-white hairy near base and short-hairy or not elsewhere, sheaths hairy near throat and on margins; **flowers** panicles terminal on ascending branches, spikelets brown, 5-7 mm long, awns round and straight, hair bristles exceed husks. **Range:** Native. TX to FL and north to NY and southwest to KY and AR, mostly in Cp and lower Pd.
Ecology: Occurs on dry to wet sites.
Synonyms: *Erianthus giganteus* (Walt.) P. Beauv., *E. giganteus* var. *compactus* (Nash) Fern., *E. laxus* Nash, *E. saccharoides* Michx., *E. tracyi* Nash, beardgrass.

Other species in the SE: *S. baldwinii* Spreng., *S. brevibarbe* (Michx.) Pers., *S. coactatum* (Fern.) R. Webster, *S. ravennae* (L.) L.

Wildlife: Plumegrass has little wildlife value. The seeds are consumed sparingly by some songbirds.

Saccharum **Plumegrass** **Poaceae**

J. Miller (Oct)

Saccharum giganteum

T. Bodner (Oct)

Saccharum alopecuroidum

T. Bodner (Sep)

Saccharum alopecuroidum

J. Miller (Oct)

Saccharum alopecuroidum

Scleria Nutrush or Egg sedge Cyperaceae

Upright, annual or perennial, clump-forming, grass-like herbs, with triangular cross-section stems, from rhizomes. Leaves hairless, grass-like, lowest leaf sheaths bladeless, and upper leaf blades folded and keeled. Spikelets in terminal clusters, or both axillary and terminal clusters, within tufts of small-leaved bracts. Male and female flowers separate, and usually mixed within a cluster, both yellow-green turning brownish or purplish. Nutlets (achenes) spherical with a point and lustrous white to gray, often resembling a tiny egg (thus the common name), smooth to rough, within tufts of bracts. 10 species in the SE.

Common Species

Scleria triglomerata Michx. **- whip or tall nutrush** **SCTR**

Plant: Erect, coarse, grass-like perennial, 50-75 cm tall, in large clumps, from hard knotty rhizomes.

Stem (culm): Stout, 3-6 mm thick at base, sharply triangular in cross-section, hairless to short hairy.

Leaves: Mostly lower stem, flat with short-tapering ends, 40-60 cm long and 4-9 mm wide, rough, often somewhat hairy on reddish sheaths and lower midvein, yellow-green often with lighter parallel veins, ligules hairy or not.

Flowers: Apr-Sep. Spikelets in terminal stalked clusters and usually 1-2 long-stalked axillary clusters, within tufts of leafy bracts, the lowest bract erect, 5-15 cm long.

Seeds: May-Dec. Nutlet (achene) smooth lustrous, semi-spherical, white or rarely gray, 2.5-3.5 mm long, resembling a tiny egg nestled within the leafy bracts.

Range: Native. TX to n FL and north to s Canada and west to MN and KS.

Ecology: Occurs in open forests, forest plantations, and along forest margins and right-of-ways, on wet to dry sites. Persists and colonizes by rhizomes and spreads by animal- and water-dispersed seeds.

Synonyms: *S. laccida* Steud., *S. nitida* Muhl. *ex* Willd.

Other species in the SE: *S. baldwinii* (Torr.) Steud., *S. ciliata* Michx., *S. georgiana* Core, *S. hirtella* Sw., *S. minor* W. Stone, *S. oligantha* Michx., *S. pauciflora* Muhl. *ex* Willd., *S. reticularis* Michx., *S. verticillata* Muhl. *ex* Willd.

Wildlife: Nutrush seeds are used sparingly by Northern Bobwhite and some songbirds. No other wildlife value reported.

Scleria **Nutrush or Egg sedge** **Cyperaceae**

D. Lauer (Aug)

Scleria triglomerata

Scirpus **Bulrush** **Cyperaceae**

Upright or ascending, annual or perennial, grass-like herbs of wet soils or shallow water, having stems round or triangular in cross-sections or tubular. Leaves either basal, lower stemmed, or merely sheaths, with blades hairless, and tips 3-sided and having rough margins. Sheaths usually loose and overlapping near plant base, some membranous. Spikelets few to many flowered, in terminal clusters, appearing lateral in some species due to an erect leafy bract resembling a stem tip. Seed grains (achenes) essentially spherical with a flattened side or slightly 3-sided, within reddish to brownish scales, and 1-several tiny bristles. 20 species in the SE.

Common Species

Scirpus tabernaemontani K.C. Gmel. **- softstem bulrush** **SCTA2**

Plant: Erect, grass-like perennial, 0.5-3 m tall, tubular stemmed and often leafless, from horizontal rhizomes (3-8 mm thick) and forming dense colonies.

Stem (culm): Stout, 1-3 cm thick at base, round in cross-section and hollow, soft and compressible, hairless, light green.

Leaves: Blades absent or to 10 cm long and partitioned, basal sheath membranous.

Flowers: Jun-Sep. Terminal, single or ovoid clusters, reddish, 3-10 mm long and 2-5 mm wide, with stems and/or stemless, stems flat and rough, clusters subtended by a leafy bract, 1-7 cm long.

Seeds: Jul-Nov. Grain (achene) spherical with flattened side, 1.8-2.5 mm long, yellowish to grayish, smooth and lustrous, within reddish scales and 6 reddish bristles.

Range: Native. Throughout US and s Canada.

Ecology: Occurs as colonies or groups in open areas with seasonal or year-long standing water, such as margins of ponds, marshes, and streams. Persists and colonizes by rhizomes and spreads by water- and animal-dispersed seeds.

Synonyms: *Schoenoplectus tabernaemontani* (K.C. Gmel.) Palla, *S. validus* Vahl., great bulrush, matrush, blackrush, tule.

Wildlife: The achenes of most species of bulrushes are important food of numerous waterfowl and shorebird species. Several species also provide extensive nesting and escape cover.

Scirpus **Bulrush** **Cyperaceae**

J. Miller (Sep)

Scirpus tabernaemontani

Setaria **Bristlegrass or Foxtail grass** **Poaceae**

Upright, annual or perennial grasses, with bristly, spike-like, cylindrical seedheads. Spikelets 2-flowered, crowded into the spike, each spikelet subtended and exceeded in length by 1-several bristles either green, yellow, purple, brown, or red. Leaves mainly on stem, with blades flat, both surfaces rough, sheaths hairless, and ligules with fringed membranes. Grain dark and released within husks, bristles not falling. 14 species in the SE, many introduced, naturalized, and weedy.

Common Species

Setaria glauca (L.) Beauv. **- pearl millet or yellow foxtail** **PEGL2**

Plant: Erect annual, 0.2-1.2 m tall, single or tufted, branched at the base and often purplish near base, from very fibrous roots.

Stem (culm): Upright, or spreading and bent upward, slender, enclosed in overlapping leaf sheaths for about half the height, smooth and hairless above, except for short hairs at base of seedhead.

Leaves: Mostly alternate upward, with few dried at base, long-lanceolate, 10-30 cm long and 4-10 mm wide, blades flat and often twisted, yellowish or light green, hairless or scattered long hairy being more dense near and within the throat, flared at throat and whitish, sheaths keeled and hairless except for upper margins, ligule a white-fringed membrane, 0.5-1.5 mm long.

Flowers: May-Nov. Spike-like panicle, 3-15 cm long and 1.5-2.5 cm wide, cylindric, yellow to brownish bristled when mature, spikelets ovate, 3-3.4 mm long, subtended by 4-12 bristles, outer husk roughly ridged.

Seeds: Jun-Jan. Grain narrowly-ovoid, 0.8-2 mm long, released within husks without the bristles.

Range: Exotic and naturalized from Europe. Throughout US and s Canada.

Ecology: Occurs as scattered plants or extensive colonies along forest margins and right-of-ways, old fields, disturbed sites, and cultivated lands. Spreads by animal-dispersed seeds.

Synonyms: *Pennisetum glaucum* (L.) R. Br., *S. lutescens* (Weigel) F.T. Hubbard, *Chaetochloa glauca* (L.) Scribn., *C. lutescens* (Weigel) Stuntz, *Panicum glaucum* L., pigeongrass.

S. parviflora (Poir.) Kerguelen **- knotroot foxtail** **SEPA10**

Similar to *Setaria glauca* except **plant** perennial, erect to bent and rooting at nodes, from knotty rhizomes; **leaves** 6-25 cm long and 1-9 mm wide; **flowers** spike-like panicle 1-10 cm long and 0.5-3 cm wide, yellow or purplish, spikelets 2-3 mm long. **Range:** Native. Throughout US and s Canada.

Ecology: Occurs along with and in the same habitats of yellow foxtail.

Synonyms: *S. geniculata* Beauv., marsh bristlegrass.

Other Species in the SE: *S. adhaerens* (Forsk.) Chiov., *S. corrugata* (Ell.) J.A. Schultes [coastal foxtail], *S. faberi* Herrm. [giant foxtail], *S. italica* (L.) Beauv. [foxtail millet], *S. macrosperma* (Scribn. & Merr.) K. Schum. [coral foxtail], *S. magna* Griseb. [giant bristlegrass], *S. palmifolia* (Koenig) Stapf, *S. pumila* (Poir.) Roemer & J.A. Schultes, *S. setosa* (Sw.) Beauv. [West Indian bristlegrass], *S. sphacelata* (Schumacher) Moss [African bristlegrass], *S. verticillata* (L.) Beauv. [bristly foxtail], *S. viridis* (L.) Beauv. [green foxtail].

Wildlife: Foxtails are a very important wild grass for Northern Bobwhite, Mourning Doves, and numerous songbirds, particularly the Northern Cardinal, Brown-headed Cowbird, Blue Grosbeak, Dark-eyed Junco, and various sparrows. Giant foxtail is an important coastal waterfowl plant. Foxtail leaves are eaten sparingly in spring by Wild Turkey, but are considered a poor White-tailed Deer forage.

Setaria **Bristlegrass or Foxtail grass** **Poaceae**

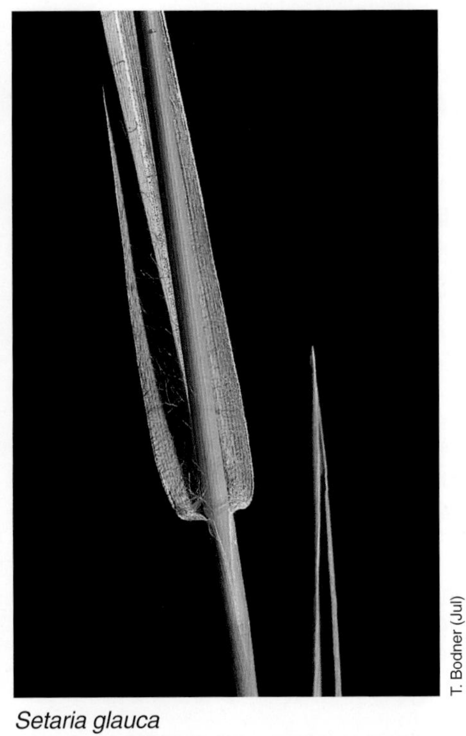

T. Bodner (Jul)

Setaria glauca

T. Bodner (Jul)

Setaria glauca

J. Miller (Nov)

Setaria parviflora

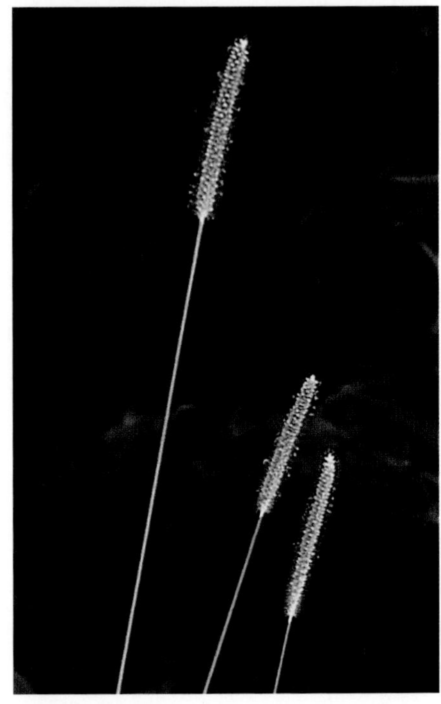

T. Bodner (Jul)

Setaria glauca

Sorghastrum Indiangrass Poaceae

Erect and rather tall, tufted, perennial grasses, from rhizomes or hardened bases. Mainly basal leaves having long-narrow flat blades and stiff ligules. Panicles terminal, narrow, brownish or tan, flowering in late summer. Spikelets spindle shaped, hairy on one side and a hairy bristle on the other side, with a thread-like terminal awn. 3 species in the SE.

Common Species

Sorghastrum nutans (L.) Nash **- yellow Indiangrass** SONU2

Plant: Tall, erect, perennial grass in loose tufts, 1-2.5 m tall, branched from base, with erect leaves, and brownish hairy panicles in late summer to fall, from scaly rhizomes.
Stem (culm): Stout, round in cross-section, tan, nodes slightly swollen and appressed hairy, internodes hairless and smooth.
Leaves: Mostly lower stem, blades to 60 cm long and 5-10 mm wide, mostly erect or close to stem, rough and sometimes whitish waxy, sheaths scattered long hairy or hairless, ligules membranous, 4-6 mm long, appearing eared when split.
Flowers: Aug-Oct. Panicles narrow-oblong and hairy, 10-40 cm long and 3-4 cm wide, bronze to golden brown and shining, often nodding at tips, with hairy lanceolate spikelets, 6-8 mm long, tipped with thread-like twisted awns, 1-1.5 cm long.
Seeds: Sep-Feb. Grain reddish, flat, 3-3.5 mm long, released within hairy husks and tipped with a thread-like awn.
Range: Native. AR to n FL and north to s Canada and west to ND, WY, and UT.
Ecology: Infrequently scattered plants in forest plantations and open forests, more commonly occurs along forest margins and right-of-ways, especially upland sites and Sandhills. Common in prairies of Blackbelt and Midwest. Persists by rhizomes and spreads by animal-dispersed seeds.
Synonyms: *S. avenaceum* (Michx.) Nash, *Andropogon nutans* L., woodgrass.

Sorghastrum secundum (Ell.) Nash **- lopside Indiangrass** SOSE5

Similar to *Sorghastrum nutans* except **plant** ascending, 1-2 m tall; **stem (culm)** slender, nodes and internodes hairless; **leaves** blades and sheaths often hairy; **flowers** (Sep-Oct) panicle branches on one side of a straight stalk, spikelet yellow-brown, awns 1.6-4.3 cm long, many twists. **Range:** Native. TX to FL and north to SC, only in Cp, especially in Sandhills and Blackbelt portions.
Synonyms: *Andropogon secundus* Ell.

Other species in the SE: *S. elliottii* (C. Mohr) Nash [slender indiangrass].

Wildlife: Indiangrass seeds are consumed by songbirds. Indiangrasses are not commonly important wildlife food plants due to their infrequent distribution, however, they can become abundant in open, burned pine forests.

Sorghastrum **Indiangrass** **Poaceae**

T. Bodner (Sep)

Sorghastrum secundum

T. Bodner (Oct)

Sorghastrum nutans

T. Bodner (Oct)

Sorghastrum nutans

Sorghum — Sorghum grass — Poaceae

Erect, large perennial and annual grasses. Only stem leaves, broad, long, and flat, having dense hairs on the collar and behind a fringed-membrane ligule. Terminal panicle, with many spike-like branches, having only 1-5 spikes per branch, and spikelets paired. One spikelet stemless and ovoid, the other stemmed and lanceolate, both shiny, pale green to purplish to gray, sometimes tipped with a thread-like awn. 3 species in the SE.

Common Species

Sorghum halepense (L.) Pers. - **Johnsongrass** — **SOHA**

Plant: Erect perennial, 1-3 m tall, branched from base with broad and long green leaves with prominent white midveins, from scaly sharp rhizomes and forming dense stands. Plant toxic, if fertilized heavily or drought stricken.
Stem (culm): Stout, hairless, upward branching somewhat.
Leaves: Alternate, long-lanceolate, 20-60 cm long and 1-3 cm wide, green with wide white midvein above, margins rough, hairless except on white flared throat and a hair patch behind the ligule, ligule a prominent white-fringed membrane, 2-5 mm long.
Flowers: Apr-Nov. Open spreading panicle, 15-50 cm long, with numerous whorled projecting branches being shorter in the upper portion, spikelets in pairs at the end of finer branchlets, one spikelet stemless and ovoid and the other stemmed and narrow, 4-6 mm long, husks shiny and short hairy, either green, yellow, purple, or black, tipped with a thread-like awn, 5-13 mm long or absent.
Seeds: May-Mar. Grain dark reddish-brown, released within the husks.
Range: Exotic and naturalized from Mediterranean area. CA to FL and north to MA and west to NE and OK.
Ecology: Occurs as dense colonies in old fields and along field margins and right-of-ways, where it invades new forest plantations, open forests, and forest openings. Highly competitive with planted and natural tree seedlings. Persists and colonizes by rhizomes and spreads by seeds.
Synonyms: *Holcus halepensis* L.

Other species in the SE: *S. almum* Parodi [columbusgrass], *S. bicolor* (L.) Moench [formerly *S. verticilliflorum* (Stud.) Stapf, *S. vulgare* Pers., or *Holcus bicolor* L., shattercane, milo, grain sorghum, broom corn (cultivated crop with many subspecies)].

Wildlife: Although Johnsongrass seeds are consumed by Northern Bobwhite, Mourning Doves, and other songbirds, this species often out-competes more valuable wildlife food plants in old agricultural fields. *Sorghum bicolor* is an agricultural crop that often is planted for wildlife. Its seed heads are used heavily by White-tailed Deer, Wild Turkey, Black Bear, and various songbirds and mammals.

Sorghum Sorghum grass Poaceae

T. Bodner (Jun)

Sorghum halepense

T. Bodner (Jun)

Sorghum halepense

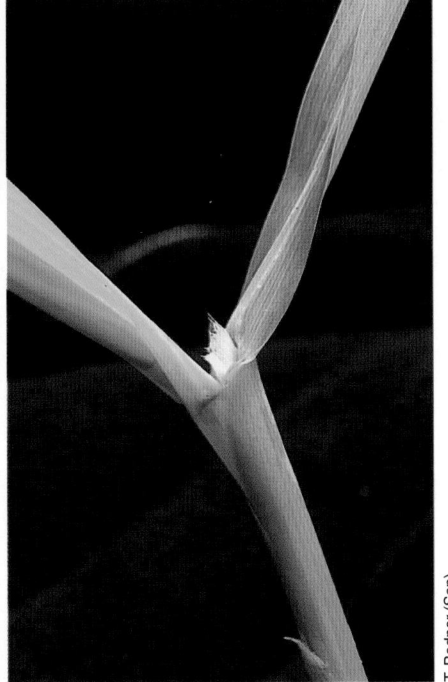

T. Bodner (Sep)

Sorghum halepense

J. Miller (Jul)

Sorghum halepense

Sporobolus Dropseed Poaceae

Upright to sprawling, tufted, perennial grasses, from rhizomes. Leaves spreading basal or lower stem, having narrow blades with margins rolled inward, appearing like pine needles. Panicles narrow or loose spreading. Grains usually falling free of husks, husks remaining on panicle, and outer coat detaching when water-soaked. 17 species in the SE.

Common Species

Sporobolus junceus (Beauv.) Kunth **- pineywoods dropseed** **SPJU**

Plant: Densely- to sparsely-tufted grass, perennial, 30-100 cm tall, dense tufts of thin leaves with upright seed stalks, from short rhizomes with fibrous roots.

Stem (culm): Erect to leaning, slender, wiry, hairless, extending well above the tufted leaves.

Leaves: Basal and lower stem, 2-25 cm long and 2-4 mm wide, margins rolled inward appearing tubular with a thickened pointed end, hairless (except occasional scattered basal hairs), upper blades short and well below the panicle.

Flowers: Sep-Oct. Panicle narrow, 10-25 cm long and 2-8 cm wide, branches brown to purple, 2-8 regular whorls, spreading or ascending, long-lanceolate spikelets, 2.5-3.5 mm long, husks green to purple to bronze, gaping husks remaining on panicle.

Seeds: Sep-Nov. Grain yellow-brown, 1.7-2 mm long, released free of outer husks.

Range: Native. TX to FL and north to VA, mainly in Cp, less frequent in the lower Pd.

Ecology: Occurs in sparse herbaceous cover forming dense clumps, or in dense grass cover or under partial shade often consisting of only a few slender tufts. Prefers open to semi-shady habitats and dry sites. Persists by rhizomes and spreads by animal-dispersed seeds.

Synonyms: *S. gracilis* (Trin.) Merr., wiregrass, pineland dropseed.

Sporobolus indicus (L.) R. Br. **- smut grass** **SPIN4**

Similar to *Sporobolus junceus* except **leaves** 15-50 cm long and 1-5 mm wide, usually folded or flat to rolled; **flowers** (Jun-Oct) spike-like panicle, with appressed or ascending branches, somewhat interrupted, 10-40 cm long and 3-6 mm wide; **seeds** spikelets black from black smut fungus (thus common name), single flowered, 1.5-2 mm long. **Range:** Naturalized from Asia. TX to FL and north to MA and west to NE.

Ecology: Common and widely-spread weedy species, occurring along forest roads and edges. Present as scattered plants or loose colonies. Plants not infected with black fungus produce seeds. Persists by rhizomes and spreads by animal-dispersed seeds.

Synonyms: *S. angustus* Buckl., *S. berteroanus* (Trin.) A.S. Hitchc. & Chase, *S. poiretii* (Roemer & J.A. Schultes) A.S. Hitchc., *Agrostis indica* L., rattail smutgrass.

Wildlife: Because of their infrequent occurrence, the dropseeds are not important wildlife plants, although the seeds are readily consumed by songbirds when available.

Sporobolus **Dropseed** **Poaceae**

T. Bodner (Sep)

Sporobolus junceus

J. Miller (Oct)

Sporobolus junceus

J. Everest (Sep)

Sporobolus indicus

J. Miller (Oct)

Sporobolus junceus

Tridens Sandgrass or Tridens Poaceae

Upright perennial grasses, tufted, or sprouting in a line from rhizomes. Leaves long narrow, basal or lower stem, hairless, blades flat or margins slightly rolled inward, and ligule a fringe of hair. Terminal panicle, much exceeding the leaves, spreading or condensed spike-like, with somewhat flattened spikelets of 3-10 flowers, often purplish. Grains yellow and ellipsoid or ovoid. 5 species in the SE.

Common Species

Tridens flavus (L.) A.S. Hitchc. - **purpletop tridens** **TRFL2**

Plant: Erect, tufted, perennial, 0.6-1.6 m tall, 1-few stems from base, from short rhizomes and forming dense colonies.

Stem (culm): Moderately stout, smooth, sticky near top and axes of panicle, green to pale green or yellowish.

Leaves: Basal and lower stem, blades to 60 cm long and 2-12 mm wide, slightly hairy on upper surface near base, collars whitish (often with purplish tinge) and slightly hairy, ligule a hairy fringe, 0.5-1mm long.

Flowers: Aug-Nov. Terminal panicles, 15-35 cm long and 10-25 cm wide, branches ascending or spreading becoming drooping at the ends, branch bases slightly swollen and hairy, spikelets usually red or purple at maturity (infrequently whitish or yellowish), flattened, with 4-8 flowers in a herring-bone, 6-8 mm long and 1.5-2.2 mm wide.

Seeds: Sep-Mar. Grain whitish yellow, ellipsoid, 1.5-2.5 mm long, released within smooth husks.

Range: Native. TX to FL and north to NH and west to MN and KS.

Ecology: Occurs in forest plantations, open forests, and old fields, and along forest margins, right-of-ways, and levees on floodplains. Present in open to shady habitats and dry to moist sites. Persists and colonizes by rhizomes and spreads by animal-dispersed seeds.

Synonyms: *T. flavus* var. *chapmanii* (Small) Shinners, *T. flavus* var. *flavus*, *T. chapmanii* (Small) Chase, *Triodia chapmanii* (Small) Bush, *T. flava* (L.) Smyth, tall redtop.

Other species in the SE: *T. albescens* (Vasey) Woot. & Standl. [white tridens], *T. ambiguus* (Ell.) J.A. Schultes [pine barren tridens], *T. carolinianus* (Steud.) Henr., *T. strictus* (Nutt.) Nash [longspike tridens].

Wildlife: During August and September, purpletop tridens seeds are used occasionally by Wild Turkey and to a lesser extent by Northern Bobwhite. Purpletop tridens is a poor White-tailed Deer forage.

Tridens **Sandgrass or Tridens** **Poaceae**

T. Bodner (Aug)

Tridens flavus

T. Bodner (Aug)

Tridens flavus

T. Bodner (Sep)

Tridens flavus

T. Bodner (Sep)

Tridens flavus

Woody Vines and Semiwoody Plants

Ampelopsis **Peppervine** **Vitaceae**

High-climbing woody vines, with somewhat swollen nodes, and climbing by scattered branched tendrils occurring opposite to leaves. Leaves alternate, multi-stemmed compound or simple grape-like, all deciduous. Flowers tiny, in branched flat clusters (cymes) opposite leaves, with 5 yellowish-green petals, 5-lobed calyx, and 1 pistil. Fruit a small spherical berry, 6-10 mm wide, green turning blue or black, containing 2-5 seeds. 4 species in the SE, including an exotic invasive species, *A. brevipedunculata*.

Common Species

Ampelopsis arborea (L.) Koehne **- peppervine** **AMAR5**

Plant: Deciduous woody vine, high-climbing, arbor-forming, or occasionally bushy, climbing by tendrils, from a woody rootcrown.

Stem: Round in cross-section, nodes flattened and swollen, light green turning reddish to purplish, becoming light tan barked with lens-shaped lighter spots (lenticels), pith white.

Leaves: Alternate, deciduous, pinnately compound being broadly triangular in outline, 10-25 cm long and wide, twice or sometimes 3-times divided, leaflets 2-5 cm long and 1-3 cm wide, oval or diamond-shaped in outline, large serrate teeth (to deep lobes), green and lustrous above, pale with hairy raised-veins beneath, petioles short, greenish to reddish to purplish, and sparsely hairy, branched tendrils opposite some leaves (may bear flowers on some tendril branches).

Flowers: Jun-Oct. Inconspicuous, flat clusters with opposite branching, clusters 2-4 cm wide and shorter than leaves, occurring opposite leaves on new growth, each flower 2-4 mm wide, petals 5, yellow-green, pistil 1, on a flat disk with nectar glands beneath.

Fruit and seeds: Sep-Dec. Berry spherical, 6-10 mm wide, green turning shiny red to black, dry, containing 2-5 seeds, black, orange-slice shaped, resembling peppercorns (thus the common name). Peppery tasting.

Range: Native. TX to FL and north to s VA and west to IL and OK, most frequent in Cp, rare in Southern Appalachian Mt.

Ecology: Occurs most frequently in wet to moist bottomland forests, along streams and rivers, but will persist in moist forest plantations and forest margins. Persists by woody rootstocks and spreads by vine growth and animal-dispersed seeds.

Synonyms: *A. bipinnata* Michx.

Other species in the SE: *A. aconitifolia* Bunge, *A. brevipedunculata* (Maxim.) Trautv. [porcelain berry - exotic invasive], *A. cordata* Michx. [cordate cissus]

Wildlife: Peppervine is a poor White-tailed Deer forage, although more extensive use has been reported at some locales. During summer, the fruits are used extensively by Raccoon and likely other wildlife species.

Ampelopsis **Peppervine** **Vitaceae**

Ampelopsis arborea

Ampelopsis arborea

Ampelopsis arborea

Berchemia Supplejack or Rattanvine Rhamnaceae

Twining woody vine, with alternate thick leaves having straight parallel veins and no tendrils. Vines wiry becoming large and stout. Flowers tiny and greenish-white, in terminal and axillary clusters. Fruit a blue drupe. 1 species in the SE.

Common Species

Berchemia scandens (Hill) K. Koch **- Alabama supplejack** **BESC**

Plant: Deciduous to tardily deciduous, woody vine, high climbing somewhat by twining or scrambling over trees as well as trailing or arching upward shrub-like, from a woody rootcrown.

Stem: Round in cross-section, tough and wiry, smooth and hairless, dull green turning to shiny drab green, becoming reddish- to grayish-brown barked having lighter streaks, growing up to 3-8 cm wide, frequent alternate branching, branch scars round and protruding, pith solid and white.

Leaves: Alternate, ovate to oval with pointed tips, 3-5 cm long and 2-4 cm wide, shiny green above and whitish to pale green beneath, parallel veins at a 30-degree angle to the midvein, margins entire or slightly wavy, petioles 0.5-2 cm long.

Flowers: Apr-May. Terminal and axillary panicles, 1-5 cm long, petals 5, each about 2 mm long and wide, whitish to greenish yellow, on a nectar disc.

Fruit and seeds: Aug-Nov. Berry-like drupe, ovoid, 5-8 mm long, blue-black and white waxy when ripe, with a 2-celled stone.

Range: Native. TX to FL and north to s VA and west to TN, s IL, and MO, mainly in Cp of the SE.

Ecology: Occurs frequently on a wide range of sites, including swamps and wet forests of Cp, high climbing in bottomland forests, and low cover in forest plantations. Also inhabits upland mixed forests and fencerows. Persists by woody rootstocks, colonizes by vine growth (not rooting at buried nodes), and spreads by animal-dispersed seeds.

Synonyms: rattanvine.

Wildlife: The fruits of Alabama supplejack are eaten by Wild Turkey, Northern Bobwhite, Raccoon, and Gray Squirrel during fall and early winter. The fruits are low in crude protein and phosphorus and high in crude fiber and calcium. It is a high preference White-tailed Deer forage, particularly in spring and summer.

Berchemia **Supplejack or Rattanvine** **Rhamnaceae**

T. Bodner (May)

Berchemia scandens

Bignonia — Crossvine — Bignoniaceae

High-climbing woody vine, slender, and climbing by twining and clinging axillary tendrils. Opposite compound leaves, having pairs of 2-leaflet leaves at a node, appearing as 4 leaflets at a node, margins entire and unlobed. Showy orange and yellow flowers appearing in spring, short and wide tube having 5 flared lobes, somewhat irregular or 2-lipped, with the lowest lip the larger. Seeds many, transversely winged, enclosed in a long-slender, dry, flattened capsule. 1 species in the SE.

Common Species

Bignonia capreolata L. **- crossvine** — **BICA**

Plant: Semi-evergreen, high-climbing, woody vine, 20-30 m long, climbing by paired tendrils, from trailing woody root. Cross-section of stem revealing a cross-shaped pith (thus the common name).

Stem: Slender, round to slightly angled in cross-section, smooth and green, becoming golden-brown to dull gray-brown barked, sloughing in thin films appearing as ridges, slightly rough, nodes slightly flattened (not swollen), with paired fine wiry tendrils, very curly, attaching to bark or twigs.

Leaves: Opposite, thin and semi-evergreen, compound with only 2 leaflets (appearing as 4 leaves per node), leaflets oblong-lanceolate with rounded or flared bases, 5-15 cm long and 2-6 cm wide, surfaces hairless and smooth, margins entire, outermost veins paralleling the margin, midvein whitish and all veins slightly protruding beneath, between leaflets a branched tendril for both twining and clinging, with a pair of small accessory leaves in each axil, resembling stipules.

Flowers: Apr-May. Axillary clusters of 2-5 flowers, showy, corolla narrowly tubular at base and flaring to about 2 cm wide with 5 spreading lobes, lobes rounded, slightly irregular, tube dull red to orange outside and yellow and/or red within, calyx small and cup-like, stalks 2-3 cm long.

Fruit and seeds: Jun-Sep. Long-flattened capsule with tapering ends, 10-20 cm long and 2-2.5 cm wide, splitting open to release numerous seeds, elliptic, flat and thin papery, attached together in pairs, each 2 winged, 2.5-3 cm long and 8-10 mm wide including wings.

Range: Native. TX to n FL and north to MD and VA and west to IL and OK, rare in Mt.

Ecology: Occurs on a wide range of sites as infrequent scattered plants, usually high in the canopy, with only vines nearby for identification. Common in the canopy of bottomland forests, but present in lowland forest plantations and upland mixed forests, and along fencerows. Persists by woody rootstocks and spreads by vine growth and wind- and water- dispersed seeds.

Synonyms: *Anisostichus capreolata* (L.) Bureau, *A. crucigera* (L.) Bureau., trumpet-flower.

Wildlife: Crossvine is a preferred food item of the Swamp Rabbit, and a moderate preference White-tailed Deer forage. Because it flowers early, crossvine is a good nectar source for Ruby-throated Hummingbirds at a time when other sources may be scarce.

Bignonia **Crossvine** **Bignoniaceae**

T. Bodner (Mar)

Bignonia capreolata

T. Bodner (Jun)

Bignonia capreolata

Campsis Trumpetcreeper Bignoniaceae

Deciduous woody vine, trailing or climbing by rootlets on the stem, with opposite pinnately-compound leaves. Tubular flowers, bright orange to red. Causes skin rash like poison-ivy on sensitive individuals. 1 species in the SE.

Common Species

Campsis radicans (L.) Seem. *ex* Bureau **- trumpetcreeper** **CARA2**

Plant: Deciduous woody vine, up to 10 m long or longer, trailing and rooting at nodes, and high climbing by rootlets at nodes, at times forming dense colonies.

Stem: Stout, up to 2-8 cm wide, round in cross-section, new growth shiny smooth and tan to brown, prior year's growth rough and gray, old vines gray to tan barked and platy-fissured, leaf scars opposite and circular to oval, with round bundle scar, climbing by clinging tufted rootlets at nodes.

Leaves: Opposite, odd pinnately compound with 7-15 leaflets, leaflets ovate, 3-5 cm long and 1-3 cm wide, margins coarsely- or finely-toothed, smooth and shiny above and lighter green with white-hairy prominent veins (or smooth) beneath, tips often sharp and twisted, blades wavy.

Flowers: Jun-Oct. Axillary branched clusters, 5-20 flowers, each tubular (6-8 cm long) with 5 squarish lobes (much shorter than tube), bright orange to red with a striped inner throat, calyx tube 1.5-2 cm, orange to red, with 5 short teeth, each 4-5 mm long. Flower buds ovoid and red-orange.

Fruit and seeds: Oct-Feb. Long narrow capsule with pointed ends, 10-18 cm long and 2-3 cm wide, green early turning light gray, splitting in halves lengthwise to release numerous seeds, 2-winged, flat with papery wings, 1.5-2.5 cm long including wings.

Range: Native. TX to FL and north to MD and west to NE and OK, most frequent in Cp and Pd (AL to NC).

Ecology: Occurs commonly in fencerows spreading into right-of-ways, fields, and forest plantations, at times invasive in the understory and forming dense colonies. Planted as an ornamental and spreads to become weedy. Persists by woody rootcrowns, colonizes by vine growth, and spreads by wind- and water-dispersed seeds.

Synonyms: *Bignonia radicans* L., *Tecoma radicans* (L.) Juss., trumpetvine, cow-itch-vine [milk from cows feeding on this plant can cause a rash on infants, thus the common name].

Wildlife: Trumpetcreeper is an excellent Ruby-throated Hummingbird plant. It is readily used by White-tailed Deer, particularly in the spring and summer.

Campsis **Trumpetcreeper** **Bignoniaceae**

J. Miller (Sep)

Campsis radicans

T. Bodner (May)

Campsis radicans

J. Miller (Jun)

Campsis radicans

Gelsemium **Jessamine vine** **Loganiaceae**

Evergreen, climbing or trailing woody vines, with thick opposite leaves, elliptical to lanceolate, and hairless. Flowers in axillary few-flowered clusters (cymes), yellow, tubular to funnelform, with 5 short and broad spreading lobes. Dry capsule, splitting into halves to release many flat, winged seeds. 2 species in the SE.

Common Species

Gelsemium sempervirens (L.) St.-Hil. **- yellow jessamine** **GESE**

Plant: Commonly-occurring woody vine, high-climbing, trailing, or carpet-forming, up to 10 m long (sometimes longer), twining by twisting rapid vine growth and rooting at litter-covered nodes, from trailing woody roots. Plant toxic to livestock and honey from plant may cause a reaction in sensitive individuals.
Stem: Slender but very tough, round in cross-section, reddish brown with frequent white waxy areas, becoming fissured with light streaks and rough, branching at infrequent swollen nodes, often becoming tangled and matted, young growth twining and also climbing trees by hugging bark fissures, rooting at nodes when covered by leaf litter.
Leaves: Opposite, lanceolate to elliptic with pointed tip, 2-6 cm long and 1-2 cm wide, thick, margins entire, lustrous green above, dull and lighter green beneath, petioles 2-7 mm long.
Flowers: Feb-early May. Fragrant, few-flowered axillary clusters (cymes) on new growth, bright yellow, 2-4 cm long, funnel-shaped with 5 broad lobes, sepals 5, rounded to blunt.
Fruit and seeds: Oct-Jun. Dry capsule, broadly oblong, 1.5-2.5 cm long and 1-1.5 cm wide and 0.4-0.8 cm thick, tip pointed, splitting into halves to release many seeds, flat and winged, 6-9 mm long. Extended release period over winter into the following early summer.
Range: Native. TX to c FL and north to se VA and west to se TN and AR.
Ecology: One of the most common woody vines, inhabiting the full range of forest types. Forms dominant ground cover in open dry forests, and less abundant but frequent in shaded bottomland forests and in seasonally wet forests. Commonly growing through and over shrubs and midstory trees, but also capable of climbing mature pine trees. Bred and planted as an ornamental. Persists by woody rootcrowns, colonizes by vine growth, and spreads infrequently by gravity- and water-dispersed seeds.
Synonyms: *Bignonia sempervirens* L., Carolina jessamine, poor man's rope, evening trumpetflower.

Gelsemium rankinii Small **- swamp jessamine** **GERA4**

Similar to *Gelsemium sempervirens* except **flowers** not fragrant, sepals pointed and persisting on capsule; **fruit and seeds** capsule elliptical with beaked tip about 3 mm long, seeds wingless. **Range:** Native. Restricted to se LA, s AL, FL panhandle, se GA, and scattered in coastal SC and NC.
Ecology: Occurs in wet forests, bogs and swamps.
Synonyms: Rankin's trumpetflower.

Wildlife: Both species of jessamine are an early spring source of nectar for Ruby-throated Hummingbirds and the Spicebush Swallowtail butterfly. *Gelsemium* are a White-tailed Deer browse of moderate preference, particularly in the Coastal Plain.

Gelsemium **Jessamine vine** **Loganiaceae**

T. Bodner (Apr)

Gelsemium sempervirens

T. Bodner (Mar)

Gelsemium sempervirens

T. Bodner (Sep)

Gelsemium sempervirens

J. Miller (Jun)

Gelsemium sempervirens

Hypericum **St. Johnswort** **Clusiaceae**

Perennial herbs, small shrubs, or annuals. Shrub species tardily deciduous or evergreen, having glands appearing as dots or short lines on the leaves, stems, sepals, and petals (requires hand lens). Leaves opposite and small, sessile, with entire margins. Flowers yellow or orange-yellow (or pink), petals 4 or 5, and sepals 4 or 5. Fruit an ovoid capsule having a persistent style tip enclosed in persistent sepals, splitting lengthwise to release many cylindric or oblong seeds. 37 species in the SE, with the annual pineweed (*Hypericum gentianoides*) in the forb section.

Common Species

Hypericum hypericoides (L.) Crantz - **St. Andrew's-cross** **HYHY**

Plant: Bushy small shrub, 0.3-1 m tall, densely and finely branched, from a woody rootcrown with fibrous roots.

Stem: Erect or ascending, 1-few stems from base, branching near base and densely branching from leaf axils above, reddish-brown finely-winged twigs, with shredding bark on older branches, often becoming smooth and glossy near base.

Leaves: Opposite, tardily deciduous to evergreen, linear to oblanceolate to elliptic, 8-25 mm long and 1-7 mm wide, sessile, margins entire, basal margins often notched or flared, round tipped, whitish-green beneath, tiny holes or black dots (glands) on lower surface (less so on upper), small leaves appearing as fascicles in leaf axils due to frequent branching.

Flowers: May-Oct. Solitary on short bracted stalks (2-3 mm long) from terminal leaf axils (appearing as pairs when from opposite axils), petals 4 (in X-shaped cross, thus common name), pale yellow, petals oblong to oblanceloate to elliptic, often irregular shaped with pointed tip offset, 6-10 mm long, quickly falling while 2 large sepals (and 2 minute sepals) persistent and becoming erect, many stamens also persistent.

Fruit and seeds: Jul-Apr. Capsule enclosed between persistent sepals, oval, greenish becoming reddish-brown, capsule ovate to elliptic and compressed, 4-9 mm long and 2.5-4 mm wide and 1 mm thick, beaked by two persistent styles, splitting lengthwise to release many seeds, oblong cylindric, 1 mm long, black.

Range: Native. TX to FL and north to MA and west to MO and e OK, mainly in Cp and Pd in the SE.

Ecology: Occurs as scattered plants on a wide range of sites and in open to partially shaded stands. Present in upland forests, sandhills to bottomlands, cypress-gum depressions, and bogs. Persists by woody rootcrowns and spreads by animal-, insect-, and water-dispersed seeds.

Synonyms: *H. stragulum* P. Adams & Robson, *Ascyrum hypericoides* L., *A. linifolium* Spach.

Hypericum **St. Johnswort** **Clusiaceae**

T. Bodner (Sep)

Hypericum hypericoides

T. Bodner (Mar)

Hypericum hypericoides

T. Bodner (Sep)

Hypericum hypericoides

Hypericum **St. Johnswort** **Clusiaceae**

Hypericum nudiflorum Michx. *ex* Willd. **- early St. Johnswort** **HYNU**
Similar to *Hypericum hypericoides* except **stem** wing-angled; **leaves** elliptic to ovate, 1.5-8 cm long and 0.5-3 cm wide, base rounded or pointed, tip rounded sometimes with tiny tip (mucronate); **flowers** (May-Sep) petals 5, yellow, obovate, 6-8 mm long; **fruit and seeds** capsule ovoid, 4.5-6 mm long and 3.5-5.5 mm wide. **Range:** Native. MS to FL and north to VA and west to TN, mainly Pd and Hilly Cp.
Ecology: Occurs in moist forests and open to shady habitats.
Synonyms: *H. apocynifolium* Small., naked-flower St. Johnswort.

Hypericum crux-andreae (L.) Crantz **- St. Peterswort** **HYCR3**
Similar to *Hypericum hypericoides* except **plant** a shrub 0.3-1 m tall; **leaves** larger, 2-3 cm long, margins slightly rolled inward, bases rounded to slightly clasping, often dull green and white waxy; **flowers** (May-Oct) petals 4, bright yellow, obovate, 10-18 mm long, round tipped, sepals 4, 2 ovate to heart-shaped (10-18 mm long and 8-15 mm wide) and 2 lanceolate (6-15 mm long and 2-3 mm wide), styles 3; **fruit and seeds** capsule ovate to oval, 7-10 mm long, beaked by 3 persistent diverging styles, enclosed within 2 persistent sepals, green or brown, splitting to release many seeds, oblong, 0.7-0.8 mm long, brown. **Range:** Native. TX to c FL and north to NJ and west to se KY, n AL, and e OK.
Synonyms: *H. stans* (Michx. *ex* Willd.) P. Adams & Robson, *Ascyrum stans* Michx. *ex* Willd., *A. crux-andreae* L., *A. cuneifolium* Chapman.

Hypericum cistifolium Lam. **- roundpod St. Johnswort** **HYCI**
Similar to *Hypericum hypericoides* except **plant** a shrub 1-2 m tall, crown bushy upward and spreading branched with short branches in upper axils; **stem** pale-green to yellowish green within flower and seed stalks, becoming reddish-brown below with 4 longitudinal wings, sloughing in flakes on main stem; **leaves** lanceolate to oblong, 2-3 cm long and 6-7 mm wide, broadest near the base, bases rounded appearing somewhat clasping, margins rolled under, main stem leaves with short branchlets in axils; **flowers** (Jun-Oct) terminal branched clusters (cymes), petals 5, bright yellow, obovate with a small tooth on one side of a broad tip (flower pin-wheel shaped), sepals 5, oblong, 2-3 mm long; **fruit and seeds** capsule ovate, 4-5 mm long, dark brown, beaked with 3 persistent styles (appearing as one or divergent at tips), 2 mm long, within 5 persistent spreading sepals and tufts of brown persistent stamens, seeds oblong, 0.4-0.5 mm long, dark brown. **Range:** Native. TX to s FL and north to NC, within the Cp.
Ecology: Occurs as scattered plants or small colonies, on wet to moist sites, such as floodplain forests, pine savannas, seasonally wet flatwoods, river and stream banks, shores of ponds and lakes, and ditches. Persists by woody rootstocks and spreads by seeds.
Synonyms: *H. opacum* Torr. & Gray.

Wildlife: The woody *Hypericum* (not *H. gentianoides*) generally are poor White-tailed Deer browse plants. Seeds of some species are consumed in low amounts by Northern Bobwhite and seed-eating songbirds.

Hypericum St. Johnswort Clusiaceae

J. Miller (Oct)

Hypericum crux-andreae

J. Miller (Oct)

Hypericum nudiflorum

D. Lauer (Aug)

Hypericum cistifolium

Lonicera Honeysuckle Caprifoliaceae

Twining or trailing woody vines and shrubs, having opposite leaves with entire margins, except some lobed or finely serrate on spring shoots. Flowers 2 to several in axillary or terminal radiating clusters on bracted stalks, corolla tubular having 5 lobes in 2 lips; white, pink, yellow or red; and a nectar gland at the base within. Fruit a spherical berry, containing few flattened-ellipsoid seeds. 10 species in the SE, with several exotic and invasive, replacing native flora.

Common Species

Lonicera japonica Thunb. **- Japanese honeysuckle** **LOJA**

Plant: Semi-evergreen woody vine, high climbing and trailing, from long woody rhizomes.
Stem: Slender becoming stout, round in cross-section, brown and hairy becoming tan barked, having fissures and sloughing with age, opposite branched, rooting at low nodes.
Leaves: Opposite, mostly semi-evergreen or evergreen, ovate to elliptic to oblong, 4-6.5 cm long and 2-3.5 cm wide, base rounded, tips blunt-pointed to rounded, margins entire with fine hairs but often lobed in the spring, smooth to rough hairy both surfaces.
Flowers: Apr-Jun. Axillary pairs on a bracted stalk, fragrant, white (or pink) and pale yellow, 2-3 cm long, thin tubular flaring to 5 lobes in 2 lips (upper lip 4-lobed and lower lip 1-lobed), longest lobes about equaling tube, 5 stamens and 1 exserted pistil.
Fruit and seeds: Aug-Mar. Berry black, glossy, nearly spherical, 5-6 mm wide, stalks 1-3 cm long with persistent sepals, seeds several, oblong, 3-3.3 mm long, black.
Range: Exotic and naturalized from Asia. TX to FL and north to MA and west to IA and OK.
Ecology: Common invasive exotic, overwhelming and replacing native flora on a wide range of sites. One of the most common vines in pine plantations and bottomland forests in mid-region. Occurs as dense infestations along forest margins and right-of-ways. Persists by rootstocks and spreads by rooting at nodes and animal-dispersed seeds.

Lonicera sempervirens L. **- trumpet honeysuckle** **LOSE**

Similar to *Lonicera japonica* except **plant** initially erect, twining, or trailing, not exceeding about 5 m long; **stem** smooth to finely hairy at first, infrequently branching, new growth purplish; **leaves** tardily deciduous, 3-7 cm long and 1-4 cm wide, variable being oblong to elliptic to rounded, with uppermost 1 or 2 pairs joined around the stem, white waxy below; **flowers** (Mar-Sep) in terminal spikes, 2-4 per node with subtending bract, red slender tube and yellow lobes, 3-5 cm long; **fruit and seeds** (Jul-Sep) berry red and seeds golden-brown. **Range:** Native. TX to FL and north to ME and west to IA, mainly Cp and Pd in the SE.
Ecology: Occurs on moist to wet sites, open forests and their borders. Often present as infrequent plants or few in a group. Bred and planted as ornamental.
Synonyms: *Phenianthus sempervirens* (L.) Raf., coral honeysuckle, mailbox vine.

Other species in the SE: *L.* x *bella* Zabel, *L. canadensis* Bartr. *ex* Marsh., *L. dioica* L., *L. flava* Sims, *L. fragrantissima* Lindl. & Paxton, *L. maackii* (Rupr.) Herder, *L. morrowii* Gray, *L. tatarica* Lindl. & Paxton.

Wildlife: Japanese honeysuckle is an important year-round browse plant for White-tailed Deer and Eastern Cottontail, particularly in the Piedmont. Both fruit and vegetative growth are eaten by Wild Turkey, Northern Bobwhite, and songbirds, particularly during winter. However, fruit production is low and inconsistent. Japanese honeysuckle also provides dense escape cover. Trumpet honeysuckle is a low preference browse. Its flowers are visited by Ruby-throated Hummingbirds and it is a larval plant for the Spring Azure butterfly.

Lonicera Honeysuckle Caprifoliaceae

T. Bodner (May)

Lonicera japonica

T. Bodner (May)

Lonicera japonica

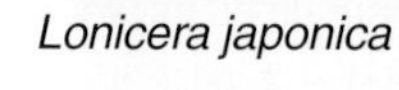

T. Bodner (Apr)

Lonicera sempervirens

T. Bodner (Oct)

Lonicera japonica

Mitchella Partridgeberry Rubiaceae

Low-prostrate evergreen perennial, with a vine-like stem, trailing below the litter layer and rooting at nodes, forming patches in semi-shady and shady habitats. Leaves opposite, rounded, and leathery. Minute axillary flowers in pairs, white, and 4 lobed. Scarlet-red berry, spherical, usually paired or fused having 2 tips. 1 species in the SE.

Common Species

Mitchella repens L. - partridgeberry MIRE

Plant: Trailing, evergreen perennial, vine-like, stem beneath litter-layer and paired leaves above, rooting at nodes and forming small colonies in shady forests.

Stem: Slender, horizontal, patchy hairy or hairless, nodes spaced 2-6 cm, slightly swollen, rooting at nodes.

Leaves: Opposite, evergreen and slightly leathery, ovate to broadly heart-shaped, 1-2 cm long and wide, shiny dark green with whitish midvein above and pale green beneath, smooth and hairless, margins entire, tips rounded, bases rounded to flared, petioles 5-12 mm long, slightly hairy, with minute stipules.

Flowers: Apr-Jun. Stalked pair in leaf axil, fused at base, corolla white or tinged with pink, funnelform, 10-15 mm long, 4 lobed, each 3-4 mm long, spreading or recurved.

Fruit and seeds: Jun-Jan. Berry-like drupe, scarlet red (rarely white), spherical or 2 lobed, 7-10 mm wide, paired and often fused, persistent calyx on upper surface in pairs (thus common name, two-eyed-berry), containing 8 seeds.

Range: Native. TX to c FL and north to s Canada and west to MN, all physiographic provinces in the SE.

Ecology: Common low, inconspicuous perennial in most moist to dry forests, on semi-shady to shady habitats. Occurs commonly as scattered patches in the understory of dense pine forests and plantations as well as in multi-canopied mixed forests. Persists and colonizes by stem growth at the soil surface and spreads by animal-dispersed seeds.

Synonyms: twin-flower, two-eyed-berry.

Wildlife: Partridgeberry is not a prolific seeder, and thus the fruits do not comprise a major portion of the diet of any wildlife species. In the Southern Appalachian Mountains, partridgeberry fruits are consumed readily by Ruffed Grouse during fall and winter, and the leaves are eaten consistently year-round. The fruits also are consumed by Northern Bobwhite, Wild Turkey, Gray Squirrel, Raccoon, and various rodents. The stems and leaves are eaten sparingly by White-tailed Deer.

Mitchella **Partridgeberry** **Rubiaceae**

J. Miller (Oct)

Mitchella repens

J. Miller (Oct)

Mitchella repens

Parthenocissus **Virginia creeper or Woodbine** **Vitaceae**

High-climbing or trailing woody vine, climbing by branched tendrils with adhering disks, or trailing and rooting along leaf-littered vines. Leaves alternate and palmately compound with 5 leaflets (3-7), long petioled. Flowers tiny and inconspicuous in terminal or axillary cymes. Fruit a drupe, black or dark blue, with few seeds. 1 species in the SE.

Common Species

Parthenocissus quinquefolia (L.) Planch. **- Virginia creeper** **PAQU2**

Plant: High climbing or trailing woody vine, to 30 m long, with many branched tendrils tipped with adhesive disks, capable of climbing trees and rock faces, or forming abundant ground cover by rooting of trailing vines.

Stem: Round in cross-section, nodes frequent and flattened, infrequently branched, brownish-green and fine hairy becoming purplish-brown and rough with pale lenticel dots, branched tendrils (pinnately compound) occurring opposite leaves of growing shoots, tendril tips develop adhesive pads when contacting suitable surfaces (or shrivel), leaf scars protruding and circular.

Leaves: Alternate, palmately compound with 5 leaflets (3-7), leaflets ovate to elliptic to obovate, 4-15 cm long and 5-8 cm wide, margins coarsely (or finely) serrate above the middle, pale and often soft hairy beneath, larger petioles 15-20 cm long, pale tan to reddish and finely hairy, long petioles provide height to leaves of trailing vines.

Flowers: May-Aug. Tiny flowers in terminal and axillary red branched clusters (cymes), petals 5, yellowish green, 2-3 mm long, calyx 5 lobed, red.

Fruit and seeds: Oct-Feb. Drupe black to dark blue and white waxy, spherical, 5-9 mm wide, containing 1-3 seeds, lustrous brown, 3.5-4 mm long.

Range: Native. TX to FL and north to s Canada and west to MN.

Ecology: Occurs as scattered plants on moist to dry sites, and as ground cover in forest plantations, or high climbing in mature forests and their margins. Most abundant in open mixed upland forests. Persists and spreads by rapid vine growth, colonizes by rooting at nodes, and spreads by animal-dispersed seeds.

Synonyms: *P. hirsuta* (Pursh) Graebn., *Vitis quinquefolia* (L.) Lam., woodbine.

Wildlife: The persistent fruits are eaten by numerous songbirds including the Carolina Chickadee, Northern Mockingbird, Gray Catbird, Tree Swallow, American Robin, along with numerous woodpeckers and thrushes. Gray Squirrel likewise consume the fruit. It is a low preference White-tailed Deer browse.

Parthenocissus **Virginia creeper or Woodbine** **Vitaceae**

T. Bodner (Apr)

Parthenocissus quinquefolia

J. Miller (Jun)

Parthenocissus quinquefolia

J. Miller (Oct)

Parthenocissus quinquefolia

Pueraria **Kudzu** **Fabaceae**

Exotic semiwoody vine, legume, twining and trailing to form dense infestations covering ground and trees. Large 3-leaflet leaves, finely hairy, slightly lobed or entire, alternate on rope-like vines. Flowers in reddish-purple spike-like racemes. Fruit a flattened legume-type capsule, brown hairy, splitting to release few seeds, oval and hard-coated. 1 species in the SE.

Common Species

Pueraria montana (Lour.) Merr. **- kudzu** **PUMO**

Plant: Twining and trailing, semiwoody vine, 10-30 m long, from semiwoody tuberous roots weighing up to many kilograms and reaching up to 1-5 m deep.

Stem: Stout, round in cross-section, frequent unswollen nodes, rooting at nodes when on ground or buried, trailing or climbing by twining on objects less than 10 cm in diameter, infrequent branching, yellow-green and dense-erect golden hairy, becoming silver matted hairy aging to light gray barked and rope-like, hairless, eventually rough-barked and usually dark brown, rigid, up to 10 cm diameter and larger.

Leaves: Alternate, 3 leaflet, leaflets 8-18 cm long and 6-20 cm wide, usually slightly lobed (or unlobed in shade), middle-leaflet symmetric and 2-lobed, side leaflets 1-lobed on lower sides, tip pointed, margins golden fine-hairy, dark green and golden hairy above and densely-matted silver hairy beneath appearing whitish-green, swollen brown-hairy petioles subtended by 2 thin bracts, about 1 cm long, leafstalks 15-30 cm long, long hairy with a swollen base and 2 stipules.

Flowers: Jun-Sep. Axillary spike-like dense racemes, 5-30 cm long, flowers in pairs (or 3's) from raised nodes spiraling up the stalk, opening from base to top on bracted short hairy pedicels, pea-type flowers, upper (hood) petal lavender-rose to wine-colored with an inner yellow spot, lower jutting wing and keel petals darker wine to purple, calyx short-tubular, greenish turning dull purple, flowers insect pollinated.

Fruit and seeds: Sep-Jan. Dry legume (pod), flattened and bulging above the seeds, 3-5 cm long and 8-10 mm wide, tan and stiff golden-brown hairy, splitting and twisting on 1-2 sides to release 1-few seeds, ovoid, 3 mm long, light brown with tan dots, small percentage filled or viable.

Range: Exotic and naturalized from China. TX to FL and north to CT and west to IL, NE, and OK.

Ecology: Nitrogen fixer. Vines root at nodes when in contact with ground (stoloniferous) or buried by leaf-litter (somewhat rhizomonous) yielding new plants when connecting vine dies, with 1-3 plants per square meter in infested patches. Twining capabilities limited to supports less than 10 cm in diameter, mainly climbing larger trees by twining on other kudzu vines or other vines. Forming dense mats over the ground, debris, shrubs, and trees, excluding most other plants except evergreen shrubs and vines, and blackberries. Planted widely for erosion control and livestock feed on about 1 million hectares from 1920-50, occurs in old infestations, along right-of-ways and stream banks, and spreading outward.

Synonyms: *P. montana* var. *lobata* (Willd.) Maesen & S. Almeida, *P. lobata* (Willd.) Ohwi, *P. thunbergiana* (Sieb. & Zucc.) Benth., kuzu.

Wildlife: Kudzu is an excellent White-tailed Deer forage until killed by frost. The dense thickets provide excellent cover for young fawns, and are a preferred location for Woodchuck dens. Kudzu is a larval host of the Silver-spotted Skipper butterfly.

Pueraria | Kudzu | Fabaceae

T. Bodner (Jun)

Pueraria montana

J. Miller (Jul)

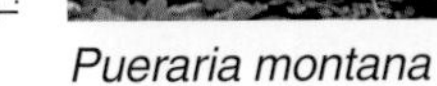

Pueraria montana

T. Bodner (Nov)

Pueraria montana

T. Bodner (Jul)

Pueraria montana

Rubus **Brambles** **Rosaceae**

Erect, arching, or trailing woody plants, often with thorny or bristly stems. Shoots arise from a perennial rootcrown in cycles with two different growth forms. The first form, the primocane, not bearing flowers or fruits, unbranched and with larger stems and larger leaves. The second form, the floricane, with smaller leaves, less elongated stems and forming short lateral branches, most bearing flowers and fruits. Leaves pinnate or palmately compound (except simple leaves with the Appalachian Mountain species, *Rubus odoratus*), and leaflet numbers and shape often differing between the two growth forms. Petals 5, white or pink, both stamens and pistils numerous. Fruit a cluster of drupelets commonly termed a "berry." About 17 species in the SE, commonly hybridizing with species delineations unclear and only examples presented.

Common Species

Rubus argutus Link **- highbush blackberry** **RUAR2**

Plant: Most commonly occurring blackberry, erect or arching cane-like shrub, 1-3 m tall or long, from rhizomes and forming dense colonies.
Stem: Ascending, arching when fruiting, often cluttered in patches, armed with broad-based and flattened hooked thorns, primocanes coarse with grooved-ridges being finely hairy when young and hairless when older, floricanes more slender and less grooved with branching from upper axils and sparsely hairy (often mottled green).
Leaves: Alternate, thin, lower surfaces short soft-white hairy and hairy or not above, floricanes with 3-leaflet leaves and primocanes with 5-(4-)leaflet leaves with the terminal leaflet largest, leaflets alike in shape, blades elliptic to lanceolate with long tapering tips, larger terminal leaflets on floricanes 3-6 cm long and on primocanes 7-12 cm long, margins serrate to twice serrate, terminal leaflet petioles longest (6-10 cm) and lower lateral petioles sessile to 5-6 mm long, petioles and lower-surface midveins with thorns.
Flowers: Apr-Jul. Loose racemes on axillary branchlets, having leafy bracts, petals 5, white (pink in bud), 2-2.5 cm long, numerous stamens and pistils, sepals 5, hairy, 4-6 mm long, flower stalks soft hairy with thorns.
Fruit and seeds: Apr-Aug. Blackberry (aggregate fruit of drupelets), 1-2.5 cm long and 1-1.5 cm wide, bright red maturing to blue-black, juicy and favorable when ripe.
Range: Native. TX to c FL and north to MD and west to KS.
Ecology: Very common early invader of harvested forests (often increasing in density with hardwood control or pine thinning treatments). Occurs on wet to dry sites and open to semi-shady habitats. Common along right-of-ways and forest edges. Persists and colonizes by rhizomes, and spreads by long-viable seeds continually brought in by birds and other animals.
Synonyms: *R. abundiflorus* Bailey, *R. betulifolius* Small, *R. floridensis* Bailey, *R. floridus* Tratt., *R. incisifrons* Bailey, *R. louisianus* Berger, *R. rhodophyllus* Rydb., sawtooth blackberry.

Rubus cuneifolius Pursh **- sand blackberry** **RUCU**

Similar to *Rubus argutus* except **stems** 0.3-1.5 m tall, densely hairy (sloughing with age); **leaves** grayish green above and grayish- to whitish-woolly hairy beneath, leaflets less than 6 cm long, oblanceolate, tips rounded, widest above middle and often straight sided; **flowers** petals white, 1-1.5 cm long, obovate to oval, sepals hairy; **fruit** 1-2.5 cm long and 0.8-1.5 wide. **Range:** Native. MS to FL and north to CT, mainly in Cp and lower Pd.
Ecology: Occurs with *R. argutus* on dry sandy soils, in open to semi-shady habitats.
Synonyms: *R. chapmanii* Bailey.

Rubus **Brambles** **Rosaceae**

J. Miller (Jul)

Rubus cuneifolius

J. Miller (Sep)

Rubus argutus

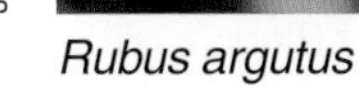

J. Miller (Jul)

Rubus cuneifolius

T. Bodner (May)

Rubus cuneifolius

Rubus Brambles Rosaceae

Rubus flagellaris Willd. - northern dewberry RUFL

Plant: Commonly-occurring, long trailing and vine-like, or initially erect and then low arching perennial, 2-5 m long, often rooting at tips, forming patches by rooting rhizomes, rooting tips, and perennial rootcrowns.
Stem: Primocanes erect initially or low arching and groove-angled, floricanes trailing vine-like, round in cross-section, slender, light-green becoming reddish later, hairless to short hairy eventually with scattered stemless red-glands (requires hand lens), few thorns initially to many fine broad-based thorns, both hooked and straight.
Leaves: Alternate, 5 leaflet on primocanes, 3 leaflet on floricanes, with lateral leaflets unequal and having 1 or more with basal lobes, blades 2-7 cm long and floricane's smallest, primocane leaflets all ovate with margins serrate to dentate, floricane leaflets ovate to obovate with margins irregular (often serrate and lobed), finely hairy to hairless and rarely densely hairy beneath although grayish green, thorns on petioles and under midveins.
Flowers: Apr-Jul. Solitary or racemes with 5-6 flowers, on branches in leaf axils of the floricane, flower stems and stalks slender and long, finely hairy with thorns, petals 5, white, 1-2 cm long, obovate, sepals triangular, 4-8 mm long, densely hairy and with red gland tips.
Fruit and seeds: Jun-Jul. Blackberry (aggregate fruit), red turning to black, 1-2 cm long.
Range: Native. TX to n FL and north to Que and west to MI.
Ecology: Commonly occurring in open to densely-shaded forests, on dry to moist sites. Present in early forest plantations and common along right-of-ways, forest margins and stream banks, and in pastures. Occurs as frequent scattered plants in all ages of pine plantations. Persists by perennial rootcrowns, colonizes by rhizomes and vine tip rooting, and spreads by animal-dispersed seeds.
Synonyms: *R. arundelanus* Blanch., *R. ashei* Bailey, *R. enslenii* Tratt.

Rubus trivialis Michx. - southern dewberry RUTR

Similar to *Rubus flagellaris* being long trailing or only initially erect or low arching often in patches except **stem** bristling with straight red-purplish hairs along with reddish thorns; **leaves** 5-leaflet on primocanes, hairless, elliptic or lanceolate, 2-6 cm long, tip pointed, margins coarsely serrate or doubly serrate, leaves persistent over winter turning red, persisting until floricane branchlets and flowers develop; **flowers** (Mar-Apr) solitary or rarely in 2-3's, petals white or white tinged with pink; **fruit and seeds** (Apr-May) aggregate berry 1.5-3 cm long and 1-1.5 cm wide, red turning to black. **Range:** Native. TX to s FL and north to se VA and west to IL and MO, mainly in Cp and Pd in the SE.
Synonyms: *R. continentalis* (Focke) Bailey, *R. mississippianus* Bailey.

Wildlife: Collectively, the various species of *Rubus* arguably are the most important group of plants to wildlife in the Southeast. The heavily used soft mast is available from spring (dewberry) through summer (blackberries and raspberries, *R. occidentalis*). Users include a long list of game mammals, gamebirds, and songbirds. Blackberry leaves appear early in spring and persist late into fall. They are an important browse source for White-tailed Deer and Eastern Cottontail. Blackberry thickets provide excellent escape cover for birds, rabbits, and other small mammals as well as nesting sites for numerous songbirds.

Rubus **Brambles** **Rosaceae**

J. Miller (Aug)

Rubus flagellaris

T. Bodner (Apr)

Rubus trivialis

T. Bodner (Jun)

Rubus flagellaris

Smilax Greenbrier or Catbrier Smilacaceae

Mostly woody, with a few herbaceous vines, many armed with thorns, most climbing by paired tendrils from leaf petioles. Woody species mostly from rhizomes, some from tubers. Leaves alternate, deciduous or evergreen, often with 3 main veins and netted interveins. Small trumpet-shaped flowers in small axillary clusters (umbels), greenish to yellowish to brownish, with male and female flowers on separate plants (dioecious). Fruit a berry, green turning red to blue or black, containing 1-3 seeds. 14 species in the SE, with woody species difficult to separate because leaves vary greatly on the same and different plants.

Common Species

Smilax glauca Walt. - **cat greenbrier** SMGL

Plant: Climbing or scrambling woody vine, 2-20 m long, having thorns, from underground runners and knotty tuberous rhizomes.

Stem: Slender to moderately stout, round in cross-section, new growth densely white waxy, becoming smooth green and often dark mottled with age, lower portion usually with many slender prickles, upper portions having flattened thorns (both straight and hooked) on nodes and scattered on internodes, thorns green with dark tips, older vines frequently alternate branched.

Leaves: Alternate, thin and semi-evergreen, broadly ovate to lanceolate, 5-9 cm long and 2-6 cm wide, distinctly whitish beneath (fine short wax hairs) and shiny green above (often with lighter mottles), margins entire, base square to rounded, 3 or 5 main veins, petioles 5-10 mm long, often with 2 twining tendrils mid-petiole extending from tiny stipule ridges. Leaves turning reddish to purplish in late fall and winter.

Flowers: Apr-Jun. Flat-top clusters (umbels) in upper leaf axils, flattened stalks 1-3.5 cm long, 1.5-3 times as long as the subtending petioles, petals yellowish or brownish, flowers 6-8 mm wide, male and female flowers on different plants.

Fruit and seeds: Sep-Oct. Spherical berry, 5-8 mm wide, green and whitish waxy ripening to shiny bluish-black, 1-3 seeds in sticky pulp.

Range: Native. TX to c FL and north to NJ and west to IL, MO, and OK.

Ecology: Very common, occurs on a wide range of sites, from open new forest plantations to open mature forests, dry to seasonally-wet habitats. Persists by tuberous rhizomes and perennial vines, and spreads by vine growth and bird-dispersed seeds, long viable.

Synonyms: *S. glauca* var. *genuina* Blake, *S. glauca* var. *leurophylla* Blake, wild sarsparilla, cat sawbrier.

Smilax bona-nox L. - **saw greenbrier or catbrier** SMBO2

Similar to *Smilax glauca* except **plant** from round tuberous rhizomes having prickles; **stem** usually 4-angled, not whitish, basally brownish scurfy and prickly, only scattered thorns; **leaves** very variable in shape, broadly ovate to slender heart-shaped, shiny green above, often with lighter mottles, and paler green beneath, margins thickened and sometimes with prickles.

Ecology: Commonly associates with *Smilax glauca* on the same sites.

Smilax rotundifolia L. - **roundleaf greenbrier** SMRO

Similar to *Smilax glauca* except **plant** high-climbing with branching stems, from long slender rhizomes; **stem** round in cross-section lower and 4-angled above, hairless and green, not whitish, thorns green with darker tips; **leaves** thick, round to broadly ovate, to 14 cm long and wide, leaves of climbing branchlets small ovate; **fruit and seeds** in rounded clusters, 1+ years to mature.

Smilax **Greenbrier or Catbrier** **Smilacaceae**

T. Bodner (May)

Smilax bona-nox

J. Miller (Jul)

Smilax glauca

T. Bodner (May)

Smilax bona-nox

T. Bodner (Apr)

Smilax rotundifolia

Smilax **Greenbrier or Catbrier** **Smilacaceae**

Smilax smallii Morong - **lanceleaf greenbrier** **SMSM**

Plant: Evergreen, high-climbing and arbor-forming woody vine, seldom thorny and much branched, from branching tuberous rhizomes to 60 cm long.
Stem: Round in cross-section, basally stout (to 2 cm thick) and whitish waxy at first with scattered thorns, becoming slender higher and dull green without thorns, densely branching into extensive festoons.
Leaves: Alternate, evergreen and thin leathery, lanceolate to ovate, 5-10 cm long and 1-4 cm wide, shiny green above and dull green beneath, 3-veined (rarely 5), margins entire, tip pointed to long tapering, base rounded, petioles short, paired tendrils from some petioles.
Flowers: Apr-Jul. Spherical clusters (umbels) in upper leaf axils, slightly-flattened stalks, 0.1-1 cm long, as long as the subtending petioles, petals green, flowers 6-8 mm wide.
Fruit and seeds: Jun-Aug (1+ year to mature). Berry spherical, 5-7 mm wide, pale green turning reddish ripening to black, with 1-3 seeds in sticky pulp.
Range: Native. TX to c FL and north to NC (rare in VA and not in TN) and west to s AR, mainly Cp and scattered in lower Pd.
Ecology: Scattered plants in moist forests, in the canopy of mid-story and overstory trees, and persisting in harvested sites and new plantations. Common along forest edges, fencerows, and sand dune areas. Persists and colonizes by rhizomes and spreads by rhizomes and bird-dispersed seeds.
Synonyms: *S. lanceolata* L., *S. domingensis* Willd., Jackson-brier, Jacksonvine.

Smilax pumila Walt. - **sarsparilla vine** **SMPU**

Plant: Low growing, often trailing, woody vine without thorns, less than 50 cm tall, from a slender tan runner, often below-litter, having triangular scale leaves and fibrous roots.
Stem: Upright or trailing from nodes on the runner, slender, round in cross-section, reddish-brown soft hairy, no thorns.
Leaves: Alternate, evergreen, ovate to broadly heart-shaped to oblong, 4-10 cm long and 3-5 cm wide, 3-veined (rarely 5), with the two lateral veins forming an ellipse around the midvein, dark green often with lighter mottles and scurfy hairy above, grayish-green and densely hairy beneath, margins entire, petioles shorter upward and hairy, tendrils on petioles few or none.
Flowers: Oct-Nov. Small clusters (umbels) in upper leaf axils, stalks 1-12 mm long, as long as petioles, petals yellow.
Fruit and seeds: Jan-Mar. Berry bright red, ovoid and pointed, 5-8 mm long, with 1 seed.
Range: Native. TX to LA and southern halves of MS, AL, GA, and SC, and to c FL, only in Cp.
Ecology: Occurs in colonies and patches in open dry to moist forests, along stream banks and forest edges. Persists and colonizes by runners and spreads by runners and seeds.
Synonyms: hairy greenbrier, wild sarsparilla, dwarf smilax.

Wildlife: The greenbriers are important wildlife plants. The fruits are consumed by Ruffed Grouse, Wild Turkey, Northern Bobwhite, along with at least 40 or more species of songbirds. The foliage is a preferred browse of White-tailed Deer throughout the Southeast. It also is important in the winter diet of the Ruffed Grouse. Swamp Rabbits, Marsh Rabbits, and Eastern Cottontails readily consume the leaves and young shoots. Tubers of some species are consumed by Beaver.

Smilax **Greenbrier or Catbrier** **Smilacaceae**

T. Bodner (Apr)

Smilax smallii

J. Miller (Oct)

Smilax pumila

J. Miller (Oct)

Smilax pumila

Toxicodendron Toxicodendron Anacardiaceae

Trailing or climbing woody vines and low shrubs, with alternate and pinnately compound leaves, having an odd-number of leaflets. Small flowers in axillary panicles on twigs of the current season. Fruit a small dry drupe, ovoid, tan to gray or ivory. Contact with any plant parts by susceptible persons, or smoke from burning plant parts, may cause a blistering rash. 3 species in the SE.

Common Species

Toxicodendron radicans (L.) Kuntze **- poison ivy** **TORA2**

Plant: High-climbing woody vine, often trailing with erect stems, and at times a low weak shrub, from vine runners, rooting extensively. May cause rash when contacted.
Stem: Older to 5-10 cm in diameter, climbing by hairy aerial rootlets, bark rough gray to brownish, twigs gray-brown and soft grayish-brownish hairy becoming hairless with age, leaf scars crescent-shaped with several bundle scars, infrequent branching.
Leaves: Alternate, 3 leaflet, thin, leaflets ovate to elliptic, 5-20 cm long and 2-12 cm wide, entire to serrate to shallowly lobed, softly hairy and lighter green beneath, lateral leaflets sessile and often asymmetric, terminal leaflet symmetric with petiole 1-4 cm long, stalk slender, 5-15 cm long, swollen and reddish at base. May turn yellow, orange, or reddish in the fall.
Flowers: Apr-May. Axillary panicles of 2-6, 2-7 cm long, petals 5, white and purplish, about 2 mm long, functionally unisexual.
Fruit and seeds: Aug-Feb. Dry drupe, spherical, 4-7 mm wide, grooved and tan, hairless.
Range: Native. AZ to s FL and north to s Canada and west to NE.
Ecology: Commonly occurs on a wide range of habitats, moist to dry sites, open to shady. Seeds widely spread by birds, colonizes by vines becoming runners when leaf covered.
Synonyms: *Rhus radicans L.*

Toxicodendron **Toxicodendron** **Anacardiaceae**

J. Miller (Sep)

Toxicodendron radicans

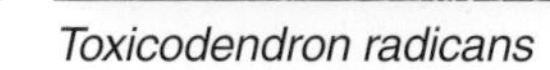

T. Bodner (Apr)

Toxicodendron radicans

T. Bodner (Oct)

Toxicodendron radicans

J. Miller (Oct)

Toxicodendron radicans

Toxicodendron **Toxicodendron** **Anacardiaceae**

Toxicodendron pubescens P. Mill **- poison oak** **TOPU2**

Similar to *Toxicodendron radicans* except **plant** low shrub, 0.3-1 m tall, from runners; **leaves** thick, rarely thin, 3 leaflet, margins 1-3 undulating to incised lobes (oak-like), often only on 1 side of lateral leaflets, rounded tips, finely to densely hairy; **fruit and seeds** drupe densely hairy, 6-8 mm wide. **Range:** Native. Same as *T. radicans.*
Ecology: Occurs as small colonies in dry and open forests, and right-of-ways.
Synonyms: *T. quercifolium* (Michx.) Greene, *T. toxicarium* Gillis, *T. toxicodendron* (L.) Britt., *Rhus toxicarium* Salisb., *R. toxicodendron* L., Atlantic poison oak.

Toxicodendron vernix (L.) Kuntze **- poison sumac** **TOVE**

Plant: Shrub to 5 m tall, sparsely branching, often from the base. Causes rash.
Stem: Ascending, twigs hairless and reddish, later tan with lenticels, smooth gray bark.
Leaves: Alternate, 7-15 leaflet, leaflets elliptic to oblong, 5-10 cm long and 2-5 cm wide, paler green below, margins entire, tips sharp to tapering, petioles and stalk reddish.
Flowers: May-Jun. Long-stalked axillary panicles, drooping or spreading.
Fruit and seeds: Aug-Jan. Drupe, 5-7 mm wide, hairless.
Range: Native. TX to s FL and north to Que and west to MN.
Ecology: Occurs on wet sites, swamps, bogs, pocosins, and depressions.
Synonyms: *Rhus vernix* L., *R. toxicodendron* Kuntze, swamp sumac, poison elder.

Wildlife: Poison ivy and poison oak fruits are consumed by many species of songbirds, primarily during winter. Woodpeckers and the Northern Flicker and Yellow-bellied Sapsucker appear particularly fond of the fruits. Poison ivy is a moderate to high preference White-tailed Deer forage, and also is an important component in the diet of the Swamp Rabbit.

Toxicodendron Toxicodendron Anacardiaceae

T. Bodner (May)

Toxicodendron vernix

T. Bodner (May)

Toxicodendron pubescens

T. Bodner (May)

Toxicodendron vernix

J. Miller (Oct)

Toxicodendron pubescens

Vitis **Grape** **Vitaceae**

High-climbing, trailing, or scrambling, deciduous woody vines (or viney shrubs), climbing by tendrils, branched or not. Leaves simple and palmately veined, cordate or orbicular in shape, margins serrate and lobed or not. Flowers in axillary panicles, inconspicuous greenish-yellowish flowers, having nectar glands. Fruit a berry, spherical, dark purple or tan when ripe, containing 1-4 seeds, usually red or brown, and edible. 10 species in the SE, rarely hybridize.

Common Species

Vitis rotundifolia Michx. **- muscadine grape** **VIRO3**

Plant: Commonly occurring woody vine, high climbing, scrambling over shrubs, and trailing, to 40 m long, with long-reaching branches extending outward during rapid mid-summer growth, rooting at litter-covered nodes.

Stem: Slender becoming stout to 5 cm in diameter, infrequent alternate branched, climbing and often binding together by unbranched tendrils, tendrils opposite to leaves at 2 consecutive nodes, twigs pale green to tan to purplish-brown, slightly angled or grooved when very young, hairless, nodes slightly swollen with a raised band, bark tight gray and sloughing in plates only on large stems, young bark with many light dots (lenticels) becoming lengthwise lines later, leaf scars half round to circular, stem pith continuous through nodes.

Leaves: Alternate, deciduous, cordate to rounded (thus the sci. name), 4-10 cm long and wide, palmately veined, with midvein short-tapering tipped (often 2 lateral veins extended tipped), margins dentate to crenate, smooth and lustrous above, shiny and smooth except short-hairy veins beneath, petioles as long as blade.

Flowers: Apr-Jun. Axillary panicles, short branching and round in shape, 3-8 cm long, inconspicuous yellow-green flowers, unisexual in function.

Fruit and seeds: Jul-Sep. Berry, a thick-skinned grape, 1-few in clusters, each spherical, 10-25 mm long and wide, green turning red to purple to black (occasionally bronze) with tan spots when ripe, containing 1-4 seeds, oval to ellipsoid, 5-8 mm long, brown.

Range: Native. TX to s FL and north to DE and west to MO.

Ecology: Common vine in mid-region, occurs on a wide range of sites, both upland and bottomland habitats. Present in open new forests and mature forests, forest edges and right-of-ways, forming arbors at all canopy heights as well as ground cover. Seeds widely spread by birds.

Synonyms: *Muscadinia rotundifolia* (Michx.) Small, scuppernong (tan skinned variety), bullace grape.

Vitis **Grape** **Vitaceae**

T. Bodner (May)

Vitis rotundifolia

T. Bodner (May)

Vitis rotundifolia

J. Miller (Feb)

Vitis rotundifolia

T. Bodner (Sep)

Vitis rotundifolia

Vitis **Grape** **Vitaceae**

Vitis aestivalis Michx. **- summer grape** **VIAE**

Plant: High-climbing and arbor-forming woody vine, often growing in the shrub and mid-story layers, climbing by branched tendrils, with golden hairy twigs and lower leaf surfaces, loose flaking bark, rooting at litter-covered nodes.
Stem: Climbing, branched tendrils bind vines together forming arbors, rarely trailing, twigs pale-green or tan or gray, densely to sparsely reddish-brown hairy, round in cross-section and swollen at nodes, younger bark sloughing in strips, light-grayish to brownish, older bark reddish-brown and shredding, having bluish-gray waxy streaks and spots, leaf scars triangular to half round, pith soft and brown, interrupted at nodes.
Leaves: Deciduous, alternate, cordate to orbicular, to 20 cm long and wide, unlobed or with 3-5 tips and lobes, some oval shaped, margins dentate to crenate, hairless to short hairy above, densely to sparsely rusty hairy and bluish-green (or whitish) beneath, petioles as long as blades.
Flowers: May-Jul. Axillary panicles, narrowly triangular, 5-16 cm long, frilly yellow-green flowers, unisexual in function.
Fruit and seeds: Jun-Oct. Berry (small soft-skinned grape), many in cylindric dangling clusters, each spherical, 5-12 mm wide, green ripening to black and white-waxy (without lenticel dots), 1-4 seeds, tan to brown, 3-6 mm long.
Range: Native. TX to s FL and north to MA and west to Ont and MN.
Ecology: Occurs in moist to somewhat dry forests of all structures, from new forest plantations to mature forests, forming arbors on shrubs and in small tree canopies. Present along forest margins, and less often along stream or river banks. Spreading by vine growth and animal- and gravity-dispersed seeds.
Synonyms: *V. argentifolia* Munson, *V. bicolor* Le Conte, *V. lecontiana* House, *V. lincecumii* Buckl., pigeon grape.

Vitis cinerea (Engelm.) Millard **- downy winter-grape** **VICI2**

Similar to *Vitis aestivalis* except **stem** angled when young and dense short hairy (lenticels absent); **fruit and seeds** berry 4-9 mm wide, black with faintly white waxy. **Range:** Native. TX to n FL and north to VA and west to IL.
Ecology: Occurs in floodplains, bottomland forests, and margins of ponds and streams.
Synonyms: *V. baileyana* Munson, *V. aestivalis* var. *cinerea* Engelm., *V. austrina* Small, *V. sola* Bailey, *V. berlandieri* Planch., *V. helleri* (Bailey) Small, sweet winter grape, graybark grape.

Other species in the SE: *V. labrusca* L. [fox or plum grape], *V. mustangensis* Buckl. [mustang grape], *V. palmata* Vahl [red, cat or catbird grape], *V. riparia* Michx. [river-bank, river-side, or november grape], *V. rupestris* Scheele [sand or sugar grape], *V. shuttleworthii* House [calusa or upland grape], *V. vulpina* L. [winter, frost, november, or chicken grape].

Wildlife: Grapes are very important soft mast producers for a variety of game species including Black Bear, White-tailed Deer, Wild Turkey, Ruffed Grouse, Raccoon, Virginia Opossum, Gray Squirrel, and Striped Skunk. Numerous songbirds also consume grapes, especially the Northern Cardinal, Northern Mockingbird, American Robin, Cedar Waxwing, and others. Muscadine grape leaves are a preferred White-tailed Deer forage, whereas the other species are of moderate preference. Gray Squirrel prefer to build leaf nests in trees supporting grapevines.

Vitis **Grape** **Vitaceae**

T. Bodner (Jun)

Vitis aestivalis

T. Bodner (May)

Vitis aestivalis

J. Miller (Sep)

Vitis aestivalis

T. Bodner (Sep)

Vitis aestivalis

Wisteria Wisteria Fabaceae

Woody vines or shrubs with twining branches, legumes, with odd pinnate leaves having stipules, and leaflets with entire margins. Known for the spring showy dangling blossoms of violet to white flowers. Fruit a flattened legume (pod), splitting in time to release several seeds. 3 species in the SE, with 1 native and 2 exotic species.

Common Species

Wisteria sinensis (Sims) DC. **- Chinese wisteria** **WISI**

Plant: High-climbing or trailing woody vine (or cultured to be a shrub), to 20 m long, with vine runners rooting at nodes, forming dense infestations around old homesite plantings.
Stem: Climbing by twining, covering shrubs and trees, 4-10 cm in diameter, branching infrequently, twigs densely short hairy, older bark tight and gray with light dots (lenticels), rooting where vines covered by leaf litter.
Leaves: Alternate, odd pinnately compound, 10-40 cm long, leaflets 7-13, each ovate to elliptic with tapering pointed tips, 4-8 cm long and 2-6 cm wide, hairless to short hairy at maturity but densely silky hairy when young, stalks with swollen bases.
Flowers: Mar-Jul. Fragrant, dangling and showy, stalked clusters (racemes) appearing when leaves emerge, dense to sparse in blossoms, all blooming at about the same time, 15-30 cm long and 7-9 cm wide, pea-type, corolla lavender to violet to pink to white, 2-2.5 cm long, calyx tube 3-3.5 mm long.
Fruit and seeds: Jul-Nov. Flattened legume (pod), irregularly oblong to oblanceolate, 6-15 cm long and 2-3 cm wide, greenish-brown to golden velvety hairy, splitting on 2 sides releasing 1-8 seeds, flat and round, 1-2 cm wide, dark brown and shiny.
Range: Exotic from Asia. LA to c FL and north to VA and west to IL, mainly in Cp and Pd of the SE.
Ecology: Nitrogen fixer. Forms dense infestations where previously planted, often with other exotic invasive plants, and only invasive after disturbance. Occurs on wet to dry sites. Colonizes by vine growth and runner sprouting, appearing not to be widely spread by seed.
Synonyms: *Rehsonia sinensis* (Sims) Stritch.

Wisteria frutescens (L.) Poir. **- American wisteria** **WIFR**

Similar to *Wisteria sinensis* except native and not forming infestations and **plant** climbing into and through shrubs and small trees, rarely running; **stem** slender, initially silky-shaggy hairy or short hairy; **leaves** 10-30 cm long, leaflets 9-15, oblong to ovate-lanceolate with pointed or blunt tips, 2-6 cm long, densely hairy becoming hairless, stalk bases only slightly swollen; **flowers** (Apr-Aug) after leaves appearing, dense racemes, 15 cm or less long, flowering from base to apex, corolla purplish blue to lilac, 1.5-2 cm long, calyx tube 5-6 mm long, with stalked glands; **fruit and seeds** (May-Sep) hairless legume (pod), linear-oblong, 5-10 cm long, seeds bean-shaped, 6-8 mm wide. **Range:** Native. TX to c FL and north to VA and west to AR, mainly Cp in the SE.
Synonyms: *W. macrostachyr*a (Torr. & Gray) Nutt. *ex* B.L. Robins & Fern., *Kraunhia frutescens* (L.) Greene, *K. macrostachya* (Torr. & Gray) Small.

Other species in the SE: *W. floribunda* (Willd.) DC. [Japanese wisteria].

Wildlife: Chinese wisteria is a low to moderate preference browse plant and a larval plant of the Zarucco Duskywing and the Long-tailed Skipper butterflies.

Wisteria Wisteria Fabaceae

T. Bodner (May)

Wisteria sinensis

J. Miller (Apr)

Wisteria sinensis

T. Bodner (Sep)

Wisteria sinensis

J. Miller (Apr)

Wisteria frutescens

Shrubs

Aesculus **Buckeye** **Hippocastanaceae**

Deciduous trees or shrubs, with opposite and palmately-compound leaves, having 5-7 leaflets, their veins pronounced and parallel from the midvein to the leaflet margin. Twigs tan to reddish brown with pale lenticels. Flowering in spring and early summer, showy red or yellow-green to white flowers, in terminal panicles on new stem growth. Fruit a large, irregularly-spherical capsule, leathery hulled, splitting at maturity to release 1-3 seeds, each rounded and tan. 5 species in the SE.

Common Species

Aesculus pavia L. **- red buckeye** **AEPA**

Plant: Deciduous shrub or small tree, 1-4 m tall, with single or multi-stems, often leaning, widely V-branched, often in small groups from rootsprouts.

Stem: Twigs greenish-gray to reddish-tan and short hairy becoming hairless, leaf scars large and U-shaped, bundle scars 3 or roughly in 3 groups around the stem, terminal buds with overlapping golden scales forming a point, bark gray and smooth with light yellow dots (lenticels), sometimes covered with white lichens. Bruised branches emit a disagreeable odor.

Leaves: Opposite, palmately compound, on long-hairy red stalks, leaflets 5 (sometimes 7), with short petioles, leaflets elliptic to oval, 6-16 cm long and 3-6 cm wide, margins finely serrate, shiny dark green above and finely hairy beneath, veins distinctly parallel from the midvein to the margin. Leaves deciduous in late summer.

Flowers: Mar-May. Many bright-red tubular flowers, projecting from a terminal red stalk (10-20 cm long), petals 4 (sometimes 5) of varying length, longest 20-30 mm, stamens 6-8, red, extending from petals, calyx tubular and 5-lobed, red, 1-2 cm long.

Fruit and seeds: Aug-Oct. Large nut-like capsule, spherically lobed, 2-6 cm wide, hull leathery, smooth and green becoming rough and golden, splitting at maturity to reveal 1-3 seeds, shiny chestnut-brown, spherical, 1.5-2 cm wide.

Range: Native. TX to FL and north to VA and west to se MO, mainly Cp.

Ecology: Common shrub in fertile, well-drained and moist forests, including low forests, protected sites on upland forests, and swamp margins. Often persists after site preparation to occur in new and older forest plantations, at times in large colonies, especially on moist sites. Flowering and fruiting occurs even in shady conditions. Colonizes by rootsprouting and spreads by animal- and gravity-dispersed seeds that are viable usually for only 1 year.

Synonyms: fire-cracker plant, scarlet buckeye.

Other species in the SE: *A. sylvatica* Bartr. [painted buckeye], *A. flava* Ait. [formerly *A. octandra* Marsh. - sweet buckeye], *A. glabra* Willd. [Ohio buckeye], *A. parviflora* Walt. [bottle-brush buckeye]. Species hybridize freely where their ranges overlap.

Wildlife: Young foliage and seeds of all species are considered poisonous to humans and livestock. Toxic materials are especially prevalent in young shoots and seeds. Gray Squirrel and feral hogs occasionally eat the seeds. Because buckeyes flower early, they are an important nectar source for Ruby-throated Hummingbirds, especially *A. pavia*.

Aesculus **Buckeye** **Hippocastanaceae**

T. Bodner (Apr)

Aesculus pavia

T. Bodner (Apr)

Aesculus pavia

T. Bodner (May)

Aesculus pavia

Aralia Aralia Araliaceae

Perennial herbs, shrubs, or small trees, with alternate spreading leaves, once to twice pinnately compound, somewhat aromatic. Flowers white or greenish-white, in spherical-shaped branched clusters, panicles or racemes or umbels. Fruit a small drupe, purple or purplish-black, 5 seeded. 4 species in the SE.

Common Species

Aralia spinosa L. - **devil's walkingstick** ARSP2

Plant: Deciduous shrub or small tree, 1-5 m tall, usually unbranched, bristling with sharp thorns, occurring at times in small groups from rootsprouts.
Stem: Straight, upright, unbranched until the season after the first flowering, afterwards with few ascending branches, bark tan and generally smooth, short sharp thorns in horizontal rows, often most numerous at prominent U-shaped leaf scars around the stem.
Leaves: Alternately spiraling, pinnately compound, very large, 60-150 cm long (including petiole), nearly triangular in outline, arising from the main stem, leaflets elliptic or ovate, 5-13 cm long and 4-7 cm wide, dark green above often with prickles on the midvein and whitish-green beneath, leafstalk stout and often prickly, base flattened and clasping the stem obliquely.
Flowers: Jul-Aug. Showy large terminal panicles, 60-120 cm tall and wide, numerous small creamy-white flowers, petals 5, stamens 5.
Fruit and seeds: Sep-Oct. Abundant purple-black drupes, 5-8 mm wide, each with 5 seeds.
Range: Native. TX to FL and north to NJ and west to IA.
Ecology: Occurs in upland and lowland forests, thickets, and swamp margins. Present in dense woods, but needs some direct sun to flower. Aromatic leaves turn a pleasing maroon-red in autumn. Persists and colonizes by rhizomes and spreads by bird-dispersed seeds.
Synonyms: angelica tree, prickly ash, toothache tree.

Other species in the SE: *A. nudicaulis* L. [wild sarsparilla], *A. racemosa* L. [American spikenard], *A. hispida* Vent. [bristly sarsparilla].

Wildlife: The small, fleshy fruits and seeds of *A. spinosa*, along with the other species of this genus are eaten by some songbirds, such as the Wood Thrush, Northern Cardinal, Northern Mockingbird, Brown Thrasher, Blue Jay, Eastern Bluebird, and White-throated Sparrow. Fruit also reportedly eaten by Red Fox, Striped Skunk, and Eastern Chipmunk, although this species is never abundant enough to compose a significant portion of their diets. Flowers are very attractive to bees, wasps, and Tiger Swallowtail butterfly. Foliage is a moderate preference White-tailed Deer browse.

Aralia Aralia Araliaceae

T. Bodner (Oct)

Aralia spinosa

T. Bodner (Aug)

Aralia spinosa

T. Bodner (Mar)

Aralia spinosa

T. Bodner (Sep)

Aralia spinosa

Aronia **Chokeberry** **Rosaceae**

Deciduous tree-like shrubs, or small trees, with alternate elliptical leaves having finely-serrated margins, teeth red or black tipped (stalked glands). Midveins with tiny reddish-brown hairs. Rose-type white flowers, in flat-topped terminal clusters on branchlets (corymbs). Fruit a small pome (tiny apple), red or black at maturity, less than 1 cm long. 3 species in the SE.

Common Species

Aronia arbutifolia (L.) Pers. **- red chokeberry** **PHPY4**

Plant: Upright, deciduous, tree-like shrub, 2-3 m tall, single or less often multi-stemmed, somewhat open and round topped, often forming loose colonies by rootsprouting.

Stem: Erect with few ascending branches, twigs slender, brownish-black and densely grayish hairy becoming hairless with age, bark light gray and not fissured, leaf scars crescent-shaped, with 3 bundle scars.

Leaves: Alternate, leathery, elliptic to oblong with tapering bases and short pointed or rounded tips, 4-10 cm long and 1.5-4.5 cm wide, petioles hairy, margins finely serrate with teeth tipped with tiny reddish to black hairs (requires hand lens), hairless above with small dark reddish-brown hairs along the midvein and dense whitish hairy beneath, green above often becoming rich crimson or reddish-purple in fall.

Flowers: Mar-May. Flat-topped clusters of up to 25 at branchlet terminals of current season, petals 5, rose-type, white to pinkish, 4-7 mm long, sepals 5, with red gland tips, becoming thick and persisting on the fruit, stamens 10-12, extended, white and pink.

Fruit and seeds: Sep-Feb. Bright red pome when ripe (resembling a tiny apple in winter), green during summer turning shiny red and cherry-like in fall, 6-9 mm long and 6-10 mm wide, firm and glossy, fleshy and bitter tasting (thus the common name).

Range: Native. TX to FL and north to Nova Scotia and west to MI, mainly Cp, infrequent in Pd and Mt.

Ecology: Common in swamps, bogs, and wet to moist pine forests, but tolerant of many sites from wet to dry. Will fruit in shady conditions, but best fruit production occurs in full sun. Colonizes by rhizomes and spreads by bird-dispersed seeds.

Synonyms: *Photinia pyrifolia* (Lam.) Robertson & Phipps, *Sorbus arbutifolia* (L.) Heynhold., dogberry tree.

Other species in the SE: *A. melanocarpa* (Michx.) Ell. [black chokeberry], *A. xprunifolia* (Marsh.) Redd. (pro sp.) [*arbutifolia x melanocarpa*].

Wildlife: The persistent fruit is used during winter by numerous birds, particularly Woodpeckers, Cedar Waxwing, Brown Thrasher, Eastern Bluebird, American Robin and the Northern Flicker. Black Bear consume the fruit. Although widely distributed, chokeberries likely are of minor importance due to their infrequent occurrence.

Aronia **Chokeberry** **Rosaceae**

T. Bodner (Sep)

Aronia arbutifolia

J. Miller (Jul)

Aronia arbutifolia

J. Miller (Oct)

Aronia arbutifolia

T. Bodner (Mar)

Aronia arbutifolia

Asimina Pawpaw Annonaceae

Deciduous shrubs or trees, having arching branches, and alternate velvety leaves, obovate to long linear, entire margins, and short petioles. Crushed leaves emit a foul odor. Flowers solitary, with 6 maroon or purple-green petals in two series, the outer set longer than the inner, stamens numerous in a globular mass. Fruit a large oblong, pulpy berry, containing few to many flattened seeds, oblong and brown. 8 species in the SE.

Common Species

Asimina parviflora (Michx.) Dunal **- dwarf pawpaw** **ASPA18**

Plant: Deciduous shrub, commonly 2-3 m tall, with one to few stems from the rootcrown, branches infrequent, low arching and horizontal.
Stem: Twigs reddish-brown and rusty hairy, becoming gray with frequent leaf scars on lateral sides, buds spherical and golden-brown hairy, pith segmented with partitions.
Leaves: Alternate, thin, obovate, large, 6-18 cm long and 8-10 cm wide, some with twisted pointed tips (acuminate), margins entire, green and hairless above, whitish-green with short rusty-hairs beneath when young, becoming sparsely red-hairy on the veins, petioles stout and curved, 4-7 mm long. Crushed leaves emitting a strong odor.
Flowers: Apr-May. Solitary in leaf axils of previous year's growth, present prior to or during emergence of leaves, small with foul aroma, greenish-purple to maroon, 7-15 mm wide, petals 6, fleshy with recurved tips, inner 3 shorter than outer 3, stalks less than 1 cm long being so short that flowers appear stalkless.
Fruit and seeds: Jul-Sep. Hard-pulpy berry, oblong, with 2 often growing together appearing pear-shaped, 4-7 cm long and 3-4 cm wide and 1-3 cm thick, greenish yellow, slightly rough becoming smooth, aromatic when mature, containing several seeds, flattened oblong, 12-15 mm long.
Range: Native. MS to n FL and north to se VA, mainly Cp and Pd.
Ecology: Occurs on moist to dry sites, open to shady habitat, from coastal hammocks to upland forests. Occurs as infrequent scattered plants or small groupings. *A. triloba* is similar but generally taller and infrequent in Cp, primarily in Mt and Pd.
Synonyms: small-fruited pawpaw, smallflower pawpaw.

Asimina angustifolia Raf. **- narrowleaf pawpaw** **ASAN6**

Similar to *Asimina parviflora* except **plant** to 1.7 m tall; **leaves** leathery, long linear or narrow spoon-shaped, 5-20 cm long and 1-4 cm wide, margins often rolled under; **flowers** (Apr-Jun) singly in leaf axils of new shoots, yellowish-white but sometimes tinged with purple, fragrant and remaining unchanged with age; **fruit** (Jul-Sep) berry, oblong, 4-10 cm long, yellow-green. **Range:** Native. Southern AL and west to n FL, only in Cp.
Ecology: Occurs in dry to moist sites of slash pine or longleaf pine flatwoods, longleaf pine sandhills, old fields, pastures, and right-of-ways.
Synonyms: *A. longifolia* Kral., slimleaf pawpaw.

Other species in the SE: *A. incarna* (Bartr.) Exell [flag pawpaw], *A. obovata* (Willd.) Nash [flag pawpaw], *A. pygmea* (Bartr.) Dunal, *A. reticulata* Shuttlw. *ex* Chapman [netleaf pawpaw], *A. tetramera* Small, *A. triloba* (L.) Dunal [common pawpaw]

Wildlife: Dwarf and narrowleaf pawpaw are poor fruit producers, although common pawpaw, abundant in the Mississippi River bottoms, is a prolific fruit producer. Wild Turkey, Gray Fox, Virginia Opossum, Gray Squirrel, and Raccoon eat the fruit. Larvae of the Zebra Swallowtail butterfly feed on the foliage.

Asimina **Pawpaw** **Annonaceae**

T. Bodner (Apr)

Asimina parviflora

T. Bodner (Jun)

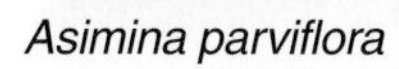

Asimina parviflora

T. Bodner (Sep)

Asimina angustifolia

T. Bodner (Jun)

Asimina parviflora

Baccharis Groundseltree or Baccharis Asteraceae

Tardily deciduous or evergreen, shrubs or small trees, with densely branched stems, often resinous. Small alternate leaves, fleshy, toothed or entire. Male and female flowers on different plants (dioecious), female flowers with long silky white pappus-bristles, bristles of male flowers shorter. Nutlets (achenes) more or less cylindric, with 5 to 10 ribs, released with pappus-bristles for wind dispersal. 4 species in the SE.

Common Species

Baccharis halimifolia L. - eastern baccharis BAHA

Plant: Erect shrub, 1-4 m tall, one to few stems from rootcrown, with numerous erect and ascending branches, bushy branched and commonly oval in profile, often forming dense colonies on open sites in eastern portions of its range.

Stem: Slender, twigs dark green and ribbed with slightly sticky resin, becoming light-green and round in cross-section, bark green with tan scaly stripes, becoming tan and slightly fissured, pith white and solid.

Leaves: Alternately spiraling, tardily deciduous to evergreen, somewhat leathery, diamond-shaped becoming obovate to elliptic upward, 2-6 cm long and 1-4 cm wide and smallest upward, coarse blunt teeth above mid-leaf, with margins becoming entire upward, surfaces green to grayish-green and shiny resin-dotted with whitish midvein (and whitish 2 lateral veins on larger leaves), bases V- or U-shaped and tapering along short petioles, 5-12 mm long.

Flowers: Sep-Oct. Male and female flowers on separate plants (dioecious), both in terminal branched clusters of small heads, female flowers with tufts of silky pappus bristles about 1 cm long and bristles of male flowers about 0.5 cm long, enclosed tightly in overlapping involucral bracts, 4-6.5 mm long, light-green.

Fruit and seeds: Oct-Nov. Cylindric nutlet (achene), 1.2-1.5 mm long, pale ribbed, topped with white bristles, about 1 cm long.

Range: Native. TX to FL and north to MA and west to WI, mainly Cp of the SE.

Ecology: Occurs as single plants or small colonies along right-of-ways and in open forests and new forest plantations as well as shore hammocks, sea beaches, salt marshes, and low grounds inland. Common along coast because of its high salt tolerance (thus common name, saltbush). Becoming noxiously weedy with an eastward expansion of the traditional range along right-of-ways and disturbed areas. Burning usually increases abundance. Seeds wind dispersed soon after ripening.

Synonyms: consumptionweed, groundseltree, manglier, saltbrush, seamyrtle, silverling.

Other species in the SE: *B. angustifolia* Michx. [falsewillow, saltbush], *B. dioica* Vahl [only in tropical FL], *B. glomeruliflora* Pers. [groundseltree].

Wildlife: Poor to moderate White-tailed Deer browse species. Small seeds may be consumed by some songbirds, but likely in very small amounts.

Baccharis Groundseltree or Baccharis Asteraceae

T. Bodner (Oct)

Baccharis halimifolia

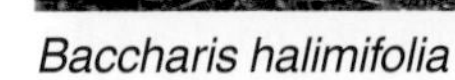

T. Bodner (Oct)

Baccharis halimifolia

T. Bodner (Oct)

Baccharis halimifolia (♀)

T. Bodner (Oct)

Baccharis halimifolia

Befaria **Tarflower** **Ericaceae**

Evergreen shrub, often growing in colonies, in a restricted range in FL and southern AL and GA. Crowded alternate leaves in a spiraling arrangement, with thick blades. Showy long-petaled, white flowers, sticky with resin (thus common name), in erect terminal spike-like racemes, on slender branches with golden-brown hairs. Fruit a spherical capsule, sticky, splitting to release numerous elongated seeds. 1 species in the SE.

Common Species

Befaria racemosa Vent. **- tarflower** **BERA2**

Plant: Upright, evergreen shrub, 1-2 m tall, with few erect branches, often forming dense colonies by rootsprouting.

Stem: Slender, erect or ascending, irregularly branched, twigs pale-green to golden-brown and long-spreading hairy, except hairless light-green ellipses where leaves attach, becoming gray-brown and hairless.

Leaves: Alternately spiraling, elliptic to ovate to oblanceolate, 2-5 cm long and 0.6-2.5 cm wide, sessile, acute to rounded tips, margins entire, midvein whitish and lateral veins not distinct, shiny often whitish waxy above, dull light-green beneath, often with few to many long hairs along midvein, leaves typically erect and often twisted thereby revealing the underside.

Flowers: May-Jul. Showy, sticky (thus common name), fragrant, terminal racemes (occasionally panicles), 3-23 flowers on red stalks (8-24 cm long), extending above current year's growth, petals 7, white and pink-tinged, long-linear, 2-3 cm long and 4-10 mm wide, stamens 12 -14, about 1/2 to 2/3 the length of the petals, blooming from bottom to top with spherical buds, each flower on a stem (1-3 cm long) with a subtending bract at base.

Fruit and seeds: Capsule spherical, sticky, 6-8 mm wide, 7-chambered, splitting to release numerous seeds, oblong, 1.3-1.8 mm long, amber.

Range: Native. Southeast AL, s GA to s FL, only in Cp.

Ecology: Occurs in patches on poorly-drained Lower Coastal Plain sites and dry sandy sites, open to semi-shady habitats. Persists by rootcrowns, colonizes by rootsprouting, and spreads by seeds.

Synonyms: flycatcher.

Wildlife: No wildlife value reported.

Befaria **Tarflower** **Ericaceae**

T. Bodner (Jun)

Befaria racemosa

T. Bodner (Jun)

Befaria racemosa

T. Bodner (Jun)

Befaria racemosa

Callicarpa Beautyberry Verbenaceae

Deciduous shrubs, much branched and bushy, with opposite leaves, being short-scurfy hairy on both surfaces, and having coarsely serrate margins. Young twigs brown and rough hairy. Small flowers in dense axillary clusters (cymes), petals lavender to pinkish. Fruit a drupe, in clusters, distinctly bright lavender to purple (rarely white), fleshy and 4 seeded. 3 species in the SE.

Common Species

Callicarpa americana L. - American beautyberry CAAM2

Plant: Multi-trunked, shade-tolerant, deciduous shrub, 2-3 m tall, with many spreading branches, bushy, often forming loose colonies. Leaves and twigs with pungent aroma when crushed or broken.

Stem: Ascending to spreading, opposite branched, young twigs light-green and scurfy hairy, slightly 4-angled at the nodes, becoming brown to gray with vertical lighter fissures, 1 bundle scar per leaf scar.

Leaves: Opposite, deciduous, ovate to broadly lanceolate, 7-17 cm long and 3-9 cm wide, margins coarsely serrate to crenate except near base, tapered tips and bases, scurfy hairy above becoming less so with age, whitish-woolly hairy beneath with prominent veins, petioles stout and scurfy hairy, 1-3 cm long. Leaves slightly aromatic when crushed.

Flowers: Jun-Jul. Dense axillary clusters, many branched and many flowered, on current growth, shorter than the subtending petioles, corolla pinkish-white, 5 lobed, flowers on short stalks.

Fruit and seeds: Aug-Jan. Spherical drupe, purple to violet (rarely white), 4-5 mm long and wide, fleshy, containing 4 seeds, flattened-ellipsoid, yellow-brown. Showy fruit clusters encircling the stem at regular intervals in late summer to early winter.

Range: Native. TX to s FL and north to MD and west to MO and OK.

Ecology: Commonly occurs on a wide variety of sites, moist to dry, open to shady habitats. Most frequent and abundant on moist sites under open pine canopies. Persists after harvesting and site preparation, and can become abundant in new forest plantations. Present along forest margins, right-of-ways, and fencerows. Fairly fire tolerant, often increasing in abundance in periodically burned pine stands. Spreads by bird-dispersed seeds.

Synonyms: French mulberry, beautybush.

Other species in the SE: *C. dichotoma* (Lour.) K. Koch. [bogs in NC and VA], *C. japonica* Thunb. [NC].

Wildlife: American beautyberry fruit consumed by more than 40 species of songbirds, particularly the American Robin, Northern Cardinal, Gray Catbird, Northern Mockingbird, Brown Thrasher, Purple Finch, and Eastern Towhee. Although considered a medium preference browse species, leaves are commonly browsed by White-tailed Deer and it can be a major part of the summer diet when highly preferred foods are scarce. Protein content ranges from 18 percent in spring to 8 percent in fall. Fruit is heavily used by White-tailed Deer in late November after leaf fall. Fruit is also consumed by Northern Bobwhite, Raccoon, Virginia Opossum and Nine-banded Armadillo as well as numerous small rodents.

Callicarpa **Beautyberry** **Verbenaceae**

T. Bodner (Jun)

Callicarpa americana

T. Bodner (Sep)

Callicarpa americana

T. Bodner (Aug)

Callicarpa americana

Calycanthus **Sweetshrub** **Calycanthaceae**

Deciduous upright shrub, with opposite leaves, oval to elliptic, having entire margins, and emitting a pleasant aroma when crushed. Flowers terminal on short branches, solitary, many petals, purplish-brown. Fruit fleshy, a large swollen floral tube, detached whole, with many seeds. 1 species in the SE.

Calycanthus floridus L. **- sweetshrub** **CAFL22**

Plant: Upright, deciduous, aromatic shrub, commonly 1-2 m tall, with spreading opposite branches, from shallow roots and often forming loose colonies by rootsprouting.

Stem: Slender, opposite and sparsely branched, twigs dark red-brown and somewhat quadrangular in cross-section, enlarged and flattened at nodes, becoming round with tiny light lenticels, leaf scars raised, horseshoe-shaped with 3-5 bundle scars, terminal bud absent.

Leaves: Opposite, aromatic, oval to elliptic with tapering tips, 5-12 cm long and 2-6 cm wide, dark green and lustrous above and white waxy beneath with raised veins, margins entire, slightly wavy, petioles stout, curved, to 1.5 cm long.

Flowers: Mar-Jun. Solitary at terminal of leafy shoots of new growth, petals many, linear and curved inward, maroon to purple-brown, calyx of many sepals, united below into a fleshy conical cup, colored like the petals, stamens numerous, short and stout, on the inner rim of the floral tube. Crushed flowers yield a strong fragrance suggestive of strawberry or banana.

Fruit and seeds: Aug-Apr. Enlarged floral tube, fleshy, cylindric with an ovoid top, 4-8 cm long and 2-5 cm wide, green in summer turning dark brown to black over winter, detaching whole in spring, containing numerous nutlets, hard-coated, about 1 cm long and 5 mm wide, brown.

Range: Native. MS to n FL and north to PA and west to OH, most frequent in Pd and Mt in the SE.

Ecology: Occurs as infrequent colonies, on moist to somewhat dry sites, and open to shady habitats. Most abundant in bottomland forests, sheltered forests, and along streams, and more abundant in the Southern Appalachian Mt. Persists after harvesting to occur in young forest plantations. Frequently cultivated as an ornamental, with yellow-flowered cultivars available. Persists by rootcrowns, colonizes by rootsprouting, and spreads by gravity- and animal-dispersed seeds.

Synonyms: *C. fertilis* Walt., *C. mohrii* Small, *C. nanus* Loisel., Carolina allspice, smooth allspice, spicebush, strawberry-shrub.

Wildlife: Sweetshrub is a high quality White-tailed Deer browse, particularly in the Southern Appalachians Mountains. Other wildlife value is limited.

Calycanthus **Sweetshrub** **Calycanthaceae**

T. Bodner (Apr)

Calycanthus floridus

T. Bodner (Apr)

Calycanthus floridus

T. Bodner (Jul)

Calycanthus floridus

Ceanothus Redroot Rhamnaceae

Low growing, deciduous or evergreen shrubs, sparsely branched, with alternate leaves. Flowers in both terminal and axillary dense clusters on stalks, panicle- or corymb-like. Petals 5, white, small, distinctly curved upward at tips to form tiny hoods, from a calyx cup with 5 lobes. Fruit a capsule-like drupe, 3 lobed, splitting forcefully to eject 3 seeds. 4 species in the SE.

Common Species

Ceanothus americanus L. - **New Jersey tea** **CEAM**

Plant: Small bushy, deciduous shrub, up to 1 m tall, with few to many slender and upturned branches, having stalked clusters of tiny floral disks persisting during winter, from stout woody rootstocks, often swollen and gnarled, deep red, with deep taproots. Used as a tea substitute during the American Revolution (thus the common name).
Stem: Twigs herbaceous near tips, often dying back during winter, tan with long spreading hairs and short curly hairs, bark reddish brown and thin, slightly rough and sparsely hairy, leaf scars small and half elliptical, somewhat raised with 1 bundle scar.
Leaves: Alternate, lanceolate to ovate, 3-8 cm long and 1-4 cm wide, distinctly 3 veined, with the outer 2 paralleling the margins (then intersecting margins above mid-leaf), margins finely serrate, often dark green and sparsely hairy above, grayish green and densely to slightly hairy beneath, petioles short and hairy.
Flowers: May-Jun. Long-stalked clusters in upper axils and many from the terminal, panicle-like, petals 5, creamy white to grayish white, about 1 mm long, hooked upward and hooded, odorless.
Fruit and seeds: Jun-Sep. Capsule-like drupe, 4-5 mm wide, 3 round lobes, splitting and ejecting 3 seeds, about 2 mm long, dark reddish brown to black, fruit base persisting through winter, saucer-shaped.
Range: Native. TX to c FL and north to Que and west to Ont and MN, most abundant in Cp, but in all physiographic provinces in the SE.
Ecology: Non-legume nitrogen fixer. Occurs as scattered plants or small groups, in well-drained open hardwood forests and sandy pine forests, including tree plantations. Inhabits gravelly or rocky banks, roadsides, and clearings. Persists by rootcrowns and spreads by ejected and animal-dispersed seeds. Burning promotes seed germination.
Synonyms: redroot.

Other species in the SE: *C. microphyllus* Michx. [small-leaved redroot], *C. herbaceus* Raf. [ovate-leafed redroot], *C. serpyllifolius* Nutt.

Wildlife: White-tailed Deer browse New Jersey tea during the growing season and to a lesser extent during winter. This species has been used as a sensitive indicator plant for identifying White-tailed Deer overabundance. Wild Turkey and Northern Bobwhite occasionally eat the seeds, as likely do numerous species of songbirds. Foliage consumed by the larvae of the Spring Azure and Mottled Duskywing butterflies.

Ceanothus **Redroot** **Rhamnaceae**

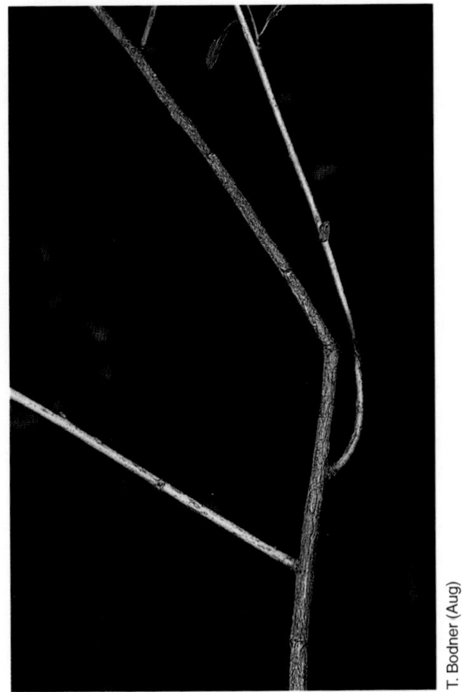

T. Bodner (Aug)

Ceanothus americanus

T. Bodner (Aug)

Ceanothus americanus

B. Craft (Jun)

Ceanothus americanus

Cephalanthus **Buttonbush** **Rubiaceae**

Deciduous, densely branching, and leafy shrub, or occasionally a small tree, growing near water or wet sites. Flowers white or yellowish-white, densely aggregated in spherical heads on long stalks, singly and in terminal clusters. Shiny opposite leaves on gray-brown stems. 1 species in the SE.

Common Species

Cephalanthus occidentalis L. **- buttonbush** **CEOC2**

Plant: Round-shaped or sprawling, bushy shrub (or small tree), 1-4 m tall, often forming dense colonies by rootsprouting, along margins of wet places.

Stem: Ascending or arching, stout, twigs gray-brown with raised linear corky lenticels, finely hairy or not, branches weakly squarish, leaf scars U-shaped with crescent-shaped bundle scars, older stems reddish brown and developing prominent ridges and furrows, pith brown or tan, solid, thick.

Leaves: Deciduous, opposite or some in whorls of 3 or 4, ovate to oblong, 6-15 cm long and 3-10 cm wide, margins entire, tip pointed, pinnately veined, bright green above and somewhat lighter green beneath, hairless or softly hairy often on upper midvein, petioles stout, light-green to reddish-brown, 2-4 cm long, stipules triangular, 2-3 mm long, often with red to black glandular tips.

Flowers: Jun-Aug. Numerous sessile flowers, white, radiating in dense spherical heads, either terminal or axillary, later developing into clusters (corymbs), on long stalks from current year's growth, each head pincushion-like, 2-3.5 cm wide, stamens 4, short, stigma often covered with masses of pollen.

Fruit and seeds: Aug-Jan. Round mass of nutlets persisting well into winter, nutlets elongate, 4-7 mm long, splitting into 4 separate nutlets.

Range: Native. Eastern US to Mexico and north to s Canada.

Ecology: Occurs along margins of streams, rivers, ponds, and marshes. Widely distributed in wet meadows, ditches, and in freshwater swamps. Highly tolerant of prolonged submergence. Grows in full sun or shade, usually late leafing out in spring. Persists by rootcrowns, colonizes by rootsprouts, and spreading by bird- and water-dispersed seeds.

Synonyms: button willow, globe-flowers, honey-balls.

Wildlife: Low preference White-tailed Deer browse, receiving only light browsing even at high deer densities. Waterfowl, particularly Wood Ducks and Mallards, feed extensively on the seeds. Ponds or swamps with an abundance of buttonbush provide excellent roost and brood rearing habitat for Wood Ducks. Seeds eaten by many species of songbirds. Flowers are an excellent nectar source for butterflies and other insects. Leaves of buttonbush poisonous to livestock.

Cephalanthus **Buttonbush** **Rubiaceae**

T. Bodner (Jun)

Cephalanthus occidentalis

J. Miller (Jun)

Cephalanthus occidentalis

J. Miller (Oct)

Cephalanthus occidentalis

T. Bodner (Jun)

Cephalanthus occidentalis

Clethra **Pepperbush** **Clethraceae**

Deciduous shrubs or trees, growing on moist sites, leafy and bushy. Leaves alternate, rather large, oblanceolate to obovate, and shiny. Branches, flower stalks, and sepals densely hairy with branched hairs (requires hand lens). Fragrant white flowers in long, narrow, spike-like racemes. Fruit a spherical capsule, 3-celled, splitting to release many seeds. 2 species in the SE.

Common Species

Clethra alnifolia **L. - sweetpepperbush** **CLAL3**

Plant: Bushy shrub, 1-3 m tall, densely leafy and branched, often oval in outline, forming colonies near streams and ponds by rootsprouting.
Stem: Ascending, twigs tan and scurfy hairy, becoming hairless and gray to red-gray, bark sloughing in areas to reveal a shiny gray inner bark, leaf scars heart-shaped with a slightly protruding bundle scar, pith spongy, whitish, and continuous.
Leaves: Alternate, oblanceolate to obovate (widest above mid-leaf), blades 5-12 cm long and 2-6 cm wide and smallest upward, veins straight and nearly parallel until curving towards margins, margins serrate above mid-leaf, scurfy above and finely white hairy (to hairless) with protruding veins beneath, petioles 5-15 mm long, whitish and scurfy hairy.
Flowers: May-Jul. Slender erect terminal and axillary racemes, with densely hairy stalks and deciduous bracts, petals 5, white (occasionally pink), 7-9 mm long, stamens 10 (in 2 whorls), very fragrant.
Fruit and seeds: Sep-Feb. Hairy capsule, spherical, 2-3 mm long, 3-celled, within persistent calyx, splitting to release many irregular seeds, about 1 mm long, pinkish to tan. Fruit stalks persisting until spring.
Range: Native. TX to FL and north to ME, mainly Cp of the SE.
Ecology: Common on wet savannas, flatwoods, pocosins, and bays, and along margins of swamps and shrub bogs. Intolerant of prolonged flooding, and increased by drainage and fire. Colonizes by rootsprouting and spreads by animal- and water-dispersed seeds.
Synonyms: *C. tomentosa* Lam., white alder, summersweet.

Other species in the SE: *C. acuminata* Michx. [only in Mt - mountain pepperbush or cinnamon clethra].

Wildlife: Sweetpepperbush is considered a preferred White-tailed Deer browse plant in the Lower Coastal Plain of the Southeast. Flowers are highly attractive to insects.

Clethra **Pepperbush** **Clethraceae**

J. Miller (Oct)

Clethra alnifolia

T. Bodner (Oct)

Clethra alnifolia

T. Bodner (Aug)

Clethra alnifolia

Cliftonia **Buckwheat tree** **Cyrillaceae**

Evergreen, thicket-forming shrub, sometimes reaching tree size, having dark scaly bark. Leaves alternate, thick and elliptic. Flowers and fruit in racemes, having winged capsules dangling along a stalk, resembling those of buckwheat (thus common name). 1 species in the SE.

Common Species

Cliftonia monophylla (Lam.) Britt. *ex* Sarg. **- buckwheat tree** **CLMO2**

Plant: Large, evergreen shrub to 15 m tall, multiple trunked and much branched, often forming dense colonies by rootsprouting.

Stem: Upright, new twigs golden brown, branches mostly crooked and dark brown to gray scaly, bark thin and weakly fissured to form small persistent elongated scales, leaf scars shield-shaped, with a slit-shaped bundle scar.

Leaves: Alternately spiraling, evergreen (may persist for 2 years) and thick, elliptic to oblanceolate, 3-6 cm long and 1.2-2 cm wide, sessile or short petioled, margins entire and rolled slightly under, surfaces glossy and bright-green often white-waxy, veins inconspicuous except for midveins.

Flowers: Feb-Mar (before new growth). Terminal racemes, 3-9 cm long, 10-30 flowers, fragrant, petals 5-8, pink-white, 5-8 mm long, sepals 5-8, stamens 10, anthers orange. New vegetative shoots arising later as a whorl at the base of the raceme.

Fruit and seeds: Aug-Feb. Three-winged capsule, ovoid, 6-7 mm long, pale lime-green becoming yellow then brown at maturity, with 2-4 seeds.

Range: Native. Southeast LA to s GA and panhandle FL, mainly Middle and Lower Cp.

Ecology: Prefers open sites along forest edges; wet sandy peat soils of alluvial swamps, bays, and acid bogs; along streams; and in flatwoods depressions. Colonizes by rootsprouting and spreads by water-dispersed seeds.

Synonyms: buckwheat-bush, black titi, ironwood.

Wildlife: White-tailed Deer browse buckwheat tree year-round, but greatest use is during winter when it composes a significant portion of the diet in the Florida flatwoods. Nutritive quality is moderate. Thickets frequently grow above the reach of White-tailed Deer and must be managed to induce sprouting. Attractive to butterflies, and it is also an excellent honey plant.

Cliftonia Buckwheat tree Cyrillaceae

T. Bodner (Mar)

Cliftonia monophylla

J. Miller (Oct)

Cliftonia monophylla

T. Bodner (Jun)

Cliftonia monophylla

Corylus Hazel Betulaceae

Multi-stemmed, deciduous shrub, with alternate leaves, doubly-serrate, and straight-veined. Male and female flowers separate on the same plant. Male catkins long, cylindric and dangling, appearing in fall, but flowering the following spring. Female flowers small, in axillary buds, ovoid, with 2 protruding red anthers. Female flowers subtended by minute bracts and 2 bractlets, enlarging at maturity to enclose a brown, hard-shelled nut, edible.
3 species in the SE.

Common Species

Corylus cornuta Marsh. **- beaked hazelnut** **COCO6**

Plant: Deciduous shrub, 0.5-1.5 m tall, multi-stemmed and often much branched in the upper plant, from a persistent rootcrown and forming dense colonies by rootsprouting.
Stem: Multiple from rootcrown, often highly branching towards extremities, twigs brown hairy.
Leaves: Deciduous, alternate, ovate and irregular in shape, 3-10 cm long and 1-7 cm wide, soft hairy beneath, margins doubly-serrate, base heart-shaped or rounded, petioles 1-2 cm long, non-glandular hairy (*C. americana* with glandular hairy petioles, requires hand lens).
Flowers: Feb-Apr (before leaves). Male catkins, cylindric and dangling, yellowish-brown, sessile, clusters of 1-3, each 1.5-2.5 cm long, on 1-year-old lateral twigs, female flowers inconspicuous, with 2 protruding anthers emerging from axillary buds, red.
Fruit and seeds: Sep-Jan. Hard nut, ovoid, 5-10 mm long, brown, covered by a hairy hull with a slender beak, 2-5 cm long.
Range: Native. KS, MO, n MS, n AL, n GA and north to Que and west to Saskatchewan, only Pd and Mt in the SE.
Ecology: Prefers slightly acidic, well-aerated, sandy loam and loam soils and rocky slopes. Forms dense thickets in open forests, pastures, and clearings. Colonizes by rootsprouting and spreads by animal-dispersed seeds.
Synonyms: beaked hazel.

Other species in the SE: *C. americana* Walt. [American hazel], *C. avellana* L. [European filbert - introduced]

Wildlife: Hazels are not a preferred White-tailed Deer browse species in the Southeast, but may be an important winter browse in northern states. Male catkins provide a rich protein source for Ruffed Grouse, especially in late winter. Nuts are readily used by Gray Squirrel, Wild Turkey, Eastern Chipmunk, and other birds and small mammals. Hazelnuts are high in protein and fats, and low in carbohydrates. Hazel thickets provide cover for American Woodcock, Ruffed Grouse, Eastern Cottontail, and other small mammals.

Corylus **Hazel** **Betulaceae**

J. Seiler (Jun)

Corylus cornuta

J. Seiler (Aug)

Corylus cornuta

J. Seiler (Jun)

Corylus cornuta

J. Seiler (Oct)

Corylus cornuta

Crataegus **Hawthorn or Haw** **Rosaceae**

Deciduous shrubs or small trees, most armed with slender spines on at least some branches. Leaves alternate, toothed and often lobed, pinnately or palmately veined, with small stipules at base of petioles (soon deciduous). Twigs usually slender and more or less zig-zagged, some with sharp spines in leaf axils. Spring flowers solitary or in terminal simple clusters (cymes), petals 5, white or lavender, sepals 5, persistent on the fruit, and stamens 5-20. Fruit a berry-like pome (small apple), containing 2-5 seeds. About 56 species in the SE, with some distinctions unclear.

Common Species

Crataegus flava Ait. **- yellow hawthorn** **CRFL2**

Plant: Common spiny shrub or small squat tree, usually 1-4 m tall, arching branched and rounded crowned.

Stem: Much branched, twigs arching, reddish-brown and scurfy hairy (or not), becoming shiny gray with long sharp spines (2-6 cm long) in some axils, bark gray with darker fissures, becoming densely blocky.

Leaves: Alternately spiraling, thick and leathery, often clustered on short lateral twigs, highly variable in shape and margins, mainly oval or obovate, 2-5 cm long and 1-3 cm wide, margins finely crenate with black gland tips (requires hand lens), margins occasionally toothed or lobed (some 3-lobed), tips rounded or pointed, bases short tapering along the petioles (with black glands), shiny green to yellowish-green above with yellow impressed veins and dull green beneath, stipules infrequent, small and lobed.

Flowers: Apr-May. Solitary in axils or terminal clusters (cymes), 3-7 flowers, each 1.6-2 cm wide, petals 5, white (pink in the bud), stamens 15-20, with red tips, sepals 5, stalks and sepals densely hairy.

Fruit and seeds: Aug-Oct. Pome (small apple), 1-2 cm wide, fleshy, green to yellow to orange (sometimes red), tipped with persistent tiny sepals, containing 3-5 nutlets.

Range: Native. MS to FL and north to VA and west to TN, mainly upper Cp and Pd, rare in Mt.

Ecology: Most common hawthorn species in mid-SE, occurs on moist to dry sites, from new forest plantations and right-of-ways to open forests, often on sandy or rocky soils. Spreads by animal-dispersed seeds.

Synonyms: summer haw, yellow haw.

Crataegus spathulata Michx. **- littlehip hawthorn** **CRSP**

Similar to *Crataegus flava* except **plant** crooked trunked shrub or often small tree, to 10 m tall; **stem** twigs slender (not as zig-zagged) and hairy or not, spines straight, 1-4 cm long, bark pale gray-brown in plates over orange-brown inner bark; **leaves** thin and hairless, narrowly oblanceolate, 2-5 cm long and 1-3 cm wide including tapering base and winged petiole, margins crenate with few end teeth, larger leaves with 3-7 lobes (no black glands), veins extend to both teeth and sinuses, stipules arching and often 2 lobed; **flowers** (Apr-May) few to many in branched clusters (cymes) on slender stalks, flowers 0.7-1 cm wide, petals white to pale pink, sepal margins entire, stamens 20, with yellow tips; **fruit and seeds** (Oct-Dec) bright red pome when ripe, 4-6 mm wide, dry pulp. **Range:** Native. TX to FL and north to VA and west to s MO, mainly Cp and Pd and along MS Valley.

Ecology: Occurs along streams, bottomlands, and moist slopes.

Synonyms: small-fruited hawthorn, red haw, sugar haw.

Crataegus **Hawthorn or Haw** **Rosaceae**

T. Bodner (Apr)

Crataegus spathulata

J. Miller (Jul)

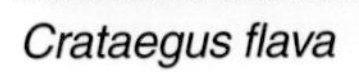

Crataegus flava

T. Bodner (May)

Crataegus flava

Crataegus **Hawthorn or Haw** **Rosaceae**

Crataegus uniflora Muenchh. **- dwarf or one-flowered hawthorn** **CRUN**
Similar to *Crataegus flava* except **plant** shrub, 1-4 m tall, with slender spreading branches; **stems** twigs shaggy hairy becoming hairless, but rough with lenticel dots, spineless or with spines 2-6 cm long; **leaves** thick when mature, ovate to oval with mostly straight tapering bases, 1-3 cm long and 1-2 cm wide, tips mostly rounded, margins serrate to serrate-crenate mostly above mid-leaf, sometimes slightly lobed, rough and shiny above and hairy at least on veins beneath; **flowers** (Mar-May) solitary (or rarely 2-3), 1-1.6 cm wide, on short hairy stalks, petals 5, white, sepals elliptic to lanceolate, 4-7 mm long and 1.5-4 mm wide, coarsely serrate with black gland tips, stamens with yellow or whitish tips; **fruit** (Aug-Oct) greenish yellow to reddish to brownish pome, 0.8-1.2 cm wide, hairy or not, thin and hard flesh, persistent sepals raised, mostly 5 nutlets. **Range:** Native. TX to FL and north to NY and west to MO.
Ecology: Occurs on mostly dry sites, open forests and margins, especially sandy ridges.

Crataegus marshallii Egglest. **- parsley hawthorn** **CRMA5**
Similar to *Crataegus flava* except **plant** shrub or small tree, 1-7 m tall, with smooth gray bark; **stems** twigs hairy, spineless or spines 2-3 cm long, bark similar to *Crataegus spathulata*; **leaves** broadly ovate in outline, deeply and finely dissected or lobed (5-7) and again divided near the tip, 1.5-5 cm long and often as wide or wider, base flat-angled, margins serrate, veins end in both sinuses and points, densely gray hairy beneath when young becoming hairless, petioles very slender and often as long as blade, hairy; **flowers** 1-1.2 cm wide in many-flowered cymes, stamens 20, with red tips; **fruit** (Oct) oblong to oval pome, 4-7 mm long, green turning bright red when ripe, nutlets 1-3. **Range:** Native. TX to FL and north to VA and west to MO, mainly Cp and lower Pd.
Ecology: Occurs along swamps and streams, and in open moist forests, often in loose colonies.

Other species in the SE: This is a variable group of species and identification to species often is difficult.

Wildlife: Hawthorns are considered a fair to medium choice White-tailed Deer browse throughout most of the Southeast. The fruit is used, but not extensively. However, fruit retained into winter can be of particular benefit to game birds (Ruffed Grouse and Wild Turkey); game animals (Raccoon, Gray Squirrel, Eastern Cottontail, and White-tailed Deer); a number of species of songbirds (Cedar Waxwing, Northern Flicker, Northern Mockingbird, Blue Jay, and Northern Cardinal); and rodents. Hawthorn thickets provide excellent nesting habitat for many species of songbirds, and excellent coverts for American Woodcock.

Crataegus **Hawthorn or Haw** **Rosaceae**

T. Bodner (Jun)

Crataegus uniflora

T. Bodner (May)

Crataegus marshallii

Cyrilla Cyrilla Cyrillaceae

Shrubs or small trees, often forming dense thickets along stream margins and in low forests. Leaves alternate and leathery, evergreen or tardily deciduous. Flowers and fruit along narrow racemes in whorls at base of current year's twigs. Flowers 5 petaled, white. Fruit an ovoid drupe. 2 species in the SE.

Common Species

Cyrilla racemiflora L. - swamp or white titi CYRA

Plant: Common shrub or small tree along streams and seasonally wet forests, 2-4 m tall, short trunks and crooked stems divided into several wide-spreading branches, from shallow roots and often forming dense colonies by rootsprouting.

Stem: Upright, twigs slender and often 3-angled, lustrous reddish-brown, leaf scars raised and shield-shaped with 1 bundle scar, younger bark reddish brown and thin, older bark divided into thin shreddy scales, pith continuous.

Leaves: Evergreen or tardily deciduous, alternately spiraling, oblanceolate to elliptic, 3-10 cm long and 0.5-2.5 cm wide, leathery when mature, both surfaces smooth and hairless with prominent veins, margins entire and often rolled under, petioles none to 1-cm long. Leaves turn scarlet red, singly, from tip to base, during fall and winter.

Flowers: May-Jul. Narrow, cylindric racemes, 2-15 cm long, clustered in whorls at the end of twigs of the previous season and below the current season's growth, petals 5, white, fragrant, 2-3.5 mm long, sepals 5, lanceolate and folded, 0.8-1.5 mm.

Fruit and seeds: Sep-Oct. Ovoid-conic drupe, 2-3 mm long, 2-celled, grey to yellow-brown, persistent pistil tipped and with persistent calyx at base, nonsplitting, containing 2 minute brown seeds in each cell.

Range: Native. TX to FL and north to se VA, mainly Cp.

Ecology: Commonly occurring along stream margins, in flatwoods depressions, acid shrub bogs, bays, and other shallow wet areas, often in pure stands. Tolerant of prolonged flooding. Colonizes by rootsprouting and spreads by animal- and water-dispersed seeds.

Synonyms: black titi, leatherwood, he-huckleberry.

Other species in the SE: *C. parviflora* Raf. [small-leaved variant only in FL panhandle - little-leaved cyrilla].

Wildlife: Young shoots are preferred browse for White-tailed Deer, while mature leaves are eaten in winter. Titi swamps offer good escape cover for White-tailed Deer and Black Bear. Flowers are attractive to butterflies, and an excellent source of nectar.

Cyrilla Cyrilla Cyrillaceae

T. Bodner (Jul)

Cyrilla racemiflora

J. Miller (Jun)

Cyrilla racemiflora

J. Miller (Jun)

Cyrilla racemiflora

T. Bodner (Oct)

Cyrilla racemiflora

Elaeagnus Elaeagnus or Silverberry Elaeagnaceae

Bushy shrubs or small trees, characterized by silvery and/or brown scaly stems, petioles, leaves, flower stalks, and fruit. Leaves alternate, with smooth margins although often rolled inward or wavy surfaced. Flowers in short racemes (infrequently solitary), in leaf axils of shoots of the season, fragrant, no petals, calyx tubular with 4 lobes, white or yellow-white. Fruit a drupe, berry-like, single seeded. 4 species in the SE, all exotic.

Common Species

Elaeagnus pungens Thunb. **- thorny olive or silverthorn** **ELPU2**

Plant: Evergreen, bushy shrub, 1-5 m tall, densely leafy, with long limber projecting shoots, often found as escaped single plants from roadside plantings.
Stem: Ascending, twigs brown and densely scaly, short shoots with small leaves becoming sharp thorns (2-4 cm long), thorns producing leafy lateral branches in second year, bearing floral racemes in fall, lateral branches distinctly long and extending beyond bushy crown.
Leaves: Alternate, oval to elliptic, 4-10 cm long and 2-5 cm wide, margins irregular and wavy, surfaces silvery scaly initially, becoming dark green or brownish-green above with a silver scaly midvein, densely silver scaly with scattered brown scales beneath.
Flowers: Nov-Dec. Axillary clusters, 1-3, tubular with 4 lobes, silvery-white to brown, 1 cm long, fragrant.
Fruit and seeds: May-Jun. Drupe (berry-like), oblong, 1-1.5 cm long, fleshy, red and brown scaly when ripe, persistent calyx tube at tip.
Range: Exotic, introduced from Asia. TX to FL and north to NC, scattered escapes from ornamental plantings, mainly Pd and Hilly and Middle Cp.
Ecology: Fast-growing, weedy ornamental. Frequently planted for hedgerows and on highway right-of-ways. Tolerant of drought and salt. Spreads by animal-dispersed seeds.
Synonyms: bronze elaeagnus.

Elaeagnus umbellata Thunb. **- autumn olive** **ELUM**

Plant: Deciduous, bushy shrub, 1-5 m tall, leafy, often with thorny branches.
Stem: Twigs slender and silver scaly, spur twigs common, with some lateral branches becoming pointed like thorns, bark light gray and smooth.
Leaves: Alternate, elliptic, 2-8 cm long and 1-3 cm wide, petioles short, margins entire, bright green to gray-green above with silver scaly midvein, densely silver scaly beneath.
Flowers: Feb-May. Axillary clusters, 5-10, fragrant, silvery-white to yellow.
Fruit and seeds: Aug-Oct. Drupe, ellipsoid, 10-15 mm long and 5-9 mm wide, red and slightly silvery to silvery-brown scaly, containing 1 nutlet, ellipsoid.
Range: Exotic and widely planted. LA to FL and north to ME and west to MI.
Ecology: Often planted in reclamation areas, tolerating a wide variety of soils, producing abundant wildlife food and cover. Escapes plantings and spreads by bird-dispersed seeds.
Synonyms: silverberry.

Other species in the SE: *E. angustifolia* L. [Russian olive, oleaster], *E. multiflora* Thunb.

Wildlife: Fruit of all species of elaeagnus provide food for Northern Bobwhite, Mourning Dove, Ruffed Grouse, Wild Turkey, and numerous songbirds. Fruit also is readily consumed by Raccoon, Striped Skunk, Virginia Opossum, Black Bear, and other mammal species. Dense hedgerows provide nesting and escape cover. Autumn olive often is recommended for conservation plantings because of its excellent wildlife value. However, it may become a pest in areas due to bird dissemination of seed.

Elaeagnus **Elaeagnus or Silverberry** **Elaeagnaceae**

T. Bodner (Mar)

Elaeagnus pungens

J. Miller (Jan)

Elaeagnus pungens

J. Miller (Oct)

Elaeagnus umbellata

J. Miller (May)

Elaeagnus umbellata

Euonymus Strawberry bush Celastraceae

Erect or trailing shrubs, evergreen or tardily deciduous, with opposite branching and opposite leaves. Unopened capsules pinkish and winged or resembling a green to red strawberry (thus common name). Stems green and slightly 4-sided. Leaves with finely serrate margins having minute gland-tipped teeth. Small flowers, opposite, with 4-5 petals and a flat center disk bearing 4-5 short stamens. 6 species in the SE, 3 exotic invasive.

Common Species

Euonymus americana **L. - strawberry bush** **EUAM7**

Plant: Erect or ascending, tardily-deciduous shrub, 0.5-2 m tall, open branched with stiffly divergent twigs.

Stem: Opposite branched, slender, twigs green and slightly 4-sided, terminal buds 3-8 mm with 4 or 5 pairs of scales in 4 ranks, lateral buds similar but smaller.

Leaves: Opposite, ovate to lanceolate, 3-10 cm long and 1-3 cm wide, thin, bright green, margins finely serrate, serrations gland tipped (requires hand lens), petioles 2-3 mm long or less.

Flowers: Mar-May. Solitary (or 2-3 in cyme), at the end of stalks arranged over leaf blades, inconspicuous, petals 5, yellowish-green to purplish-green, round to broadly triangular, 2.5-4 mm wide, on a flat floral disk.

Fruit and seeds: Sep-Dec. Warty and leathery capsule, lime green in summer turning pink and bright crimson when ripe (resembling a strawberry), 1-2 cm wide, splitting into 3-5 lobes in late-summer or fall, each lobe tipped with a dangling seed, ovoid, orange-red, thin-fleshy, persisting into winter.

Range: Native. TX to FL and north to NY and west to s IL and MO.

Ecology: Occurs in moist forests, including bottomland hardwood forests and adjacent to small streams. Persists after harvesting to occur in new forest plantations. Fruits best in partial to full sun, but is very shade tolerant. Spreads by animal-dispersed seeds.

Synonyms: *E. americanus* L., hearts-a-busting, hearts-bursting-open-with-love, burning bush, bursting-heart.

Other species in the SE: *E. alata* (Thunb.) Sieb., *E. atropurpurea* Jacq. [eastern wahoo], *E. fortunei* (Turcz.) Hand.-Maz., *E. japonica* Thunb., *E. obovatus* Nutt.

Wildlife: Seeds reportedly consumed (sparingly) by Wild Turkey and some songbirds. Leaves are readily consumed by White-tailed Deer and Eastern Cottontail, although it is not abundant enough to be important in the total diet. Because it is browsed at low White-tailed Deer densities, strawberry bush may be considered an important indicator of deer presence. It virtually disappears on overstocked range.

Euonymus **Strawberry bush** **Celastraceae**

T. Bodner (Sep)

Euonymus americana

T. Bodner (Sep)

Euonymus americana

T. Bodner (May)

Euonymus americana

Frangula Buckthorn Rhamnaceae

Deciduous shrubs or small trees, with alternate, thick leaves, having parallel veins, and finely-serrate margins. Inconspicuous greenish flowers, appearing after the leaves, petals 5, shorter than the 5-lobed calyx, stamens 5. Fruit a black drupe when ripe, with 2-3 seeds. 2 species in the SE, 1 exotic invasive.

Common Species

Frangula caroliniana (Walt.) Gray **- Carolina buckthorn** **FRCA13**

Plant: Shrub or small tree, usually 1-3 m tall (up to 14 m), with a spreading crown.
Stem: Upright base, twigs reddish-brown and hairy becoming gray and hairless, leaf scars raised with 3 bundle scars, no terminal bud and most-upper lateral buds scaleless and hairy (other lateral buds inconspicuous), bark smooth brown.
Leaves: Alternate, deciduous, narrowly elliptic to oblong to obovate, 5-15 cm long and 3-5 cm wide and largest towards branch tip, margins minutely and irregularly serrate, bases rounded, tips rounded to pointed, lustrous dark green and hairless above and hairy to hairless beneath, veins parallel with 8-10 pairs (ends following margins), petioles 1-3 cm long.
Flowers: May-Jun. Tiny axillary clusters on the current growth, petals 5, whitish or greenish-white, shorter than 5-lobed calyx, 3-4 mm long.
Fruit and seeds: Aug-Jan. Berry-like drupe, spherical, 5-10 mm wide, fleshy, red ripening to dark black, persistent in winter, containing 3 nutlets.
Range: Native. TX to n FL and north to sw VA and west to NE.
Ecology: Scattered plants occurring on moist to dry sites, open to semi-shady habitats. Occurs in new forest plantations to mature forests, and in open hardwood and bottomland forests. Often abundant on shallow calcareous rocky bluffs and shell mounds. Spreads somewhat by animal-dispersed seeds.
Synonyms: *Rhamnus caroliniana* Walt., Indian cherry, pole-cat-tree.

Other species in the SE: *F. alnus* P. Mill. (exotic).

Wildlife: Very little wildlife use, although White-tailed Deer infrequently browse buckthorn in the winter. Drupes persistent and sparingly consumed by songbirds.

Frangula **Buckthorn** **Rhamnaceae**

T. Bodner (Jun)

Frangula caroliniana

J. Miller (Sep)

Frangula caroliniana

T. Bodner (Aug)

Frangula caroliniana

Gaylussacia **Huckleberry** **Ericaceae**

Deciduous or evergreen, low shrubs, often spreading by underground stems. Leaves small, alternate, margins entire, and covered on underside with yellow resin droplets (except *G. brachycera*). Twigs slender, commonly pink to red, greenish on underside and *not* with warty dots as in some species of Vacciniums. Fruit a berry-like drupe, black or blue (rarely white), containing 10 seed-like nutlets. 6 species in the SE.

Common Species

Gaylussacia dumosa (Andr.) Torr. & Gray **- dwarf huckleberry** **GADU**

Plant: Deciduous or semi-evergreen shrub, 10-50 cm tall, branched from an underground stem.

Stem: Twigs red-brown to gray, densely short-curly hairy, with short-stalked glandular hairs.

Leaves: Alternate, oblanceolate to obovate, 2-4 cm long and 0.6-2 cm wide, scattered resinous glands beneath (tiny amber dots turning yellow to brown to black, requires hand lens), margins entire to very finely crenate and hairy, tip rounded with a short point, glossy dark green above and pale green beneath.

Flowers: Mar-Jun. Axillary leafy-bracted racemes, 4-10 flowers, corolla white or pinkish, bell-shaped with 5 tiny lobes, 5-10 mm long and wider than long, stalks and floral tubes short-curly hairy and often short-stalked glandular, bracts persistent until fruits mature.

Fruit and seeds: Jun-Oct. Berry-like drupe, spherical, 7-10 mm long, shiny black, short-stalked glandular hairy, containing 10 nutlets. Fruit edible but not sweet.

Range: Native. LA to c FL and north to Newfoundland, mainly Cp in the SE, local in Pd and Mt.

Ecology: Commonly occurs in sandy soils of well-drained pine and pine-oak forests as well as in transitional areas from high elevation pine forests to shrub-tree bogs. Responds vigorously to prescribed fire at 2 to 3 year intervals. Spreads by animal-dispersed seeds.

Synonyms: *Lasiococcus dumosus* (Andr.) Small, gopherberry, bush huckleberry.

Gaylussacia frondosa (L.) Torr. & Gray *ex* Torr. **- dangleberry** **GAFR2**

Similar to *Gaylussacia dumosa* except **plant** widely branched shrub, to 2 m tall; **stem** twigs hairy (or not), with sessile glands (not stalked); **leaves** deciduous, elliptic to oblanceolate, 3-7 cm long and 1.5-3.5 cm wide, hairless above and often densely short hairy with resinous dots beneath; **flowers** greenish-white to pinkish, 2-5 mm long, slightly longer than broad, bracts small and not persistent, stalks short hairy and sparsely glandular dotted; **fruit and seeds** (Jun-Aug) blue and white waxy, 5-8 mm long, sweet and juicy. **Range:** Native. LA to c FL and north to NH and MA, mainly Cp in the SE, local in Pd.

Ecology: Occurs in wet and well-drained flatwoods, sandhills, and along margins of bays, shrub-bogs, and cypress depressions.

Synonyms: *G. nana* (Gray) Small, *G. tomentosa* (Gray) Pursh *ex* Small, tangleberry, blue huckleberry, small dangleberry.

Other species in the SE: *G. baccata* (Wangenh.) K. Koch [black huckleberry], *G. brachycera* (Michx.) Gray [box huckleberry], *G. mosieri* Small, *G. ursina* (M.A. Curtis) Torr. & Gray *ex* Gray [buckberry].

Wildlife: Huckleberry fruits are eaten by Northern Bobwhite, Wild Turkey, Ruffed Grouse, and Gray Squirrel, along with several songbirds including Eastern Bluebird, Blue Jay, Northern Mockingbird, Northern Cardinal, Eastern Towhee, and Gray Catbird. In Florida, huckleberry leaves have been reported to be consumed in small quantities by White-tailed Deer.

Gaylussacia **Huckleberry** **Ericaceae**

T. Bodner (Sep)

Gaylussacia dumosa

D. Lauer (Aug)

Gaylussacia frondosa

D. Lauer (Aug)

Gaylussacia frondosa

Hamamelis **Witchhazel** **Hamamelidaceae**

Tree-like deciduous shrubs or small trees, often with leaning stems, sparsely branched. Alternate leaves, broadly ovate, having undulating margins, and unequal bases. Flowers in axillary clusters on twigs of previous season, with 4 linear petals, yellow (to red), and 4 short-triangular sepals. Fruit a woody capsule, ovoid, splitting and ejecting seeds, hard and black. 2 species in the SE.

Common Species

Hamamelis virginiana L. **- witchhazel** **HAVI4**

Plant: Multi-stemmed shrub or small tree, 1-7 m tall, with a flattened and spreading crown, forming small colonies by rootsprouting.

Stem: Young twigs slender, zigzagged, reddish-brown, hairy or not, bark smooth and light-brown and flaky, inner bark purplish, terminal buds in pairs without scale covering (1 stalked and curved), 0.6-1.2 cm long, and densely yellowish-brown hairy.

Leaves: Alternate, deciduous, ovate to obovate with pointed or rounded tips, 6-15 cm long and 4-8 cm wide, base unequal, margins wavy-toothed, surfaces hairy becoming hairless with age, straight veined, petioles 5-15 mm long. Leaves emitting a slight fragrance when crushed. Leaves turn bright yellow in fall.

Flowers: Sep-Dec. Showy axillary clusters, usually in 3's, appearing after leaf fall and before prior year's dried capsules open, petals 4, yellow, linear and curled, 1.5-2 cm long, sepals 4, triangular, dull yellow above, stamens 4.

Fruit and seeds: Oct-Nov. Woody capsule, densely hairy, ovoid, 6-15 mm long, splitting at beaked end and expelling 4 seeds (up to 10 m away), shiny brown-black. Empty capsule with 4 sharply recurved points, persistent.

Range: Native. TX to n FL and north to s Canada and west to MI.

Ecology: Occurs on moist sites, open to shady habitat, as an understory plant in mixed forests, bottomland forests, and along forest margins and streams. Vigorous resprouter on harvested areas. Colonizes by rootsprouting and spreads by ejected and animal-dispersed seeds.

Other species in the SE: *H. vernalis* Sarg. [AR, OK, MO, and LA - springtime witch-hazel].

Wildlife: Witchhazel is browsed by White-tailed Deer to varying degrees. In the more northern range it may be a significant part of the diet, whereas in the southern range it is browsed sparingly. Seeds are used by Wild Turkey, Northern Bobwhite, and Gray Squirrel. Seeds, buds and flowers are consumed by Ruffed Grouse. Witchhazel thickets provide escape cover. Witchhazel is pollinated by Cuculiid moths at night.

Hamamelis — Witchhazel — Hamamelidaceae

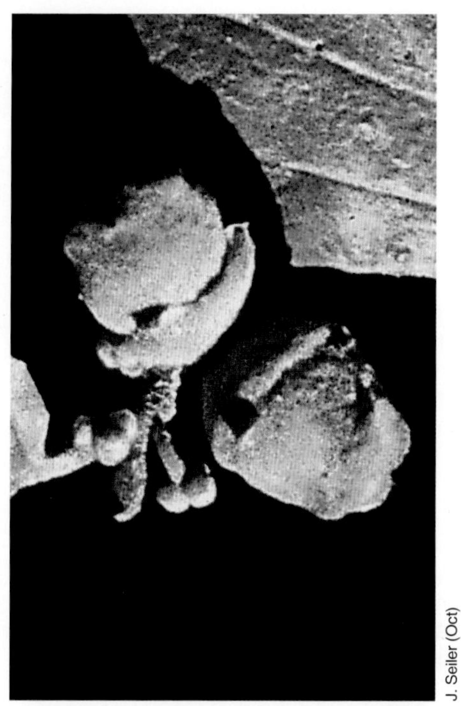

J. Seiler (Oct)

Hamamelis virginiana

T. Bodner (Jul)

Hamamelis virginiana

J. Miller (Oct)

Hamamelis virginiana

Hydrangea Hydrangea Hydrangeaceae

Deciduous, soft-woody shrubs, having somewhat stout branches with thin flaky bark. Leaves opposite or sometimes whorled, with long petioles. Flowers in dense-branched terminal clusters, having many small fertile flowers, along with longer-stemmed showy sterile flowers of only 3-4 large white or colored sepals. Many spindle-shaped capsules, splitting at the top to release numerous tiny seeds. 5 species in the SE.

Common Species

Hydrangea quercifolia Bartr. - **oakleaf hydrangea** HYQU3

Plant: Multi-stemmed, deciduous shrub, 1-2 m tall, sparsely branched, often forming small colonies or loose groupings.
Stem: Young twigs reddish-brown hairy, larger branches and stems with thin reddish-brown bark peeling in films, often hanging in shreds and revealing the tan inner-bark, leaf scars crescent-shaped, with 3-11 bundle scars in an indented curve, pith spongy and tan.
Leaves: Opposite, deciduous, oak-shaped with 4-6 lobes, blades 20-30 cm long and 20-30 cm wide, margins irregularly pointed-toothed, dark green and uneven surfaced above, whitish softly hairy with protruding whitish veins beneath, petioles 5-15 cm long, finely-white to brownish hairy.
Flowers: May-Jul. Showy elongated panicles, 15-30 cm long, each lateral branch bearing cymes of tiny fertile flowers (white, petals 5), terminating in a larger single sterile flower with 4 greatly enlarged sepals, white and papery becoming pink-purplish and finally tan, persisting into winter, fertile flowers about 8 mm wide, sepals 5.
Fruit and seeds: Oct-Feb. Cup-shaped capsule, about 1 mm long, 8-10 ribbed, brown, splitting to release few small seeds, brown, strongly-ribbed, 0.5-1 mm long.
Range: Native. LA to nw FL and GA and north to c TN and west to Mississippi River.
Ecology: Occurs in shady to semi-shady habitats, on moist slopes and in bottomland forests, and along ravines and streams, increasing in frequency on calcareous soils. Persists after harvesting to inhabit new forests and plantations. Colonizes by rootsprouting and spreads by bird-dispersed seeds.
Synonyms: ninebark, sevenbark, graybeard.

Hydrangea arborescens L. - **wild hydrangea** HYAR

Similar to *Hydrangea quercifolia* except **stems** twigs sparsely hairy, bundle scars 3; **leaves** not deeply lobed, ovate to oval, 10-18 cm long and 6-12 cm wide, margins finely to coarsely serrate, green to gray-green above and green or grayish- to whitish-hairy beneath, veins straight and parallel to margin, petioles 6-12 cm long, hairy; **flowers** terminal flat-topped clusters, white fertile flowers, with sterile large white flowers only around the edges.
Range: Native. LA, s AL, sw GA, and c FL and north to NY and west to IA, mainly Mt and Pd, local in Cp in the SE.
Ecology: Occurs in shady habitat, mostly rocky-sloped forest stands, often on calcareous soils.
Synonyms: sevenbark, smooth hydrangea, mountain hydrangea.

Other species in the SE: *H. cinerea* Small [ashy hydrangea], *H. radiata* Walt. [snowy hydrangea], *H. paniculata* Siebold. [a naturalized exotic].

Wildlife: *Hydrangea arborescens* is considered a preferred spring-summer White-tailed Deer browse; intensive use is indicative of an overpopulated range. On northern ranges it is an important winter browse. Flowers and seeds reportedly consumed by Wild Turkey and White-tailed Deer. Seeds consumed by some songbirds.

Hydrangea **Hydrangea** **Hydrangeaceae**

T. Bodner (Jun)

Hydrangea arborescens

T. Bodner (May)

Hydrangea quercifolia

J. Miller (Oct)

Hydrangea quercifolia

Ilex **Holly** **Aquifoliaceae**

Shrubs or small trees, with alternate leaves, simple, evergreen or deciduous, with minute stipules at petiole bases, often soon deciduous. Leaves often toothed, with teeth spine-tipped or thickened in some species. Flowers in axillary clusters, small, white or greenish, usually with male and female flowers on separate plants. Fruit a small drupe, yellow-white or red or black, sometimes persistent in winter, containing 4-7 nutlets. 13 species in the SE.

Common Species

Ilex glabra (L.) Gray **- gallberry** **ILGL**

Plant: Upright evergreen shrub, 1-3 m tall, soft-woody, densely ascending branched, from rhizomes and often forming dense and extensive colonies.
Stem: Twigs light green (often dark mottled) and finely hairy, becoming hairless and light brown, lenticel dots round with a center slit, leaf scars half-round to oval with 1 bundle scar, bark slightly ridged and alternating dark- and light-brown.
Leaves: Alternately spiraling, evergreen and leathery, elliptic with a tapering base, 1-4 cm long and 0.5-2.5 cm wide, lustrous yellowish-green above and pale green with minute reddish glands beneath (requires hand lens), margins entire with few blunt teeth on upper margin, petioles 3-8 mm long.
Flowers: May-Jun. Female and male flowers on separate plants, female flowers mostly solitary and male flowers in small clusters, both axillary, flowers about 8 mm wide, petals 5-8, white.
Fruit and seeds: Sep-Feb. Dry drupe, solitary in axils, spherical, 5-7 mm wide, green maturing to black, dull to lustrous, persistent into winter, containing 5-7 nutlets.
Range: Native. LA to FL and north to ME and Nova Scotia, mainly Lower Cp, local in Middle and Upper Cp.
Ecology: Often the most abundant understory shrub in flatwood forests of the Lower Cp, particularly on acid soils. Prolific resprouter following prescribed fire or bedding, often forming a nearly continuous understory in frequently-burned stands, becoming scattered plants in northern range. Colonizes by rootsprouts and spreads by animal-dispersed seeds.
Synonyms: inkberry, bitter gallberry, evergreen winterberry.

Ilex coriacea (Pursh) Chapman **- large gallberry** **ILCO**

Similar to *Ilex glabra* except **plant** to 9 m tall; **stems** twigs often reddish becoming golden, then gray with light lenticel dots; **leaves** darker green, ovate to elliptic, 3.5-8 cm long and 1.5-4 cm wide, tip pointed, margins entire or few minute teeth widely-spaced near tip, often bristly; **flowers** (Apr-May) female flowers clustered; **fruit and seeds** black drupes 1-5 in axils, each 7-10 mm wide, soft and pulpy, not persistent into winter. **Range:** Native. LA to FL and north to se VA, only Cp.
Ecology: Scattered plants in pocosins, bays, sandy woods, and swamps. Usually associated and intermixed with *I. glabra*.
Synonyms: sweet gallberry, shining inkberry, baygall-bush.

Ilex **Holly** **Aquifoliaceae**

T. Bodner (Oct)

Ilex glabra

J. Miller (Oct)

Ilex glabra

T. Bodner (Oct)

Ilex coriacea

T. Bodner (Oct)

Ilex glabra

Ilex **Holly** **Aquifoliaceae**

Ilex vomitoria Ait. **- yaupon** **ILVO**

Plant: Upright, tree-like evergreen shrub or small tree, to 10 m tall, single stemmed and divergently-branched, often forming small to extensive colonies.
Stem: Many protruding rigid twigs (almost spine-like), purple and downy hairy when young, becoming whitish gray and hairless, bark thin and gray to whitish gray.
Leaves: Alternate, leathery, oval to elliptic, 1-4 cm long and 0.5-2 cm wide, base and tip rounded, margins crenate with minute black tips (requires hand lens), shiny green with whitish midvein above and dull pale green beneath, petioles 2-5 mm long, finely hairy or not, pale green to purplish.
Flowers: Mar-May. Female flowers 1-3 in axillary clusters (nearly sessile), male flowers 3-9 in clusters, petals 4, whitish.
Fruit and seeds: Oct-Nov. Pulpy persistent drupe, almost spherical, 5-6 mm long, scarlet-red and shiny, containing 4 nutlets, each ovoid, 2-4 mm long.
Range: Native. Lower and Middle Cp in TX to c FL and north to coastal NC and se VA, and local in c AR.
Ecology: Commonly occurring shrub in Cp forests, increasing in abundance in the western part of the range, dominating forest understories in east Texas. Less common in interior sandhills. Planted inland by Native Americans. Bred and widely planted today as an ornamental. Colonizes by rootsprouting and readily spread by bird-dispersed seeds.
Synonyms: cassine, evergreen holly, Christmas berry.

Ilex opaca Ait. **- American holly** **ILOP**

Plant: Evergreen shrub or medium-sized understory tree, usually 3-10 m tall, leafy and bushy with branches at right angles, conical or columnar in form.
Stem: Twigs stout and rusty hairy becoming hairless, bark slightly rough to slightly warty, whitish gray.
Leaves: Alternate, evergreen and thick, broadly elliptic to oblong, 4-8 cm long and 2-4 cm wide, margins rolled under and wavy to dentate often spiny tipped (rarely entire with only a spine tip), shiny dark green with whitish midvein above and dull green beneath, petioles 8-12 mm long, often reddish.
Flowers: Apr-Jun. Female and male flowers in axillary clusters below leaves, petals 4, white, sepals and stamens 4.
Fruit and seeds: Sep-Apr. Dry drupe, spherical to ellipsoid, 7-12 mm long, dull red or orange (rarely yellow), containing 4 nutlets, rounded with flat side, 4-8 mm long, ribbed, grooved on rounded side. Fruit persistent until spring.
Range: Native. TX to c FL and north to MA and west to IN and AR.
Ecology: Common scattered understory shrub or tree in mixed upland and bottomland forests, and borders of swamps. Bred and widely planted as an ornamental. Spreads mainly by bird-dispersed seeds.

Wildlife: Fruits of all four species are consumed by several species of birds including Cedar Waxwings, Wild Turkey, Northern Bobwhite, Eastern Bluebird, Brown Thrasher, American Robin, Hermit Thrush, Red-bellied Woodpecker, Northern Mockingbird, Eastern Towhee, and Blue Jay, along with several mammalian species, especially Raccoon. Fruits of persistent species such as gallberry, yaupon, and American holly are important components of the winter diets of several songbirds. Yaupon is an important browse for White-tailed Deer. Gallberry is less palatable than yaupon and large gallberry, but is an important browse species because of its abundance. In some areas of the Florida flatwoods, gallberry may comprise more than 20 percent of the winter diet of deer.

Ilex **Holly** **Aquifoliaceae**

J. Miller (Oct)

Ilex vomitoria

T. Bodner (Oct)

Ilex vomitoria

T. Bodner (Oct)

Ilex opaca

Illicium Anise Illiciaceae

Evergreen shrubs or small trees, much branched, growing as scattered plants in moist sites along streams, ponds, and lakes. Spirally arranged leaves crowded at branch tips, elliptic to oblanceolate, evergreen but not thick. Showy early-spring flowers, dark red or white to yellow, with many petals. Dry capsules, joined at bases in a wheel-shaped aggregate, each capsule containing 1 seed. 2 species in the SE.

Common Species

Illicium floridanum Ellis. - **Florida anise** **ILFL**

Plant: Evergreen shrub or small tree, usually 2-3 m tall, multi-stemmed and crooked trunked. Plant parts emit a spicy rank odor when crushed.

Stem: Spreading, opposite branched and often crowded, twigs reddish brown and shaggy hairy near tip, becoming scattered hairy and shiny golden, leaf scars broadly heart-shaped and inconspicuous, with 1 bundle scar, branches dull reddish brown to gray and slightly fissured, bark gray-brown and essentially smooth.

Leaves: Alternately spiraling, crowded near twig tips often appearing whorled, evergreen although not thick, slightly leathery, narrowly elliptic to oblanceolate and tapering at both ends, blades 6-15 cm long and 2-6 cm wide, margins smooth with fine curved hairs (requires hand lens), midvein whitish above and beneath, softly hairy both surfaces becoming hairless above. Often emitting a pungent anise-like odor when crushed (thus the common name).

Flowers: Mar-May. Solitary and in terminal axils, stalks 2-4 cm long, flowers dark-red to maroon (rarely white or pink), foul-scented, petals 20-33, linear, 1.5-2 cm long and 4-5 mm wide, stamens 30-40, red. Flower buds spherical, whitish green, and present over winter.

Fruit and seeds: Aug-Sep. Dry wheel-shaped aggregate, 2.5-3 cm wide, yellow ripening to brown, 11-15 capsules radiating from joined bases, each splitting and ejecting 1 seed up to many meters away, seed about 8 mm long.

Range: Native. Southeast LA , s AL and c AL (Blackbelt), sw GA, and panhandle FL.

Ecology: Infrequent along wooded streams and margins of ponds and lakes. Prefers moist to wet soils. Spreads by ejected seeds.

Synonyms: purple anise, stink-bush, stinking laurel, polecat-tree.

Other species in the SE: *I. parviflorum* Michx. *ex* Vent. [only in FL - star anise].

Wildlife: None reported. Leaves are toxic to livestock.

Illicium **Anise** **Illiciaceae**

T. Bodner (Mar)

Illicium floridanum

Itea Virginia willow Grossulariaceae

Deciduous shrubs, growing in small colonies near streams, rivers, or lakes. Leaves alternate with finely serrate margins. Flowers in erect or arching terminal racemes, many small white flowers, petals and sepals 5. Slender capsules along a stalk, persisting into winter. 1 species in the SE.

Common Species

Itea virginica L. - **Virginia willow** **ITVI**

Plant: Tardily-deciduous shrubs, 1-2 m tall, much branched and leafy, rootsprouting and often forming small colonies near streams, ponds, and swamps.
Stem: Twigs hairy and green, turning reddish and hairless, branches erect or arching, terminal and lateral buds small, leaf scars half-elliptical with 3 bundle scars.
Leaves: Alternate, deciduous or tardily deciduous, elliptic to obovate, 3-9 cm long and 2-4 cm wide, tapering tips and bases, margins minutely and sharply serrate; petioles 3-10 mm long.
Flowers: Apr-Jun. Upright or arching terminal racemes, narrowly cylindric, 5-15 cm long, stalk hairy, with numerous flowers spiraling and radiating all sides, petals 5, white or pale pink, 3-6 mm long, stamens 5, calyx cup-shaped, 5 lobed.
Fruit and seeds: Jul-Feb. Cylindric capsules along stalk, capsules 3-8 mm long, flattened-ovoid, downy hairy, 2-grooved, beaked with swollen pistil, within persistent calyx, splitting to release many seeds, ellipsoid, about 1 mm long. Capsules persist into winter.
Range: Native. TX to FL and north to NJ and west to IL and OK.
Ecology: Common along the water-edge of streams and rivers, and in swamps with fertile soils. Grows best at forest edges in partial sun, but also present in dense shade. Colonizes by rootsprouting and spreads by water-dispersed seeds.
Synonyms: sweetspire, tassel-white, Washington-plume.

Wildlife: Virginia willow is a palatable, heavily used White-tailed Deer browse, particularly in Florida. However, it is not a major component of the diet of White-tailed Deer due to its scattered occurrence. Protein content generally is moderate to high (11-21 percent). Flowers commonly visited by butterflies.

Itea **Virginia willow** **Grossulariaceae**

J. Miller (Jun)

Itea virginica

J. Miller (Jun)

Itea virginica

T. Bodner (Jul)

Itea virginica

T. Bodner (May)

Itea virginica

Kalmia **Laurel** **Ericaceae**

Evergreen shrubs or small trees, most forming colonies from rhizomes. Leaves thick, alternate, whorled, or rarely opposite. Young twigs hairy and older twigs hairless. Flowers in terminal or axillary clusters, conspicuous corymbs or simple fascicles. Corolla disk- or saucer-shaped, white or pink-white, stamens 10, with anthers lodged into pockets of the corolla until "tripped" by insect pollinators. Fruit a spherical to ovoid capsule, 5 celled, splitting to release many seeds. 5 species in the SE.

Common Species

Kalmia latifolia L. **- mountain laurel** **KALA**

Plant: Upright, leafy, evergreen shrub or small tree, usually 2-4 m tall, multi-stemmed from base or near base and much branched, often forming small to extensive colonies from rhizomes. Plant toxic to livestock, especially the leaves.
Stem: Twigs green becoming reddish brown and short hairy, turning brown and rigid, leaf scars indented with a rim, with bud above and projecting outward, V-branched, bark reddish brown with narrow ridges and finely sloughing.
Leaves: Alternately spiraling and crowded near twig tips, evergreen and thick, elliptic to elliptic-oblanceolate, blades 5-12 cm long and 3-5 cm wide, short tapering tips (obtuse), margins entire, dark green with whitish midvein above and pale green sometimes with tiny reddish hairs beneath, petioles 1-2 cm long, pale-green.
Flowers: Apr-Jun. Showy terminal compound corymb, intricate corolla, cup-shaped with 5 shallow lobes and 10 pockets, white to pinkish-white to pink (sometimes purple tinged at pockets), 2-3 cm wide, stamens 10, anthers held in pockets in corolla and spring upward, pistil 1.
Fruit and seeds: Sep-Dec. Spherical capsule, 5-7 mm wide, 5 lobed and celled, red glandular hairy, within persistent sepals at base, splitting to release many seeds, irregular, 0.8-1 mm long.
Range: Native. LA to n FL and north to s Canada and west to s IN.
Ecology: Common shrub with dense thickets in the Southern Appalachian Mt. Infrequent along wooded streams and shady bluffs in the Lower Pd and Cp. Resprouts vigorously after fire or cutting. Shade tolerant, but most vigorous in full sun. Bred and planted widely as an ornamental. Colonizes by rootsprouting and spreads by seeds.
Synonyms: mountain ivy, calico bush, spoonwort.

Other species in the SE: *K. angustifolia* L. [TN, VA, and SC - sheep-laurel or lamb-kill], *K. carolina* Small [southern Blue Ridge Mt - sheep-laurel or lamb-kill], *K. cuneata* Michx. [NC and SC Cp - white wicky], *K. hirsuta* Walt. [LA to FL - hairy wicky].

Wildlife: Mountain laurel thickets provide dense cover for White-tailed Deer, Eastern Cottontail Rabbit, Snowshoe Hare, Ruffed Grouse, and Black Bear. Although toxic to domestic livestock, mountain laurel is browsed commonly by White-tailed Deer during winter and early spring. Mountain laurel leaves also are an important component of the winter diet of Ruffed Grouse.

Kalmia **Laurel** **Ericaceae**

T. Bodner (Apr)

Kalmia latifolia

T. Bodner (Apr)

Kalmia latifolia

T. Bodner (Jun)

Kalmia latifolia

Leucothoe Doghobble or Leucothoe Ericaceae

Evergreen or deciduous, solitary shrubs or colony forming. Leaves thick, alternate, elliptic to lanceolate, with short petioles, and serrate margins. Small white flowers dangling along bracted stalks in axillary spike-like clusters (racemes). Fruit a spherical capsule, 5-celled, splitting to release numerous winged or wingless seeds. 4 species in the SE.

Common Species

Leucothoe axillaris (Lam.) D. Don - **coastal doghobble** **LEAX**

Plant: Evergreen, arching-branched shrub, to 1.5 m tall, often forming dense colonies in Cp. Plant toxic to livestock, especially the leaves.
Stem: Arching with leaves horizontal, much branched, twigs green (somewhat zig-zagged), finely scurfy hairy becoming brown.
Leaves: Alternate, evergreen and thick, lanceolate to ovate, 6-14 cm long and 3-5 cm wide, margins thickened and finely serrate with sharp serrations, base rounded, tip short-pointed to tapering, shiny dark green above and pale green and finely hairy beneath, petioles 3-8 mm long, hairy, turning reddish.
Flowers: Mar-May. Axillary spike-like clusters (racemes), 2-7 cm long, each with tiny bracts, corolla 5-lobed, dangling, white to tinged with pink, 5-7 mm long, sepals broadly ovate to rounded.
Fruit and seeds: Sep-Jan. Spherical capsule (beaked by persistent pistil), 5-celled, 5 mm wide, within sepals, green ripening to dark brown with lighter seams, splitting to release many seeds, 1.5-2 mm long, reddish and shiny.
Range: Native. LA to n FL and north to se VA and west to TN, mainly Cp.
Ecology: Occurs on wet to moist sites and open to shady habitats. Inhabits swamp, floodplains, and pocosin forests. Apparently colonizes by rhizomes and spreads by water-dispersed seeds.
Synonyms: *L. axillaris* var. *ambigens* Fern., *L. catesbaei* (Walt.) Gray, *L. platyphylla* Small, fetterbush.

Leucothoe fontanesiana (Steud.) Sleumer - **highland doghobble** **LEFO**

Similar to *Leucothoe axillaris* except **plant** grows in Mt; **leaves** tips long tapering, longest petioles 8-15 mm long; **flowers** sepals tip pointed. **Range:** Native. Mt and upper Pd of n AL to n GA and north to WV and west to TN.
Ecology: Occurs in extensive colonies along streams and wet thickets.
Synonyms: *L. axillaris* (Lam.) D. Don var. *editorum* (Fern. & Schub.) Ahles, *L. editorum* Fern. & Schub., drooping leucothoe.

Other species in the SE: *L. racemosa* (L.) Gray [dog-hobble], *L. recurva* (Buckl.) Gray [recurved leucothoe or fetterbush].

Wildlife: Highland doghobble is browsed infrequently by White-tailed Deer in the Southern Appalachian Mt. Dense colonies of leucothoe along streams provide cover for various wildlife species including amphibians and small mammals. Leaves of *L. axillaris* poisonous to livestock.

T. Bodner (Jun)

Leucothoe axillaris

T. Bodner (Jun)

Leucothoe axillaris

Ligustrum **Privet** **Oleaceae**

Tall shrubs or small trees, semi-evergreen to evergreen with simple, opposite, entire leaves. White or cream-colored flowers in terminal panicles, often on short lateral branches. Fruit a small blue-black drupe in terminal clusters. 6 species in the SE, all exotics.

Common Species

Ligustrum sinense Lour. **- Chinese privet** **LISI**

Plant: Thicket-forming, semi-evergreen shrub, to 5 m tall, soft-woody, multi-stemmed with long leafy branches, from rootsprouts and often forming extensive colonies.
Stem: Opposite or whorled branched, branching increasing upward, twigs long slender and projecting outward at right-angle, brownish gray and short hairy (rusty or grayish) with light dots (lenticels), becoming gray-green, leaf scars raised half-round with 1 bundle scar, bark brownish-gray to gray and slightly rough (not fissured).
Leaves: Opposite in 2 rows at near right-angle to stem, semi-evergreen, ovate to elliptic, 2-4 cm long and 1-3 cm wide, rounded tip (tip often minutely indented), lustrous green above and pale green with hairy midvein beneath, petioles 1-5 mm long, rusty hairy.
Flowers: Apr-Jun. Terminal and upper axillary clusters on short branches forming panicles, fragrant, corolla 4-lobed, tube 1.5-3 mm long (equal or shorter than the lobes).
Fruit and seeds: Oct-Feb. Ovoid drupe, 6-7 mm long and 4 mm wide, pale green ripening to dark purple appearing almost black, hanging or projecting in dense clusters.
Range: Exotic, introduced from China in 1852. Naturalized and now common throughout the SE, even at higher elevations in Mt.
Ecology: Aggressive and troublesome exotic in most areas. Often forming dense thickets, particularly in bottomland forests, excluding hardwood regeneration. Common fencerow shrub, thus gaining access to forests, fields, and right-of-ways. Colonizes by rootsprouting and abundant seed production, spreads by abundant seeds dispersed widely by birds and other animals.

Ligustrum lucidum Ait. f. **- glossy privet** **LILU2**

Similar to *Ligustrum sinense* except **plant** larger to 6 m tall; **stem** round in cross-section, light gray and relatively smooth with no fissures, branches brownish gray with raised light dots (lenticels), twigs light green to gray and not hairy; **leaves** evergreen, leathery, larger, ovate, 8-15 cm long and 2.5-6 cm wide, tip acute, glossy black-green above and light green beneath, often with yellowish margins, petioles 1-2 cm long; **fruit and seeds** drupe, ellipsoid, 6-10 mm long and 5 mm wide. **Range:** Exotic, introduced from Japan in 1794. TX to FL and north to NC and west to TN.
Ecology: Same habitats as *L. sinense*, generally not as abundant depending upon location.

Other species in the SE: *L. amurense* Carr. [amur privet], *L. japonicum* Thunb. [Japanese privet], *L. ovalifolium* Hassk. [California privet], *L. vulgare* L. [common privet].

Wildlife: Dense thickets formed by Chinese privet provide cover for numerous wildlife species. Where abundant, young privet is an important, high-quality winter browse for White-tailed Deer. The persistent fruits are important winter and early spring food sources for Raccoon and numerous bird species including Northern Mockingbirds, Cedar Waxwings, Eastern Bluebirds, American Robins, and Hermit Thrushes. In spring, flocks of north-migrating American Robins (and likely other species) descend on privet thickets. Leaves, stems and bark are preferred Beaver forage. Privet thickets also provide important foraging sites for wintering American Woodcock.

Ligustrum **Privet** **Oleaceae**

J. Miller (Jun)

Ligustrum lucidum

J. Miller (May)

Ligustrum sinense

J. Miller (Jun)

Ligustrum lucidum

T. Bodner (May)

Ligustrum sinense

T. Bodner (Oct)

Ligustrum lucidum (left)
Ligustrum sinense (right)

J. Miller (Jan)

Ligustrum sinense

Lyonia Lyonia Ericaceae

Deciduous or evergreen rhizomatous shrubs. Leaves alternate, usually thick, with short petioles. Flowers in racemes arising from leaf axils or on leafless branches, corolla pink or white, bell-shaped or urn-shaped with 5 lobes, stamens 10. Fruit a capsule, 5-chambered, thickened along the sutures. Seeds small (1-3 mm), tan-brown. 5 species in the SE.

Common Species

Lyonia ligustrina (L.) DC. **- maleberry** **LYLI**

Plant: Deciduous shrub, usually 1-3 m tall, multi-stemmed with ascending branches, from rhizomes, occurring as scattered plants or loose colonies. Plant toxic to livestock, especially the leaves.

Stem: Young twigs flattened, finely hairy and tan, becoming reddish and light gray and hairless (variable in hairiness), numerous inconspicuous white dots (lenticels).

Leaves: Deciduous, alternate, variable, obovate to elliptic, 3-7 cm long and 1-3.5 cm wide, margins entire or with small scattered teeth (about 1 mm long) on lower half, tiny tipped end (mucronate), finely hairy when young becoming hairless above (somewhat shiny), whitish midvein and only hairy-veined beneath, petioles 1-5 mm long, hairy.

Flowers: Apr-Jun. Terminal or short lateral branching racemes, stalks hairy, corolla white, ovoid, 2.5-3 mm long, calyx lobes triangular, 0.5-1 mm long.

Fruit and seeds: Sep-Oct. Spherical capsule, hairy ribbed, 3-4 mm long, splitting to release narrow seeds, 1.5-2 mm long.

Range: Native. TX to c FL and north to ME and west to OH, KY, and OK, mainly Cp in the SE.

Ecology: Frequent scattered shrub in many forest types, especially moist to wet habitats. Plant very variable in foliage and hairiness. Colonizes by rhizomes and spreads by seeds.

Synonyms: male-blueberry, privet-andromeda, seedy-buckberry.

Lyonia lucida (Lam.) Koch **- fetterbush lyonia** **LYLU3**

Plant: Evergreen shrub, usually 1-3 m tall, from rhizome and often in small colonies.

Stem: Young twigs flattened, lime-green with loose blackish scales, mature twigs hairless, branches slightly ridged on opposite sides or 3-angled, brownish and black scaly, branching near base.

Leaves: Evergreen and firm, alternate, broadly elliptic, 3-9 cm long and 2-4 cm wide, margins thick with a paralleling vein, tips short tapering to a point, dark green and glossy above, lighter green and dull beneath with white midvein and minutely scattered black scaly, petioles 1-5 mm.

Flowers: Mar-Jun. Axillary clusters of nodding racemes from axils of persistent leaves of previous year, corolla pink or pink-white, ovoid, 6-8 mm long, calyx lobes lanceolate, pedicels hairless.

Fruit and seeds: Sep-Oct. Capsule, ovoid, 3.5-5 mm long, seeds minute, wedge-shaped.

Range: Native. MS and LA to FL and north to VA, mainly Cp.

Ecology: Occurs in wet to dry sites and open to shady habitats, including margins of cypress ponds, bogs and in flatwood forests.

Synonyms: staggerbush, hurrah-bush, shiny lyonia.

Lyonia Lyonia Ericaceae

J. Miller (Sep)

Lyonia ligustrina

D. Lauer (Aug)

Lyonia lucida

J. Miller (Sep)

Lyonia lucida

Lyonia **Lyonia** **Ericaceae**

Lyonia ferruginea (Walt.) Nutt. **- rusty staggerbush** **LYFE**
Plant: Evergreen shrub, usually 1-4 m tall, with several crooked trunks and spreading crowns, from rhizomes and often forming small and discontinuous colonies.
Stem: Irregularly branching, twigs scurfy-rusty scaly and short gray hairy, bark brown scaly in fissures.
Leaves: Alternate, elliptic to obovate, 2-7 cm long and 0.5-4 cm wide, new leaves smallest and scurfy rusty scaly (and gray scaly), becoming smooth on upper surfaces, margins entire and slightly rolled under, tip obtuse, petioles 1-7 mm, mature leaves with lateral veins impressed on the upper surface.
Flowers: Apr-May. Axillary clusters on twigs of previous year (short racemes), corolla white, 2-4 mm long, with short lobes, calyx lobes triangular, pedicels with rusty scales.
Fruit and seeds: Sep-Oct. Ovoid capsule, 5 sided, 3-6 mm long, seeds narrow, 2.7-3 mm long.
Range: Native. Central GA to n FL and SC, only Cp.
Ecology: Occurs in poorly-drained flatwoods and in oak-pine forests on well-drained sandy soils.
Synonyms: poor-grub.

Lyonia fruticosa (Michx.) G. S. Torr. **- coastal plain staggerbush** **LYFR3**
Similar to *Lyonia ferruginea* except **plant** only to 2 m tall, rigidly ascending stems, in dense to loose colonies; **stem** young twigs densely white hairy and pale green, becoming less hairy to hairless and golden brown; **leaves** whitish to very white beneath and veins protruding, only to 5 cm long, tip rounded; **fruit and seeds** capsule 3-5 mm long.
Synonyms: poor-grub.

Other species in the SE: *L. mariana* (L.) D. Don [staggerbush].

Wildlife: Flowers are an important nectar source. *Lyonia lucida* is a poor to fair White-tailed Deer browse. *Lyonia mariana* and *L. ligustrina* leaves are poisonous to livestock.

Lyonia Lyonia Ericaceae

J. Miller (Sep)

Lyonia fruticosa

J. Miller (Sep)

Lyonia ferruginea

T. Bodner (Sep)

Lyonia ferruginea

T. Bodner (May)

Lyonia ferruginea

Morella **Bayberry or Waxmyrtle** **Myricaceae**

Evergreen and semi-evergreen shrubs or small trees, with alternately spiraling leaves, crowded near branch tips, having entire margins or teeth above mid-leaf. Leaves of most species aromatic when crushed (except *M. inodora*). Flowers unisexual on the same or separate plants. Male flowers axillary in erect catkins up to 2 cm long. Female flowers axillary in rounded to cylindric catkins, 5-10 mm long. Fruit a spherical drupe, black and whitish waxy, clustered along branches. 4 species in the SE.

Common Species

Morella cerifera (L.) Small **- waxmyrtle** **MOCE2**

Plant: Evergreen shrub or small tree, usually 1-3 m tall (to 12 m tall), multi-stemmed, spreading branched, and leafy, often forming small to extensive colonies from rootsprouts.

Stem: Twigs slender, pale green becoming purplish- to grayish-brown, golden-brownish or grayish hairy, branches slightly warty with many raised lenticel dots and slightly raised leaf scars, being round to half-round with 1 bundle scar, upper bark sometimes sloughing in films, and lower bark becoming light-grayish, mostly smooth (not fissured) and often mottled with lichen.

Leaves: Alternately spiraling, evergreen and thin, narrowly oblanceolate with long tapering base, 3-8 cm long and 1-2 cm wide and smallest upward, lustrous above and whitish-green to golden beneath, golden-green especially when young due to tiny golden resin dots on both surfaces, but more beneath (requires hand lens), margins toothed mostly above mid-leaf (teeth minutely tipped). Leaves emit a fragrance of "Vicks" when crushed.

Flowers: Feb-Apr. Axillary in prior year's growth, bracted, male catkins solitary, long ovoid, 6-12 mm long and 4-6 mm wide, female catkins ovoid, 5-10 mm long, dense clusters, often reddish.

Fruit and seeds: Aug-Feb. Spherical drupes in clusters along twigs, each 2-4 mm wide, hard, dark bluish and white waxy (once used for candle wax, thus common name).

Range: Native. TX to s FL and north to NJ and west to AR and OK, mainly Cp and lower Pd.

Ecology: Nitrogen fixer (although not a legume). Occurs on wet to dry sites, open to semi-shady habitats, from dunes to bog margins to upland forests and right-of-ways. Vigorous resprouter following fire. Persists by rootstocks, spreads by rootsprouting and abundant animal-dispersed seeds.

Synonyms: *Myrica cerifera* L., *M. pusilla* Raf., *Cerothamnus ceriferus* (L.) Small, southern bayberry, candleberry.

Other species in the SE: *M. caroliniensis* (P. Mill.) Small, *M. inodora* (Bartr.) Small [odorless waxmyrtle], *M. pensylvanica* Loisel. [bayberry]. *Myrica gale* L. [rare in the SE - sweet gale; previously listed in genus].

Wildlife: Browsed sparingly by White-tailed Deer. Seeds are persistent into winter and are consumed by Northern Bobwhite, Wild Turkey, and waterfowl. Seeds also consumed by songbirds, including Gray Catbird, Brown Thrasher, Eastern Bluebird, Ruby-crowned Kinglet, White-eyed Vireo, Yellow-rumped Warbler, Eastern Towhee, Red-cockaded Woodpecker, and others. Fruits are high in fat and crude fiber but low in protein and phosphorus. They are a main food of wintering Tree Swallows in Florida.

Morella — Bayberry or Waxmyrtle — Myricaceae

T. Bodner (Oct)

Morella cerifera

T. Bodner (Jul)

Morella cerifera

T. Bodner (Apr)

Morella cerifera

Physocarpus **Ninebark** **Rosaceae**

Multi-stemmed, deciduous shrubs, having arching branches with brownish bark peeling or shredding in papery layers. Leaves alternate, ovate and mostly 3-lobed, with crenate or doubly-crenate margins. Small white to pinkish flowers, with 5 petals, sometimes in dense terminal clusters (corymbs). Capsules splitting to release few seeds. 1 species in the SE.

Common Species

Physocarpus opulifolius (L.) Maxim. **- ninebark** **PHOP**

Plant: Bushy deciduous shrub, to 3 m tall, leafy, multi-stemmed with erect or arching branches.

Stem: Ascending, twigs slender and slightly ridged from extensions of both sides of petiole bases, shiny golden- to grayish-brown with minute dark dots, smooth to densely hairy, leaf-scars slightly protruding with persistent petiole bases, branches initially erect becoming strongly arched, bark gray and peeling in papery layers to reveal smooth tan inner-bark.

Leaves: Alternate, ovate to heart-shaped in outline and often 3-lobed, with 2 prominent lateral veins and midvein, blades 4-7 cm long and 4-5 cm wide, margins crenate or doubly crenate, green and hairless above, whitish green with downy hairs and protruding veins beneath, petioles 0.5-2 cm long, petiole base slightly clinging around the stem.

Flowers: Apr-Jul. Terminal and often dense rounded clusters (racemes), on short branches of the current year, rose-type, petals 5, round, 5-7 mm long, white (to pinkish), stamens 25-30 in 2 whorls, nectar glands in the floral cup.

Fruit and seeds: Jun-Sep. Capsule smooth and shiny, pear-shaped, 2-3 mm long, tan to reddish, clustered on stalk, splitting in halves to release 2-4 seeds, each ellipsoid, 1.5-2 mm long.

Range: Native. AR and MO to n FL and north to VA and west to CO, all physiographic provinces in the SE.

Ecology: Occurs along stream and river banks and bog margins and moist cliffs. Spreads by water-dispersed seeds.

Synonyms: *P. intermedius* (Rydg.) Schneid., *Spiraea opulifolia* L.

Wildlife: Foliage consumed by White-tailed Deer.

Physocarpus Ninebark **Rosaceae**

T. Bodner (Jun)

Physocarpus opulifolius

T. Bodner (Apr)

Physocarpus opulifolius

Prunus **Plum or Cherry or Peach** **Rosaceae**

Deciduous or evergreen shrubs or small trees, with smooth shiny bark having horizontal markings. Leaves alternate, with reddish glands at base and/or along the serrate margins. Late-winter or early spring flowers, white to pinkish, rose-type, in showy clusters. Fruit a drupe, sweet or tart. 22 shrub species in the SE.

Common Species

Prunus umbellata Ell. **- hog or flatwoods plum** **PRUM**

Plant: Low branching shrub or small tree, usually 1-3 m tall, occurring in loose small colonies but not thickets.

Stem: Erect trunk, often with basal sprouts, twigs reddish and shiny (hairless), with scattered light dots (lenticels), branches dark purplish-black and dull, many short lateral branches lacking leaves, becoming somewhat sharp, bark smooth and shiny to wide fissuring with curling flakes.

Leaves: Alternate, elliptic, flat, 3-7 cm long and 2-3 cm wide, margins finely serrate to crenate, softly hairy (or hairless) above and beneath, protruding midvein beneath, often red dotted or blotched, tips tapering or pointed, leaf base rarely with 1-2 minute dark glands where it meets petiole. Leaves falling in dry summers.

Flowers: Feb-Apr. Axillary clusters of 2-5 flowers, along short twigs of prior season, on stalks 4-10 mm long, rose-type, petals 5, white, 0.8-1.2 cm long, many stamens, odorless.

Fruit and seeds: Aug-Sep. Plum, a drupe 1-1.3 cm long, green turning reddish and ripening to purplish black and whitish waxy, dangling on stems, tart, containing 1 seed.

Range: Native. TX to c FL and north to s NC and west to AR, only Cp and Pd.

Ecology: Occurs often in scattered groupings, on moist to dry sites, and open to shady habitats. Present in forest openings and along right-of-ways. Persists in new forest plantations and occurs in the understory of mature upland forests. Spreads by animal-dispersed seeds.

Prunus angustifolia Marsh. **- Chickasaw plum** **PRAN3**

Similar to *Prunus umbellata* except **plant** small upright shrub forming thickets; **stem** with small lateral twigs, rigid and thorn-tipped; **leaves** often folded and turning upward, somewhat thick and hairless, elliptic to lanceolate, 4-7 cm long 1-2.5 cm wide, margins serrate with red gland tips or not; **flowers** (Mar-Apr) 1-2 in fascicles, fragrant; **fruit** (May-Jul) plum red to yellow, 2-3 cm long, sweet. **Range:** Native. TX to FL and north to NJ and west to KS, mainly Cp and Pd in the SE.

Ecology: Occurs in thickets along right-of-ways, forest margins, and in old fields. Colonizes by rootsprouting and spreads by animal-dispersed seeds.

Other shrub species in the SE: *P. alabamensis* C. Mohr, *P. alleghaniensis* Porter, *P. americana* Marsh., *P. avium* (L.) L., *P. caroliniana* (P. Mill.) Ait., *P. cerasus* L., *P. geniculata* Harper, *P. glandulosa* Thunb., *P. gracilis* Engelm. & Gray, *P. hortulana* Bailey, *P. mahaleb* L., *P. mexicana* S. Wats., *P. munsoniana* W. Wight & Hedrick, *P. myrtifolia* (L.) Urban, *P. persica* (L.) Batsch [peach], *P. pumila* L., *P. rivularis* Scheele, *P. spinosa* L., *P. triloba* Lindl., *P. virginiana* L.

Wildlife: Plums are important soft mast producers in early to mid-spring, and eaten by White-tailed Deer, Black Bear, Gray Fox, Raccoon and Virginia Opossum. The fruit is low in protein, but moderate in phosphorus and calcium content. Thickets provide excellent escape and nesting cover for the Northern Bobwhite and songbirds such as the Northern Mockingbird, Brown Thrasher, and Gray Catbird. Foliage is a low preference White-tailed Deer browse.

Prunus **Plum or Cherry or Peach** **Rosaceae**

T. Bodner (Jun)

Prunus umbellata

J. Miller (Feb)

Prunus umbellata

J. Miller (Oct)

Prunus angustifolia

T. Bodner (Jul)

Prunus umbellata

Quercus | **Shrubby oak** | **Fagaceae**

Deciduous or evergreen trees and shrubs, with alternate leaves, often lobed or cleft, having short petioles. Male and female flowers separate on the same plant. Male flowers drooping catkins in clusters. Female flowers clustered in spikes. Fruit an acorn, with a cap, falling in the first or second year. About 35 species in the SE, with 4 being shrubby.

Common Species

Quercus pumila Walt. **- running oak** **QUPU80**

Plant: Tardily deciduous shrub, 1-2 m tall, from underground running stems and having erect branches, often forming extensive clonal colonies.

Stem: Low and spreading, densely branched, twigs brownish to grayish and hairy, eventually becoming hairless.

Leaves: Alternate, narrowly elliptic to oblong, 4-8 cm long and 0.7-2 cm wide (much larger from new sprouts), shiny green and hairless above and densely grayish hairy beneath, bristle tipped, margins mostly entire, petioles 1-3 mm long.

Flowers: Mar-Apr. Oak-type, male drooping catkins and female clustered spikes.

Fruit and seeds: Sep-Feb. Acorn about 1 cm long, hairy, cup about 1 cm wide and covering about 1/3-1/2 of the acorn, containing 1 seed.

Range: Native. MS to c FL and north to s NC, mainly Cp, rare in Pd.

Ecology: Occurs in sandy pine forests (wet or dry) as colonies in mixtures with *Q. minima*. Resprouts readily after burning. Colonizes by running underground stems and spreads by animal-dispersed seeds.

Other shrub species in the SE: *Q. minima* (Sarg.) Small [CP - creeping live oak], *Q. myrtifolia* Willd. [Cp - myrtle oak], *Q. prinoides* Willd. [Pd and Mt - dwarf chinkapin oak]

Wildlife: The acorns of the shrubby oak species are available to wildlife before most of the tree oaks. The acorns are readily used by White-tailed Deer, Wild Turkey, Black Bear, Gray Squirrel, and numerous other wildlife species. Although individual stems may produce only a few acorns, collectively the understory colonies can produce abundant hard mast. However, mast production is not consistent among years. In general, acorns are low in protein but high in crude fat or carbohydrates.

Quercus | **Shrubby oak** | **Fagaceae**

T. Bodner (Sep)

Quercus pumila

T. Bodner (Sep)

Quercus pumila

Rhododendron — Rhododendron or Azalea — Ericaceae

Evergreen or deciduous shrubs or small trees, with showy 5-lobed flowers of many bright colors in spring or early summer. Flowers in terminal or near-terminal clusters on prior year's shoots. Leaves alternate, often thick, densely hairy or not, with entire margins. Fruit an upright cylindric capsule with 5 cells, many seeds in rows. 17 species in the SE.

Common Species

Rhododendron maximum L. **- rosebay or great laurel** **RHMA4**

Plant: Upright, evergreen shrub or small tree, to 10 m tall, often in colonies from rootsprouts. Plant toxic to livestock, especially the leaves.

Stem: Erect trunk and densely branching, twigs stout and brownish hairy, becoming hairless, many slit-shaped leaf scars near twig tips, bark semi-rough and dark brown.

Leaves: Alternately spiraling, thick and leathery, narrowly elliptic to oblanceolate, 10-30 cm long and 2-8 cm wide, margins entire and rolled under, shiny dark green above and pale green (sometimes downy hairy) beneath, tip pointed, petioles 1-4 cm long, hairy.

Flowers: Jun-Aug. Terminal clusters of 10-30, each 3-4 cm wide, corolla 5-lobed (each rounded), pinkish-white with a yellowish to greenish spot in the largest lobe, stamens 10.

Fruit and seeds: Sep-Feb. Cylindric capsule, 1-1.5 cm long, downy hairy, splitting to release many seeds, each 1.5-2 mm long.

Range: Native. North AL to n GA and north to NY and west to OH, mainly Mt in the SE.

Ecology: Occurs in moist forests and along stream and river banks, often in dense thickets.

Synonyms: white laurel.

Rhododendron canescens (Michx.) Sweet **- wild azalea** **RHCA7**

Plant: Upright, deciduous shrub, to 4 m tall, often from rootsprouts.

Stem: Twigs slender, very coarsely hairy or not, brownish to gray.

Leaves: Alternately spiraling, elliptic to oblanceolate, 3-9 cm long and 2-4 cm wide, finely hairy above, velvety hairy with long hairs on the midvein beneath, tiny tipped.

Flowers: Feb-May. Terminal clusters, corolla pink or white, 1.5-2.5 cm long, hairy and glandular.

Fruit and seeds: Sep-Dec. Cylindric capsule, oblong, 1.4-1.7 cm long, densely hairy.

Range: Native. TX to FL and north to DE and west to AR, only Cp and Pd in the SE.

Ecology: Occurs widely on wet to moist sites, along streams and bogs to bluffs.

Synonyms: piedmont azalea, hoary azalea, sweet pinxter azalea, southern pinxterbloom.

Rhododendron austrinum (Small) Rehd. **- yellow or orange azalea** **RHAU**

Similar to *Rhododendron canescens* except **stem** twigs reddish-brown, flower buds gray hairy in winter; **leaves** elliptic to obovate, 3-11 cm long and 2-4.5 cm wide, densely hairy both surfaces with dark dotted glands; **flower** (Feb-Apr) yellow to orange to red, 2.5-4.5 cm long. **Range:** Native. South MS, s and c AL, sw GA, and FL.

Synonyms: Florida azalea.

Wildlife: Although a low quality forage, leaves of rosebay are an important component of the winter diet of White-tailed Deer in the Southern Appalachians, particularly in years when acorn mast is scarce. Leaves, twigs, buds and seeds are taken infrequently by Ruffed Grouse. Rhododendron thickets provide excellent escape cover for White-tailed Deer, Black Bear, Eastern Cottontail, Snowshoe Hare, Ruffed Grouse, Wild Turkey, and many songbirds. The deciduous azaleas are low preference deer browse species. Azaleas are frequented by butterflies, especially swallowtails, Gulf Fritillaries, and migrating Monarchs.

Rhododendron **Rhododendron or Azalea** **Ericaceae**

T. Bodner (Mar)

Rhododendron canescens

T. Bodner (Jun)

Rhododendron maximum

J. Miller (Oct)

Rhododendron austrinum

T. Bodner (Mar)

Rhododendron austrinum

Rhus **Sumac** **Anacardiaceae**

Deciduous and upright small shrubs, tall shrubs, or small trees, not poisonous (see *Toxicodendron* if poisonous). Leaves pinnately compound, with 3-31 leaflets, often on many long branch-like stalks. Small 4-6 petaled flowers in terminal conical panicles. Fruit a tiny dry drupe in the terminal clusters, ripening to red hairy in the fall. Most spreading by rhizomes. 5 species in the SE.

Common Species

Rhus copallinum L. **- winged sumac** **RHCO**

Plant: Upright stout-stemmed shrub or small tree, usually 1-2 m tall (to 7 m tall) with long pinnately compound leaves having a winged stalk and spirally arranged at the end of upward turned branches, forming colonies by rhizomes.
Stem: Erect trunk, twigs golden to brownish hairy, becoming hairless below, leaf scars heart-shaped, bark smooth and glossy to dull. Clear sap when injured.
Leaves: Alternately spiraling, pinnately compound, 10-30 cm long, leaflets 9-23, with leafy wings along stalk, leaflets elliptic to lanceolate and often asymmetric, 3-8 cm long and 1-3 cm wide, margins entire (or serrate), shiny green and hairless or slightly hairy above and pale green and hairy beneath, leaflet petioles short.
Flowers: Jun-Sep. Large terminal conical panicles, 50-150 cm long, often with hundreds of small greenish to white flowers having a red tinge, petals 5.
Fruit and seeds: Aug-Jan. Drooping panicles with tiny spherical drupes, dry, 4-5 mm wide, dull red hairy when ripe, containing 1 seed, 2.5-3 mm long.
Range: Native. Throughout Eastern US, all physiographic provinces in the SE.
Ecology: Commonly occurring shrub in new forest plantations and along forest right-of-ways and margins. Abundance may increase following site preparation burns. Present in open forests. Occurs on moist to dry sites and open to semi-shady habitats. Colonizes by rhizomes and spreads by seeds dispersed by birds and other animals.
Synonyms: dwarf sumac, shining sumac, flameleaf sumac.

Rhus glabra L. **- smooth sumac** **RHGL**

Similar to *Rhus copallinum* except **stem** hairless, twigs whitish waxy becoming dotted with raised dots (lenticels), milky sap; **leaves** no wing on stalk, often reddish, leaflets 11-31, each 5-15 cm long and 1-4 cm wide, margins serrate, dull green above and whitish waxy green beneath; **flowers** (May-Jul) yellowish green; **fruit and seeds** (Jul-Feb) drupe bright red, hairy and sticky, 3-4 mm long. **Range:** Native. Throughout Eastern US and s Canada.
Ecology: Less frequent than winged sumac, but occurring on same sites. Burning enhances seed germination.
Synonyms: *R. borealis* Greene, *R. calophylla* Greene.

Rhus hirta (L.) Sudworth **- staghorn sumac** **RHHI2**

Similar to *Rhus glabra* except **plant** larger, to 10 m; **stem** twigs densely hairy; **leaves** hairy stalks and beneath leaflets. **Range:** Native. Mt of GA, SC, NC, TN, VA, and north to Nova Scotia and west to MN.
Synonyms: *R. typhina* L.

Rhus **Sumac** **Anacardiaceae**

T. Bodner (May)

Rhus glabra

T. Bodner (Aug)

Rhus copallinum

T. Bodner (Jul)

Rhus glabra

T. Bodner (Jul)

Rhus glabra

Rhus **Sumac** **Anacardiaceae**

Rhus aromatica Ait. **- fragrant sumac** **RHAR4**

Plant: Low ascending shrub, 0.5-2 m tall, often in patches, resembling poison oak with 3-leaflet leaves and having toothed margins, sweet scented when crushed (unlike poison oak), sprouting from spreading rhizomes.
Stem: Upright to ascending or drooping, sparsely branched, twigs slender and brown and hairy, becoming hairless and lustrous turning dull with lenticel dots, leaf scars raised and angled.
Leaves: Alternate, leaflets 3, moderately thick, leaflets ovate to obovate (lateral leaflets asymmetric), 3-8 cm long and 2-5 cm wide, coarsely toothed above mid-leaflet, pointed or blunt tipped, lateral leaflets rounded at base and sessile, smooth above and short hairy or not beneath, petioles 2-3.5 cm long. Leaves often sweet-scented when crushed.
Flowers: Feb-May. Short dense catkin-like clusters, near twig tips, appearing late-summer to fall of year prior to flowering, each 1-2 cm long, corolla yellowish-green, opening before or during leaf expansion.
Fruit and seeds: Apr-Sep. Spherical drupe, 5-7 mm wide, densely red hairy.
Range: Native. TX to w FL and north to VT and Que and west to IL and NE, mainly Pd and Hilly Cp in the SE.
Ecology: Occurs on dry sites, in open or semi-shady habitats, often along right-of-ways and forest margins. Forms small to extensive colonies in new forest plantations and open forests, on sandy or rocky soils, and igneous intrusions. Colonizes by rhizomes and spreads by animal-dispersed seeds.

Other species in the SE: *R. michauxii* Sarg. [rare and endangered, Cp and Pd of GA, SC, and NC - Michaux's sumac].

Wildlife: Fruiting heads remain on the plant into winter and are available when more desirable foods are scarce. Fruits, particularly smooth and staghorn sumac, are consumed by White-tailed Deer, Virginia Opossum, Northern Bobwhite, Ruffed Grouse, and Wild Turkey. Fruit also consumed by numerous birds, including woodpeckers, thrushes, vireos, Eastern Bluebirds, Ruby-crowned Kinglets, Gray Catbirds, and others. The bark is an emergency winter forage for the Eastern Cottontail.

Rhus Sumac Anacardiaceae

J. Miller (Oct)

Rhus aromatica

T. Bodner (May)

Rhus aromatica

J. Miller (Oct)

Rhus aromatica

Rosa Rose Rosaceae

Deciduous or evergreen shrubs, with upright, arching, or trailing prickly stems. Leaves pinnately compound, margins serrate or crenate, and base of petioles winged (stipules). Showy rose flowers, solitary or in clusters, 5-petaled, pink or white. Fruit a somewhat fleshy hip containing several seeds. 6 native or naturalized species in the SE, and 6 more rare escapes.

Common Species

Rosa carolina L. - wild rose ROCA4

Plant: Low prickly shrub, 0.5-1.5 m tall, forming small groups by underground runners.
Stem: Erect or slightly arching, sparsely branched, twigs round slender (slightly zig-zagged), hairless, with few to many slender straight spines, 5-10 mm long, branches becoming dull green and mottled, plant base often reddish brown.
Leaves: Alternate, leaflets 3-7, leaflets elliptic to oval, 1.5-3 cm long and 0.8-2 cm wide, terminal leaflet long petioled and laterals leaflets short petioled, margins serrate (often entire near base), dull to glossy green above and whitish green (sometimes downy) beneath, stipules along base of leafstalks, narrow with flared tips.
Flowers: May-Jun. Solitary at terminal, rose-type, petals 5, pink, 2-3 cm long, numerous yellow stamens, sepals 5 (persistent on early fruit), long narrow and tapering tipped, sepals and stalk bristling with red gland-tipped hairs (requires hand lens).
Fruit and seeds: Aug-Dec. Rose hip, spherical and fleshy, about 1 cm wide, green to yellow and ripening to glossy red, sepals present or not at end.
Range: Native. TX to n FL and north to s Canada and west to MN, throughout, rare in Middle or Lower Cp in the SE.
Ecology: Occurs on moist to dry sites, and in open to semi-shady habitats. Commonly grows in the understory of open mixed forests and new forest plantations. Present along forest margins and shaded parts of right-of-ways. Colonizes by rhizomes and spreads by animal-dispersed seeds.
Synonyms: *R. serulata* Raf., *R. texarkana* Rydb., pasture rose, low rose, Carolina rose.

Rosa multiflora Thunb. *ex* Murr. - multiflora rose ROMU

Similar to *Rosa carolina* except **plant** large and arching, clump forming, with large downward curving thorns with wide bases; **leaves** leaflets 7, stalks hairy (some gland tipped), thorny, channeled, stipules pronounced, extending 3-15 mm along the stalk and out 2-6 mm, margins bristle toothed; **flowers** terminal branching clusters on lateral branchlets, petals 5, white, heart-shaped, 1-1.5 cm long. **Range:** Exotic from China. Throughout Eastern US.
Ecology: Widely planted for "living fences," escaping along right-of-ways and invading new forests and forest margins. Infests pastures and old fields. Colonizes by prolific sprouting and runners, and spreads by animal-dispersed seeds.

Other common species in the SE: *R. bracteata* J.C. Wendl. [Macartney rose], *R. laevigata* Michx. [Cherokee rose], *R. palustris* Marsh. [swamp rose], *R. setigera* Michx.

Wildlife: Wild roses are a preferred White-tailed Deer browse. The fleshy fruits (hips) remain on the shrubs through winter and are available when other more preferred foods are scarce. Ruffed Grouse and Wild Turkey feed on the fruits along with at least 38 species of songbirds. Fruits are consumed by Eastern Cottontail and several rodent species. Thickets of wild roses, especially multiflora rose, provide excellent nesting and escape cover for gamebirds and songbirds.

Rosa **Rose** **Rosaceae**

J. Miller (Sep)

Rosa carolina

J. Miller (May)

Rosa carolina

T. Bodner (Jun)

Rosa multiflora

J. Miller (May)

Rosa multiflora

Sambucus Elderberry Caprifoliaceae

Deciduous, multi-stemmed and soft-stemmed shrubs, from stolons, often in colonies. Leaves opposite and pinnately compound. Many small flowers in terminal flat or conical clusters (cymes), 5-lobed corolla and calyx. Fruit a small berry with 4 seeds. 3 species in the SE.

Common Species

Sambucus canadensis L. - elderberry SACA12

Plant: Multi-stemmed upright shrub, to 4 m tall, forming colonies by underground runners.
Stem: Erect trunks, twigs light-green and herb-like, becoming grayish brown with warty lenticel dots, long internodes, pith white.
Leaves: Opposite, pinnately compound, leaflets 5-11, 10-18 cm long, leaflets lanceolate to ovate, 5-15 cm long and 3-6 cm wide, margins finely serrate, tapering tipped, pale green beneath (sometimes downy), sometimes stalked glands on small leaflets at leaf base.
Flowers: Apr-Jul (infrequently later). Flat- or curved-topped terminal clusters (cymes),10-20 cm wide, corolla 5-lobed, white, 3-5 mm wide.
Fruit and seeds: Jul-Oct. Spherical berry, 4-6 mm long, green ripening to purplish black, juicy, containing 3-5 seeds.
Range: Native. Throughout Eastern US, all physiographic provinces in the SE.
Ecology: Occurs commonly on wet to moist sites, in open habitat. Forms small colonies around swamp and stream margins, and along ditches and disturbed areas. Colonizes by rhizomes and stolons, and spreads by animal-dispersed seeds.
Synonyms: *S. simpsonii* Rehd. *ex* Sarg., common elder.

Other species in the SE: *S. nigra* L. ssp. *canadensis* (L.) R. Bolli, *S. racemosa* L. [formerly *S. pubens* Michx. - high elevations in Mt].

Wildlife: Elderberry fruit is an important soft mast in late summer and early fall for more than 50 species of songbirds, particularly the American Robin, Northern Mockingbird, and Gray Catbird. Wild Turkey, Northern Bobwhite, Ruffed Grouse, and Mourning Dove also consume the fruit. White-tailed Deer occasionally feed on the foliage during summer although preference may vary across the Southeast.

Sambucus Elderberry Caprifoliaceae

J. Miller (Jun)

Sambucus canadensis

T. Bodner (May)

Sambucus canadensis

T. Bodner (Jul)

Sambucus canadensis

T. Bodner (May)

Sambucus canadensis

Symplocos **Sweetleaf** **Symplocaceae**

Tardily deciduous, tall shrub or small tree, with leaves simple and alternate, somewhat leathery, having a sweet taste (thus the common name). Showy spring flowers, yellow, tufted in leaf axils. Green drupes in tight axillary clusters. 1 species in the SE.

Common Species

Symplocos tinctoria (L.) L'Her. **- sweetleaf** **SYTI**

Plant: Upright, tardily-deciduous shrub or small tree, to 4 m tall, one to few stems and sparsely branched (tree-like in form).

Stem: Twigs stout, grayish hairy becoming reddish-brown and gray waxy, waxy coating decreasing with age, somewhat ridged but round in cross-section, branching appearing opposite, leaf scars half-round and tan, lateral buds small and somewhat embedded, bark smooth dark brown, pith chambered.

Leaves: Alternate, somewhat thick and persisting into winter, narrowly elliptic with tapering tips and bases, 5-15 cm long and 3-5 cm wide, petiole and midvein (and a few lateral veins) whitish or yellowish green, margins entire or very swallow crenate (slightly rolled under), dark green and short-stiff hairy above and whitish green and densely short hairy beneath, petioles about 1 cm long. Older leaves may be slightly sweet tasting (especially near midvein).

Flowers: Mar-May. Showy tuft-like clusters in leaf axils of previous year's branches, petals 5, 6-8 mm long, yellow or light yellow with many light-yellow stamens, fragrant.

Fruit and seeds: Aug-Nov. Small dry drupe, in bracted stemless clusters, green to orange brown, each 8-12 mm long and 3-6 mm wide, containing 1 seed.

Range: Native. TX to n FL and north to e VA and west to AR and OK, mainly Cp and Southern Appalachian Mt, infrequent in Pd of the SE.

Ecology: Occurs on wet to dry sites, and open to shady habitats. Widely scattered plants occur in upland to bottomland forests, and along stream margins. Does not occur on range with cattle or horses due to overgrazing. Spreads by seeds.

Synonyms: horse-sugar, yellowwood.

Wildlife: A moderate preference White-tailed Deer browse in the Southern Appalachian Mountains.

T. Bodner (Mar)

Symplocos tinctoria

T. Bodner (Oct)

Symplocos tinctoria

Vaccinium Vaccinium or Blueberry Ericaceae

Spreading-branched, leafy shrubs to small trees, deciduous or evergreen, with alternate leaves having short petioles. Small flowers in various arrangements, on current or prior-year's twigs, appearing in spring. Corolla tube white to greenish to reddish, 5 lobed, calyx 4-5 lobed and persistent on the fruit. Fruit a spherical berry, greenish to reddish to black, containing 5 to many seeds. 18 species in the SE, with many hybrids, some of commercial use.

Common Species

Vaccinium elliottii Chapman - **Elliott's blueberry** **VAEL**

Plant: Erect branchy shrub, deciduous, 1-3 m tall, bushy with small shiny leaves, may appear in patches but not from runners.

Stem: Ascending, slender twigs, round in cross-section, light-green, somewhat zig-zagged, projecting outward or arching, rough downy hairy or not, branches with transition areas being smooth green streaked and tan-brown fissured, bark on trunk grayish and shredding in thin plates.

Leaves: Alternate, small and thick, elliptic to oval, 1-3 cm long and 0.6-1.5 cm wide, shiny green above and paler with fine rough hairs beneath, margins finely serrate (each serration with tiny gland-tipped hairs when new), tips blunt to pointed, petioles short and often reddish. Leaves often turning bright red in fall.

Flowers: Mar-Apr (before and as leaves expand). Cluster of 3-5 on previous year's twigs, corolla urn- or vase-shaped, white to pinkish, 5-8 mm long and 2-4 mm wide.

Fruit and seeds: May-Jun. Spherical berry, 5-10 mm wide, green ripening to bluish-black to black, occasionally whitish waxy, with persistent calyx crown, juicy and sweet when ripe, containing many tiny seeds.

Range: Native. TX and AR to n FL and north to se VA, mainly Cp and lower Pd.

Ecology: Abundant where present, on wet to dry sites and in open to semi-shady habitats. Persists in new forest plantations and right-of-ways, and occurs in upland and bottomland open forests. Readily resprouts after burning. Often in patches or extensive colonies, but not from rhizomes. Seeds dispersed by birds and other animals.

Synonyms: huckleberry, mayberry.

Vaccinium myrsinites **Lam. - shiny blueberry** **VAMY3**

Similar to *Vaccinium elliottii* except **plant** low evergreen shrub, only 20-60 cm tall, in colonies from runners; **leaves** tiny, oblanceolate to elliptic, mostly 5-15 mm long and 2-10 mm wide, shiny green to grayish green; **fruit** (May-Jun) berry black and often whitish waxy, 4-8 mm wide, juicy and sweet when ripe. **Range:** Native. South AL, s GA, FL, and se SC. **Ecology:** Abundant where present, in wet to dry forests, usually open habitat, especially pine forests and scrub areas.

Synonyms: evergreen blueberry.

T. Bodner (Apr)

T. Bodner (May)

Vaccinium elliottii

J. Miller (Jun)

Vaccinium elliottii

J. Miller (Jun)

Vaccinium myrsinites

T. Bodner (Jun)

Vaccinium elliottii

Vaccinium **Vaccinium or Blueberry** **Ericaceae**

Vaccinium stamineum L. **- deerberry** **VAST**

Plant: Multi-stemmed, deciduous shrub, usually 1-2 m tall (to 5 m tall), leafy with many spreading branches, from runners and forming loose colonies of somewhat scattered plants.
Stem: Erect or short with spreading branches (often horizontal at one height), twigs green to gray-green, downy hairs or not, trunks brownish to grayish scaly.
Leaves: Alternate, deciduous, variable, thin to leathery, elliptic to oblanceolate, 3-8 cm long and 2-4 cm wide, blades often slightly wavy or arching, softly hairy or not, margins mostly entire or few scattered gland-tipped teeth near base, sometimes with gland-tipped hairs appearing also on stems, flower stalks, and flowers.
Flowers: Mar-Jun. Drooping clusters with leaf-like bracts, corolla vase-shaped with 5 lobes, white, enlarging with age to 5-8 mm long, stamens protruding.
Fruit and seeds: Jul-Aug. Dangling spherical berry, 1-1.5 cm wide, green to pink to yellowish or purple, calyx tipped, mostly sparsely hairy (or not), tartly sweet or bitter.
Range: Native. TX to n FL and north to NY and west to IN and MO.
Ecology: Abundant where present. Occurs in moist to dry forests, open to shady habitats. Very variable in leaf and fruit. Persists in new forest plantations and frequents open pine-hardwood forests. Spreads by runners and seeds dispersed by birds and other animals.
Synonyms: *V. neglectum* (Small) Fern., squaw-huckleberry, southern gooseberry, buckberry.

Vaccinium arboreum Marsh. **- sparkleberry** **VAAR**

Plant: Erect, tardily-deciduous shrub or small tree, usually 2-4 m tall (to 10 m tall), with a single trunk or multi-stemmed and leafy.
Stem: Twigs round in cross-section, somewhat zig-zagged, brownish-gray becoming shiny brown with light lenticel dots, bark grayish brown and sometimes slightly purplish, sloughing in irregular thin plates or flakes with patches of smooth reddish-brown inner-bark.
Leaves: Alternate, evergreen or tardily deciduous, thick, oval to obovate, 2-5 cm long and 1-3 cm wide, margins entire or finely toothed, tiny pointed tip (mucronate) or not, shiny green above and whitish green beneath, often spotted with golden rust infections. Turning golden, pinkish, or reddish in the fall.
Flowers: Mar-Jun. Leafy-bracted racemes or sprays, corolla white, rounded urn-shaped with 5 lobes, 3-7 mm long.
Fruit and seeds: Aug-Feb. Spherical berry, 5-8 mm wide, green turning purplish to lustrous black, dry and mealy.
Range: Native. TX to FL and north to VA and west to IN and KS.
Ecology: Occurs commonly in dry forests, thickets, right-of-ways, and new forest plantations, acid to calcareous soils, sandy and rocky sites.
Synonyms: farkleberry, tree-huckleberry, winter-huckleberry.

Wildlife: Vaccinium fruit are a major component in the summer diet of Ruffed Grouse and Wild Turkey. Fruit also are consumed frequently by Northern Bobwhite, Scarlet Tanager, American Robin, Northern Cardinal, Eastern Bluebird, Brown Thrasher, and others. Black Bear make heavy use of the fruits, as do several small mammal species. Sparkleberry apparently is not as preferred as other species, but persists into winter when other more preferred forages are scarce. Vaccinium leaves are a moderate preference White-tailed Deer forage, although use varies across the Southeast, with highest use occurring in the flatwoods of the Lower Coastal Plain. Vacciniums are the preferred larval food plant of the Henry's Elfin butterfly.

Vaccinium **Vaccinium or Blueberry** **Ericaceae**

T. Bodner (Aug)

Vaccinium stamineum

J. Miller (May)

Vaccinium stamineum

T. Bodner (Oct)

Vaccinium arboreum

J. Miller (Oct)

Vaccinium arboreum

Viburnum — Haw or Viburnum or Arrowwood — Caprifoliaceae

Deciduous upright shrubs or small trees, with opposite leaves and branching. Leaf scars V-shaped or 3-lobed. Tiny 5-petaled flowers, white or pinkish, in terminal flat-topped clusters (cymes). Fruit a small 1-seeded drupe. 11 species in the SE.

Common Species

Viburnum nudum L. - **possumhaw** — VINU

Plant: Erect leafy and branchy shrub, usually 1-2 m tall (to 5 m tall), multi-stemmed and much branched.
Stem: Twigs green with rusty scurfy scales, becoming grayish brown and warty, winter buds with slender sheaths, having reddish-brown scales.
Leaves: Opposite, thick, oblong to elliptic, 5-15 cm long and 2-6 cm wide, margins finely serrate to crenate (or not) and usually rolled under, shiny green with a lighter midvein having rusty scales (dotted) above, scurfy hairy along the midvein beneath (densely dotted), petioles finely winged, 5-22 mm long.
Flowers: Apr-May. Terminal, flat-topped dense clusters (cyme), 4-5 branches, 8-15 cm wide, petals 5, white.
Fruit and seeds: Aug-Dec. Ovoid drupe, dry, 6-10 mm long, yellow to pink and ripening to deep blue and whitish waxy, containing 1 seed, flattened ovoid, 5-8 mm long.
Range: Native. TX to c FL and north to Ont and southwest to KY and AR.
Ecology: Occurs on moist to wet sites and open to shady habitats. Common in bottomland and wet forests, bogs and bays, and exposed rocks in Mt. Spreads by animal-dispersed seeds.
Synonyms: *V. cassinoides* L., *V. nitidum* Ait., swamphaw, witherrod, nanny-berry.

Viburnum rufidulum Raf. - **rusty blackhaw** — VIRU

Similar to *Viburnum nudum* except **stem** light reddish gray with rusty scales, becoming gray with scattered reddish lenticel dots, bud scales maroon velvety hairy, bark increasingly fine platy fissured; **leaves** elliptic to oval, broadest above the middle, 5-8 cm long and 3-5 cm wide, shiny above, reddish scales on midvein above and beneath and on petioles (often reddish), margins finely serrate; **flowers** (Mar-Apr) clusters 5-10 cm wide, white; **fruit and seeds** (Sep-Dec) drupe 10-15 mm long. **Range:** Native. TX to c FL and north to s VA and west to IL and MO, except Mt.
Ecology: Occurs on dry sites in open to shady habitats. Common as scattered plants in upland forests in Pd and Cp, and along forest margins and right-of-ways.

Viburnum **Haw or Viburnum or Arrowwood** **Caprifoliaceae**

T. Bodner (May)

Viburnum nudum

T. Bodner (Jun)

Viburnum nudum

J. Miller (Oct)

Viburnum rufidulum

T. Bodner (Sep)

Viburnum rufidulum

Viburnum **Haw or Viburnum or Arrowwood** **Caprifoliaceae**

Viburnum acerifolium L. - **mapleleaf viburnum** **VIAC**

Plant: Erect, slender shrub, 1-2 m tall, multi-stemmed, leafy and much branched, forming colonies from underground runners.
Stem: Twigs green and densely white hairy, becoming reddish brown, bark brown and smooth.
Leaves: Opposite, mostly 3-lobed (and 3-veined), resembling a maple leaf (thus the common name), 7-14 cm long and 6-12 cm wide, margins irregularly toothed, bright green with sparse straight hairs above, pale green and hairy beneath with scattered reddish dots (sometimes hairy tufts in vein axils), bases rounded to cordate, tips pointed or blunt, petioles 1-4 cm long (often reddish), with stipules (2-3 mm long) on the base or not.
Flowers: Apr-May. Terminal flat-topped clusters (cyme), with 6-8 branches, on stalks with 1 leaf pair, clusters 2-6 cm wide, petals 5, white.
Fruit and seeds: Aug-Dec. Ovoid drupe, 8-9 mm long, green to pink and ripening to purplish black.
Range: Native. TX to w FL and north to s Canada and west to MN, infrequent in Hilly Cp and not in Lower Cp.
Ecology: Occurs on moist to dry sites and in open to shady habitats. Common to mountain forests and margins, forming colonies. Colonizes by rhizomes and spreads by seeds.
Synonyms: *V. densiflorum* Chapman, arrowwood.

Viburnum dentatum L. - **arrowwood** **VIDE**

Similar to *Viburnum acerifolium* except **plant** medium shrub or small tree, 1-3 m tall, often with single stem; **stem** twigs straight, rough shaggy hairy and becoming finely warty, bark tight gray-brown to reddish; **leaves** broadly ovate or elliptic, 6-12 cm long and 5-10 cm wide, margins wavy serrate to almost dentate, base rounded to cordate, tapering tipped, green with scattered rough hairs above and pale scurfy hairy (or not) beneath, scurfy petioles, 1-3 cm long, often red; **flower** (Mar-Jun) cymes with 5-7 branches, 4-10 cm wide, white; **fruit and seeds** ellipsoid drupe, 6-9 mm long, blue-black. **Range:** Native. TX to c FL and north to ME and west to IL, mainly Cp and infrequent Pd in the SE.
Ecology: Occurs on wet to dry sites, and open to semi-shady habitats. Inhabits river and stream banks, low flatwoods, and upland forests. Occurs as scattered plants. Spreads by animal-dispersed seeds.

Other species in the SE: *V. bracteatum* Rehd. [protected in GA], *V. lantanoides* Michx. [formerly *V. alnifolium* Marsh. - hobble-bush or witch-hobble], *V. lentago* L. [sweet viburnum], *V. molle* Michx. [MO and AR in the SE - softleaf arrowwood], *V. obovatum* Walt. [small-leaf viburnum], *V. prunifolium* L. [blackhaw], *V. rafinesquianum* J.A. Schultes [downy arrowwood].

Wildlife: Viburnum fruit are available from late summer through fall. They constitute a minor portion of the diet of numerous wildlife species, including Ruffed Grouse, Wild Turkey, various songbirds, White-tailed Deer, Gray Squirrel, and others. They generally are a moderate preference White-tailed Deer browse species in the Southeast.

Viburnum **Haw or Viburnum or Arrowwood** **Caprifoliaceae**

T. Bodner (May)

Viburnum acerifolium

T. Bodner (Jun)

Viburnum dentatum

T. Bodner (Jun)

Viburnum acerifolium

Palms and Yucca

Rhapidophyllum **Needle palm** **Arecaceae**

Shrub with erect, leaning, or reclining stems, with upright and fan-like corrugated leaves, having long 3-angled petioles, from an underground or short stem, with needle-spine projections. Flowers in large showy clubs, initially enclosed by a special sheathing bract (spathe). Fruit a spherical, hairy drupe. 1 species in the SE.

Common Species

Rhapidophyllum hystrix (Pursh) H. Wendl. & Drude *ex* Drude **- needle palm RHHY**

Plant: Slow-growing palm, stem underground for some time, eventually erect or leaning, to 1.5 m tall, having fan-shaped leaves on long petioles with sharp spines projecting from base, from stolons.
Stem: Erect or leaning, stout, 8-10 cm wide at base, brownish, bristling with long and very sharp spines, clump to 50-60 cm wide with crowded overlapping old petioles.
Leaves: Spiraling, projecting upward and outward from stem, fan-shaped and segments faintly corrugated, 50-70 cm wide, deeply divided into dagger-shaped leaflets, with a fold at midvein, shiny bluish-green above and scurfy green beneath, petioles slender, longer than leaves, triangular in cross-section (unarmed), dull grayish-silvery covering turning to green-gray mottled, leafstalk base with persistent fibrous web or fabric and long spines.
Flowers: May-Jun. Stout, dense axillary clusters, 10-25 cm long, nestled among the spines, flowers small, unisexual.
Fruit and seeds: Oct-Feb. Drupe-like fruit, fleshy, oval, 15-20 mm long, brown to red, loose golden hairy, held in cluster of needles and petioles, each contains 1 seed.
Range: Native. Southeast MS, lower half of AL and GA, c FL, and se SC, mainly Cp.
Ecology: Occurs on river bluffs, seepage areas, ravine slopes, moist to wet floodplains, and limestone hammocks.
Synonyms: blue-palmetto, vegetable porcupine.

Wildlife: No wildlife value reported.

Rhapidophyllum **Needle palm** **Arecaceae**

T. Bodner (May)

Rhapidophyllum hystrix

T. Bodner (Mar)

Rhapidophyllum hystrix

T. Bodner (May)

Rhapidophyllum hystrix

J. Miller (Mar)

Rhapidophyllum hystrix

Sabal Palmetto Arecaceae

Shrub or tree palms, without spines or prickles, from a horizontal rootstock. Leaves fan-like and corrugated with dagger-shaped segments, having long 3-angled petioles. Flowers on an elongated stalk with drooping branched clusters. Fruit a drupe, berry-like, ovoid to spherical. 3 species in the SE.

Common Species

Sabal minor (Jacq.) Pers. - **dwarf palmetto** **SAMI8**

Plant: Shrub palm, 2-3 m tall, from an underground stem with the leaf crown mostly just beneath the soil surface, often forming colonies in wet and poorly-drained areas.

Stem: Below ground (rarely emerging), straight or curved, stout.

Leaves: Fan-like and fan-shaped, variable in size, nearly to 1 m wide, 30-40 dagger-shaped segments, 30-150 cm long and 1-2 cm wide, nearly flat or slightly folded, some usually drooping, blue-green to light green and yellow green midvein below, margins usually without stringy fibers.

Flowers: May-Jul. Stalk exceeding leaves, 1-2 m tall, having tubular light-brown bracts and narrower branching at spaced intervals, petals 2-3 mm long, whitish, calyx tubular, 3 lobed, subtended by 2 unequal bracts.

Fruits and seeds: Sep-Jan. Drupe-like fruit, flat-circular, 8-10 mm wide, green maturing to black, each contains 1 seed.

Range: Native. TX and AR to FL and north to ne NC, only Cp.

Ecology: Infrequent, being most common around seasonally wet areas in swamps and poorly-drained forests. Sometimes growing on moist bluffs and ravine slopes. Persists by underground stems and spreads by animal- and water-dispersed seeds.

Synonyms: *S. adansonnii* Guersent, *S. deeringiana* Small, *S. glabra* Sarg., *S. louisiana* (Darby) Bomhard, *Corypha minor* Jacq., blue-stem, bush palmetto.

Sabal palmetto (Walt.) Lodd. *ex* J.A. & J.H. Schultes - **cabbage palmetto** **SAPA**

Similar to *Sabal minor* except **plant** a shrub or tree, to 25 m tall; **stem** 30-40 cm wide, leaf-sheaths persisting near crown, after shedding becoming slightly rough and light gray; **leaves** midvein curved downward and segments folded upward, margins with stringy fibers. **Range:** Native. FL and Lower Cp of se GA and se SC.

Ecology: Occurs in coastal forests and tidal flats, elevated areas in marshes.

Synonyms: *S. jamesiana* Small, *Corypha palmetto* Walt., swamp palmetto.

Other species in the SE: *S. etonia* Swingle *ex* Nash [only in FL, scrub palmetto].

Wildlife: Dwarf palmetto fruits are eaten by several songbirds including the Yellow-rumped Warbler and various woodpecker species, along with Raccoon, Gray Squirrel, and White-tailed Deer. Cabbage palmetto fruits are eaten by songbirds, Northern Bobwhite, small mammals, White-tailed Deer, and Black Bear. The tree provides excellent nesting sites for Gray Squirrel and the Crested Caracara. The fruits are also important in the diet of Wild Turkey during winter. The palmettos are a food plant of the Monk butterfly.

Sabal **Palmetto** **Arecaceae**

B. Sorrie (Mar)

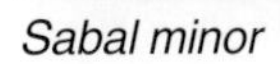

Sabal minor

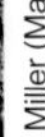

J. Miller (Mar)

Sabal minor

B. Sorrie (Mar)

Sabal palmetto

J. Miller (Mar)

Sabal minor

Serenoa | Saw palmetto | Arecaceae

Shrub or tree palms, with upright and fan-like leaves, often having short recurved prickles along the petioles, from a horizontal underground stem. Flowers on an elongated stalk, with special curved bracts (spathes) at the base of the stalk and branches. Flowers with 3 sepals and 3 petals. Fruit a drupe, ellipsoid or oval. 1 species in the SE.

Common Species

Serenoa repens (Bartr.) Small **- saw palmetto** **SERE2**

Plant: Dwarf palm, to 3 m tall, with a horizontal stem (or erect or leaning), often branched and forming extensive and dense colonies.

Stem: Stout, creeping horizontal atop or just below the soil surface (not always evident), sometimes an erect or leaning terminal portion, appearing scaly, frequently branched.

Leaves: Spiraling around stem, densely crowded with overlapping petiole bases, fan-shaped and segments corrugated, nearly circular in outline, to 1 m wide, deeply divided into dagger-shaped segments, with a fold at midvein, shiny green to greenish-yellow, petioles slender, 0.5-1 m long and 1-2 cm wide, recurved prickles along margins or not, bases with fibrous brown web or fabric.

Flowers: May-Jul. Large branched cluster of spikes, not longer than the leaves, individual flowers small and non-showy, 5-6 mm long.

Fruit and seeds: Oct-Feb. Drupe-like fruit, variable in shape, mostly ovoid, 15-25 mm long and 12-15 mm wide, shiny, fleshy, golden-green ripening to black to bluish black, each contains 1 seed.

Range: Native. Southeast LA to s FL and north to se SC, mainly Lower and Middle Cp.

Ecology: Occurs on dry to seasonally wet sites, in open to semi-shady habitats. Commonly forming colonies on sand ridges, flatwood forests, and coastal dunes, and on islands near marshes. Colonizes by branching surface stems and spreads by animal-dispersed seeds.

Synonyms: *S. serrulata* (Michx.) Nichols, *Brahea serrulata* (Michx.) H. Wendl., *Corypha repens* Bartr.

Wildlife: Saw palmetto is an inconsistent soft mast producer. When available, the fruits are used heavily by Black Bear, White-tailed Deer, and feral hogs. Fruiting is reduced temporarily by fire, although infrequent burning can promote fruit production. Important as a honey plant.

Serenoa Saw palmetto Arecaceae

T. Bodner (Sep)

Serenoa repens

D. Lauer (Aug)

Serenoa repens

T. Bodner (Sep)

Serenoa repens

T. Bodner (Sep)

Serenoa repens

Yucca Yucca Agavaceae

Woody, stout, short-stemmed plants (except for the tall-stemmed Joshua Tree), from a woody rootcrown. Crowded, evergreen leaves, spiraling around and radiating from the stem, thick and leathery, sword-like or dagger-like. Leaf tips hard and spiny with bases flared and partially overlapping. Large white or nearly-white flowers, either sessile or on long bracted stalks, in showy racemes or panicles. Fruit an oblong fleshy capsule or berry, with many seeds, triangular and flattened. 6 species in the SE, only 1 widely occurring.

Common Species

Yucca filamentosa L. **- beargrass or yucca** **YUFI**

Plant: Short, leafy-stemmed, woody plant, 20-60 cm tall with flower stalks to 3 m tall, from stout branching roots, producing rosette-like plants that very gradually fade after flowering, and new rosettes appearing near the base from new root branches.
Stem: Short or absent, covered by overlapping leaf bases.
Leaves: Numerous, spiraling rosette-like, stiff (to drooping with age), linear and dagger-shaped, 20-80 cm long and 2-7 cm wide, yellowish-green to gray-green, parallel veined and somewhat folded along the midvein, with white flaring bases, long-tapering tips to a spiny point, white margins fraying into thread-like curly fibers.
Flowers: Apr-Jun. Tall stalked panicle, 1-3 m tall, erect bracts along the stalk to 20 cm long and shorter upward, showy bell-shaped white to cream flowers, drooping on hairy or hairless branches, 6 unequal-length petals, lanceolate to ovate, 4-5 cm long and 1-3 cm wide.
Fruit and seeds: Aug-Oct. Erect capsules, oblong or cylindrical with a tip, 2-5 cm long and 2-3 cm wide, splitting to release many black seeds, flat and half-round, 6-8 mm long.
Range: Native. LA to s FL and north to MD, mainly Cp and Pd, rare in Mt.
Ecology: Occurs mainly on sandy soils and open to semi-shady habitats, but also on ridges of the Pd and Mt. Widely transported and planted during early settlement and to the present, and escaping. Pollinated by a small moth, the larvae feed upon the seeds.
Synonyms: *Y. concava* Haw., *Y. flaccida* Haw., *Y. smalliana* Fern., beargrass, silkgrass, Adam's needle, curly-hair yucca, spoonleaf yucca.

Other species in the SE: *Y. aloifolia* Engelm. *ex* Trel. [Spanish bayonet - cultivated and native on dunes], *Y. arkansana* Trel. [TX, AR, OK and MO - Arkansas yucca, soapweed], *Y. gloriosa* L. [Spanish dagger, Spanish bayonet, mound-lily yucca], *Y. louisianensis* Trel. [Louisiana beargrass], *Y. recurvifolia* Salisb. [curveleaf yucca]

Wildlife: Flowers are visited by Ruby-throated Hummingbirds. No other wildlife value reported.

Yucca **Yucca** **Agavaceae**

T. Bodner (May)

Yucca filamentosa

T. Bodner (May)

Yucca filamentosa

T. Bodner (May)

Yucca filamentosa

Cane

Arundinaria **Cane or Switchcane** **Poaceae**

Erect, woody perennial, cane-sized grass, from hard rhizomes, forming open to dense stands (brakes). Stems persistent, at first unbranched, later branching freely, forming fan-like clumps. Leaves lanceolate with hairy sheaths having bristle tufts near the throat (soon deciduous) and sheaths overlapping near the summit. Flowering branches from the main stem, or from the rhizome, with terminal panicles of spikelets. Largest native grass in the US. 1 species in the SE, with 2 subspecies, being difficult to separate.

Common Species

Arundinaria gigantea (Walt.) Muhl. **- river cane or switchcane** **ARGI**

Plant: Upright cane-sized grass, bamboo-like, 1-8 m tall, branched at upper nodes, multiple stems from a stout rhizome, and forming dense to open stands near pond and stream banks and bogs.

Stem (culm): Erect, tubular with slightly swollen nodes, to 3 cm wide, smooth, green with dark bands at nodes, hollow internodes and solid nodes.

Leaves: Alternate, grass-like, sessile, lanceolate, 10-30 cm long and 2.5-4 cm wide, base rounded and tip long tapering, hairless above and densely hairy beneath and on the outer sheath, bristles 6-9 mm long on both sides of throat and often on ears (soon deciduous), sheaths widely spaced below and overlapping on current growth, ligules firm, 1 mm long.

Flowers: Mar-Jul. Crowded terminal panicles, few to several spikelets on slender angled stalks, 2-30 cm long, hairy or not, spiklets 4-7 mm long and 6-8 mm wide, with 8-12 flowers, hairy husks, pale or purplish.

Seeds: Jun-Aug. Grain brown, ellipsoid, 7-8 mm long and 3 mm wide. Flowering stem or plant dying after seed matures.

Range: Native. TX to n FL and north to s MD and west to s IL and s MO, all provinces in the SE.

Ecology: Occurs on moist to wet sites, in open to semi-shady habitats. Common to open river and stream banks, shrub bogs, sloughs, and bayous. Scattered small plants occur on uplands adjacent to bottomlands, even in dry forests. Flowering and seeding at irregular intervals like other bamboos. Extensive cane brakes now diminished by overgrazing and clearing.

Synonyms: 3 subspecies: ssp. *gigantea* (Walt.) Muhl. [giant cane], ssp. *macrosperma* (Michx.) McClure, ssp. *tecta* (Walt.) McClure [switchcane], giant cane.

Wildlife: Stands of switchcane provide excellent cover, particularly for wetland species such as the Swamp Rabbit. It also is the preferred nesting substrate of the Hooded Warbler.

Arundinaria **Cane or Switchcane** **Poaceae**

T. Bodner (Oct)

Arundinaria gigantea

Cactus

Opuntia **Pricklypear** **Cactaceae**

Low growing, woody succulent plants, cactus, with round flattened stem segments, having scattered tufts of hair-like spines at nodes, often with long slender spines. Leaves minute scales at nodes and soon deciduous. Showy flowers, singly or several at an edge node, with many spreading green sepals gradually grading into bright-yellow spreading petals. 3 species in the SE.

Common Species

Opuntia humifusa (Raf.) Raf. **- spreading picklypear** **OPHU**

Plant: Low growing succulent plant, with segmented spiny stems, to 40 cm tall, forming carpets to 1 m across, from branching fibrous roots. Mature fruit edible.

Stem: Reclining, ascending, or prostrate, segments flattened and pad-like, round to ovoid to oblanceolate, 3-20 cm long and 2-8 cm wide and 1-2 cm thick, pale to deep green or purplish, smooth and sometimes white-waxy, segments usually joined tightly, nodes (dark spots) spaced along the pad margins and scattered on the surfaces, axillary tufts of hairs, hair-like barbed bristles at nodes, often along with 1-2 woody spines, 1-5 cm long, light brown to whitish.

Leaves: Inconspicuous, scale-like at nodes, 4-10 mm long, quickly deciduous.

Flowers: May-Jul(-Oct). Axillary, 1-3, showy yellow, 5-9 cm wide, many spreading petals grading to many green sepals, short spines in center, many yellow stamens, and sometimes a red star-shaped eye, from a swollen spiny base.

Fruit and seeds: Jul-Oct. Large berry, rounded and tapering at base, 2-5 cm long and 1.5-2 cm wide, spotty tufts of hair-like spines, green ripening to purple or reddish brown, containing many seeds, 3-5 mm long, flattened. Edible and somewhat sweet when ripe.

Range: Native. OK to s FL and north to MA and west to MI and MO, all physiographic provinces in the SE, but infrequent in Southern Appalachian Mt.

Ecology: Occurs as scattered plants or colonies on dry rocky or sandy sites, open to slight shade. Recovers rapidly after burning. Spreading by animal-dispersed stem segments and seeds.

Synonyms: *O. ammophila* Small, *O. compressa* J.F. Macbr., *O. impedata* Small, *O. lata* Small, *O. pisciformis* Small, *O. turbinata* Small, *O. turgida* Small, pricklypear cactus, eastern pricklypear, devil's-tongue.

Opuntia pusilla (Haw.) Nutt. **- cockspur pricklypear** **OPPU2**

Similar to *Opuntia humifusa* except **plant** to 20 cm tall; **stem** segments narrowly oblanceolate and readily separating, 1-5 cm long and 1-2 cm wide, nearly cylindric or only slightly flattened, bearing 2 or more long unequal spines at some nodes, 1-3 cm long.

Range: Native. MS to FL and north to NC, sand dunes and sandy forests along the coast.

Synonyms: *O. drummondii* Graham, *O. pusillus* Haw., *O. tracyi* Britt.

Other species in the SE: *O. vulgaris* P. Mill. [cultivated and rarely escapes].

Wildlife: Most species of pricklypear are native to the Southwest where they are commonly used by songbirds and small mammals. Wildlife use of *Opuntia* in the Southeast has not been researched, although Raccoon reportedly consume the fruits and Gopher Tortoise consume the pads.

Opuntia **Pricklypear** **Cactaceae**

T. Bodner (Oct)

Opuntia humifusa

T. Bodner (May)

Opuntia humifusa

J. Miller (Jun)

Opuntia humifusa

Ferns

Asplenium **Spleenwort** **Aspleniaceae**

Mostly evergreen ferns, having upright to reclining fronds, from the end of scaly underground rhizomes, either creeping or erect. Fronds clustered, long lanceolate to triangular, 1-2 pinnately compound, with wiry stalks, hairless to sparsely hairy, red-brown to black at least near the base and green above, with 2 vascular bundles uniting below the blade. Spore clusters linear at the edge of fertile leaflets that occur along the length of the frond. 12 species in the SE, many hybridize to produce plants with intermediate characteristics.

Common Species

Asplenium platyneuron (L.) B.S.P. **- ebony spleenwort** **ASPL**

Plant: Upright to reclining, slender fronds, 10-60 cm tall or long, with dark-brown stalks, from the end of horizontal rhizomes, 3-4 mm wide, having few linear dark-brown scales.

Stem: Ascending or reclining leafstalks in clusters from the soil surface, wiry, shiny and hairless, brittle, dark brown (thus the common name) to dark reddish-brown.

Leaves (fronds): Shorter reclining sterile fronds (evergreen) and taller upright fertile fronds (deciduous), long lanceolate, 10-60 cm long and 2-12 cm wide, nearly opposite arranged leaflets, semi-thick to thin, narrow triangular, sessile, flaring at base and mostly overlapping the stalk, margins coarsely to finely serrate.

Flowers: Spore clusters linear, near the edge of fertile leaflets on the taller upright fronds.

Seeds: Apr-Oct. Minute spores, wind dispersed.

Range: Native. AZ to FL and north to Canada and west to CO.

Ecology: Commonly occurring fern, often growing with *Polystichum acrostichoides*. Present in rocky and sandy forests, pine plantations, and along stream banks and right-of-ways. Persists and colonizes by rhizomes and spreads by wind-dispersed spores.

Other species in the SE: *A. bradleyi* D.C. Eat. [Bradley's spleenwort], *A. heterochroum* Kunze, *A. x heteroresiliens* W.H. Wagner (pro sp.) [*heterochroum* x *resiliens*], *A. monanthes* L. [single-sorus spleenwort], *A. montanum* Willd. [mountain spleenwort], *A. myriophyllum* (Sw.) K. Presl, *A. pinnatifidum* Nutt. [lobed spleenwort], *A. resiliens* Kunze. [black-stalked spleenwort], *A. rhizophyllum* L. [walking fern], *A. ruta-muraria* L. [wall-rue], *A. trichomanes* L. [maidenhair spleenwort].

Wildlife: No wildlife value reported, and likely of very minor importance.

T. Bodner (Oct)

Asplenium platyneuron

T. Bodner (May)

Asplenium platyneuron

Athyrium **Lady fern** **Dryopteridaceae**

Herbaceous upright ferns, with fronds upright in rows, from short creeping rhizomes having wide brown scales. Large deciduous fronds, light green, 1-3 pinnately compound, lanceolate to ovate, with simple or forking veins ending in teeth. Frond stalks light-green to yellow, and reddish towards the base, furrowed or flattened on the upper side. Spore clusters in rows along veins underneath fertile leaflets. 1 species in the SE.

Common Species

Athyrium filix-femina (L.) Roth **- southern lady fern** **ATFI**

Plant: Erect, light-green, large fronds, to 1 m tall, clustered from a thick rhizome (ascending or horizontal), having abundant persistent scales (long triangular).

Stem: Ascending leafstalks in clusters from the soil surface, stalks light-green (sometimes reddish) with basal scales, stem about as long as blades.

Leaves (fronds): Upright from the soil, 1-3 pinnately compound, lanceolate to long triangular in outline, being broadest near the base, 0.4-1 m long and 10-35 cm wide, nearly oppositely arranged leaflets, ovate- to linear-lanceolate, margins serrate.

Flowers: Spore clusters, dark brown, 2 rows, crescent-shaped, under fertile leaflets next to veins, about 1 mm long.

Seeds: May-Sep. Minute spores, wind-dispersed.

Range: Native. TX to s FL and north to MA and west to IN and MO.

Ecology: Occurs on moist to wet sites, forest openings and shady habitats, and remaining after harvest. Often growing in small colonies. Persists by rhizomes and spreads by windblown spores.

Synonyms: *A. asplenioides* (Michx.) A.A. Eaton.

Wildlife: The southern lady fern provides little wildlife value other than modest cover for small mammals, amphibians, and reptiles.

Athyrium Lady fern Dryopteridaceae

J. Miller (Jul)

Athyrium filix-femina

T. Bodner (Apr)

Athyrium filix-femina

T. Bodner (Apr)

Athyrium filix-femina

Lygodium **Climbing fern** **Schizaeaceae**

Perennial ferns, climbing and twining, vine-like, with delicate pinnately or palmately compound leaves, from rhizomes. Vines often wiry and twisted. 3 species in the SE, with 2 being exotics.

Common Species

Lygodium japonicum (Thunb. *ex* Murr.) Sw. **- Japanese climbing fern LYJA**

Plant: Climbing and twining, perennial vine, up to 7 m long, from slender rhizomes, widely creeping, black and wiry, often forming mats and becoming shrub- and tree-covering infestations.

Stem: Trailing, twining, or climbing, wiry, slender but difficult to break, green to straw-colored or reddish.

Leaves (fronds): Opposite on vine, light green, compound once or twice divided, varying in appearance according to the number of divisions, with highly dissected leaves, appearing lacy, generally triangular in outline, 8-15 cm long and 5-8 cm wide.

Flowers: Fertile fronds on upper part of blades, usually smaller segments with fingerlike projections around the margins bearing sporangia, in double rows on the under margins.

Seeds: Minute spores, wind dispersed.

Range: Exotic. Native to tropical and subtropical areas of eastern Asia and Australia, introduced in US as an ornamental for horticultural purposes. TX to FL to NC, increasingly common across the Cp, becoming scattered in the Pd.

Ecology: Spreading along highway right-of-ways (preferring under and around bridges) and invades into open forests, forest road edges, and stream margins. Scattered in occurrence, but can increase in cover to form mats, covering shrubs and trees. Spreads rapidly by wind-dispersed spores.

Synonyms: *Ophioglossum japonicum.*

Other species in the SE: *L. microphyllum* (Cav.) R. Br. [old world climbing fern, exotic with singly pinnately-dissected fronds], *L. palmatum* (Bernh.) Sw. [native to US with singly pinnately-dissected fronds].

Wildlife: No wildlife value reported.

T. Bodner (Sep)

Lygodium japonicum

J. Miller (Jul)

Lygodium japonicum

J. Miller (Jul)

Lygodium japonicum

Polystichum **Wood fern or Shield fern** **Dryopteridaceae**

Upright to reclining evergreen fronds, from the end of stout underground rhizomes. Fronds leathery textured, unbranched, with leaflet segments in 2 rows, along stalks having sloughing membranous scales. Spores from round central bodies under fertile leaflets near the tip of fronds. 1 species in the SE.

Common Species

Polystichum acrostichoides (Michx.) Schott **- Christmas fern** **POAC4**

Plant: Upright and reclining, slender, evergreen fronds, to 70 cm tall, pinnately compound, from the end of an ascending or horizontal stout rhizome, covered with old stems and fronds and filmy scales having dense fibrous black roots, often forming dense patches.

Stem: Ascending leafstalks in clusters from the soil surface, stalks slender with filmy scales.

Leaves (fronds): Upright, leathery and evergreen, once pinnately compound, lanceolate in outline, 25-60 cm long and 5-12 cm wide, lustrous above, nearly oppositely arranged leaflets, lanceolate to oblong with triangular lobes on the upper base, 2-5 cm long and 5-12 mm wide, margins serrate with soft spiny tips, with upper 12-20 leaflet pairs fertile and much smaller, this upper section becoming brown during late summer and winter.

Flowers: Spore clusters, brownish, 2 rows under fertile leaflets near the tip of fronds, nearly covering the surface.

Seeds: Jun-Oct. Minute spores, wind dispersed.

Range: Native. Eastern US and into Canada.

Ecology: Common fern, occurs on moist to wet sites, shady to semi-open habitats, especially ravines and rocky slopes, and swamp margins. Persists and colonizes by rhizomes and spreads by wind-dispersed spores.

Synonyms: daggerfern, canker-brake.

Wildlife: Christmas fern is an important winter food source for Ruffed Grouse in the Southern Appalachian Mountains. Wild Turkey infrequently feed on the fronds during summer and fall.

J. Miller (Jan)

Polystichum acrostichoides

J. Miller (Jan)

Polystichum acrostichoides

J. Miller (Jan)

Polystichum acrostichoides

Pteridium **Brackenfern** **Dennstaedtiaceae**

A perennial fern, found world-wide, having stiff upright branching stems, from creeping and forking underground rhizomes. Deciduous triangular-shaped fronds, appearing as bipinnate or tripinnate compound leaves, with many spatular-shaped leaflet-like segments, having margins rolled under. Spores produced from a narrow band of sporangia along the underside of the margin during specific times. 1 species in the SE.

Common Species

Pteridium aquilinum (L.) Kuhn **- brackenfern** **PTAQ**

Plant: Widely occuring perennial, coarse fern, with stiff upright stems, 0.3-1.5 m tall, having triangular-shaped fronds, from a black and branching underground rhizome, to 6 m long. Often forming extensive colonies and becoming brown and dry usually after mid-summer. Plant toxic to livestock.

Stem: Erect, stiff, alternate branches from the rhizome, often branched into 3 fronds at a set height, base dark brown and sparsely hairy becoming straw colored and hairless upward.

Leaves (fronds): Deciduous, 1-3, upright or horizontal from a stem, widely triangular or ovate in outline, 20-60 cm long and 20-60 cm wide, 1-3 times divided into stems having spatular-shaped lobes or leaflet-like segments, 1-7 cm long and 3-7 mm wide, becoming shorter outward, with a longer terminal segment (2-4.5 mm wide), hairless or slightly hairy beneath, margins slightly rolled under and increasingly so when sporangia (fruiting bodies) develop along the underside.

Flowers: Spore clusters, brownish, along margins on underside of leaflets at specific time, usually after fire or disturbance.

Seeds: Jul-Sep. Minute spores, wind dispersed.

Range: Native, and world-wide with different varieties. Southern bracken (var. *pseudocaudatum*) from e OK to FL and north to MA and west to IL and s MO, especially Cp in the SE. Eastern bracken (var. *latiusculum*) from e OK to nw NC and north to VA and Newfoundland and west to Que and MN.

Ecology: Widely occurs on acid soils from open pine forests to abandoned pastures, especially those frequently burned. Also occurs on moist sites under mature forests, along forest margins and right-of-ways. Leaves killed by summer drought or first frost, and remain standing. Poisonous to livestock if consumed in large quantities. Spreads rapidly by creeping rhizomes, often forming continuous cover over extensive areas. Plants produce no seeds, only minute windblown spores, barely visible.

Synonyms: *P. aquilinum* var. *pseudocaudatum* (Clute) Heller [southern brackenfern], *P. aquilinum* var. *latiusculum* (Desv.) Underwood *ex* Heller [eastern brackenfern].

Other species in the SE: *P. caudatum* (L.) Maxon [only in FL].

Wildlife: Other than minor amounts of cover, brackenfern has little to no wildlife value.

T. Bodner (May)

Pteridium aquilinum

J. Miller (Sep)

Pteridium aquilinum

Ground Lichen

Cladina **Reindeer lichen** **Cladoniaceae**

Fruticose lichens, being short shrubby- and hair-like, growing in pillow-shaped mats. About 40 species in the SE.

Common Species

Cladina subtenuis (Abbayes) Hale & Culb. **- reindeer lichen** **CLSU7**

Plant: Ground lichen, 4-10 cm tall, growing in gray-green mats, densely branched, up to 30 cm wide, on the surface of sandy soils. Often forming extensive loose colonies.
Stem: Greenish-gray, round in cross-section, branched in 2's, hollow, brittle when dry.
Leaves: none.
Flowers (apothecia): Apothecia (spore heads) very rare.
Fruit and seeds: Wind-dispersed spores.
Range: Native. TX to FL and north to Nova Scotia and southwest to IL, MO, and OK.
Ecology: Nitrogen fixer. Most common reindeer moss in mid-SE. Inhabits sandy soil sites in open pine and pine-hardwood forests, along forest edges and right-of-ways. Promotes soil formation in sterile habitats. Binds together humus, litter, and the soil surface.
Synonyms: *Cladonia subtenuis* (Abb.) Evans, reindeer moss.

Cladina rangiferina (L.) Nyl. **- graygreen reindeer lichen** **CLRA60**

Similar to *Cladonia subenuis* except **stem** ashy gray, branched in 4's. **Range:** Native. AR to n GA and north to Canada and southwest to s IN and s IL, mainly in Mt and n SE.
Synonyms: *Cladonia rangiferina* (L.) Wigg., reindeer moss.

Wildlife: No wildlife value reported.

Cladina **Reindeer lichen** **Cladoniaceae**

T. Bodner (Sep)

Cladina subtenuis

J. Miller (Oct)

Cladina rangiferina

Glossary...understanding terms

Achene - a small, dry, non-splitting fruit with a single seed, common to grasses, asters, and nutrushes.
Acute tip - terminating in a sharp or well defined point, with more or less straight sides.
Alternate leaves - one leaf at each node and alternating on sides of the stem (Fig. 3).
Alternately spiraling leaves - one leaf at each node and their points of attachment forming a spiral up the stem.
Annual - a plant that germinates, flowers, produces seed and dies within one growing season.
Anther - the pollen bearing portion of the stamen (Fig 2).
Apical - at the tip or summit.
Apothecia - the reproductive structure of lichen, usually disc- or cup-shaped.
Appressed hairs - hairs lying close to or flattened against stems, leaves, or other structures.
Arbor - vine entanglement within the crowns of shrubs or trees.
Armed - bearing spines, thorns, or prickles.
Aromatic - having a pleasant and/or strong odor, sometimes requires crushing of plant parts to detect.
Ascending - tending to grow upward, slightly leaning to somewhat erect.
Aster-type flower - compound flowers of Asteraceae in heads with a circular center portion of disk flowers surrounded by radiating ray flowers, such as a sunflower (Fig. 2).
Attenuate - gradually tapering to a long point at the apex or base.
Awn - a stiff bristle at the tip of a part, such as on grass husks.
Axil - the angle formed between two structures, such as between a leaf and the stem.
Axillary - located in an axil or along a side.

Barbed - with minute, backward projecting points, as on some awns and bristles.
Bayou - a slowly draining body of water, often formed from ox-bow lakes on floodplains.
Beaked - bearing a prolonged thick tip, as on some seeds and fruiting structures.
Bearded - having a cluster or tuft of hairs or bristles.
Bee guides - markings in flowers that attract and direct bees.
Berry - a fleshy or pulpy fruit from a single ovary with one to many embedded seeds, such as tomato and grape.
Biennial - a plant that lives for about two years, typically forming a basal rosette in the first year, flowering and fruiting in the second year, and then dying.
Bipinnately compound - twice pinnately compound; a pinnately compound leaf being again divided (Fig. 3).
Blade - the expanded part of a leaf (Fig. 4).
Bract - a small leaf or leaf-like structure often located at the base of a flower, inflorescence, or fruit.
Branchlet - a small branch.
Branch scar - a characteristic marking on a stem where there was once a branch.
Bristle - a short, stiff, slender, hair-like structure.
Browse - young twigs, leaves, and fruit of woody plants, vines, or herbs that are consumed by herbivores; typically refers to deer forages.
Bud - an undeveloped flower, flower cluster, stem, or branch, often enclosed within tiny specialized leaves termed bud scales.
Bulb - an underground, short, vertical stem with fleshy scales or leaves, such as an onion; swollen segments along a rhizome, as with *Cyperus rotundus*.
Bunch grass - a grass species with a cluster-forming growth habit; a grass that grows in upright large tufts.

Bundle scar - tiny raised area(s) within a leaf scar, from the broken ends of the vascular bundles, found along a twig.

Callous-tipped - hardened or thickened at the tip(s), such as the teeth on leaves of *Lobelia brevifolia*.

Calyx - the collective term for all of the sepals of a flower, commonly green, but occasionally colored and petal-like or reduced to absent (Fig. 2).

Calyx tube - a tube-like structure formed by wholly or partially fused sepals.

Cane - very tall grasses, such as switchcane and bamboo; tall, stiff stems, as with some species of *Rubus*.

Capsule - a dry fruit that splits into two or more parts at maturity; for example, the fruit of *Hypericum*.

Catkin - an inflorescence consisting of a dense spike (upright or drooping) of small, unisexual flowers, such as the inflorescences of *Ambrosia*, *Corylus*, and *Quercus*.

Channeled leaves - having one to a few, lengthwise grooves or furrows and ridges.

Clasping - base that partly or wholly surrounds another structure, such as a leaf base surrounding a stem (Fig. 3).

Cleft - incised or cut through more than halfway (such as a cleft leaf margin, Fig. 3).

Coastal Plain (Cp) - the physiographic province extending from the Fall Line to the Atlantic Ocean or Gulf of Mexico, characterized by thick soils derived from relatively recent sediments of loose, unconsolidated material, divided into Hilly, Middle, and Lower (Fig.1).

Collar - the area of a grass leaf blade where it attaches to the sheath (Fig. 4).

Colony - a stand or group of one species of plant, from seed origin or those connected by underground structures such as rhizomes.

Concave - having a surface that is rounded and curved inward.

Convex - having a surface that is rounded and curved outward.

Cordate - heart-shaped (Fig. 3).

Cordate base - a leaf base resembling the double-curved top of a heart shape (Fig. 2).

Corm - a short, swollen, underground stem base from which roots form, similar to bulbs but not scaly.

Corolla - the collective name for all of the petals of a flower (Fig.2).

Corona - a crown-like extension of flower petals consisting of a hood and horns, as with *Asclepias*.

Corrugated leaves or leaflets - having tiny lengthwise ridges and furrows, zig-zagged in cross-section.

Corymb - a flat- or curve-topped inflorescence with the lower pedicels becoming increasingly longer (Fig. 2).

Crenate - margin with shallow, rounded teeth; scalloped (Fig. 3).

Culm - the flowering stems of grasses, sedges, and rushes.

Cultivar - a form or variety of plant originating under cultivation.

Cyme - a flat or rounded inflorescence, with the central flowers blooming first (Fig. 2).

Deciduous - falling off or shedding; not persistent; refers to leaves, bracts, stipules, and stipels.

Dentate - margin with sharp outward-pointing teeth (Fig. 3).

Dioecous - plants with unisexual flowers and having male and female flowers on separate plants.

Disk (disc) - circular, center portion of an aster-type flower, containing disk (disc) flowers (Fig. 2).

Doubly serrate - margins with small serrations on larger saw-like teeth; twice serrate.

Drupe - a fleshy fruit, surrounding a stone (endocarp) that contains a single seed.

Drupelet - small drupes, usually clustered like those forming a blackberry.

Ellipsoid - a three-dimensional ellipse; narrow or narrowly-rounded at ends and widest in the middle.

Elliptic - oval shaped; broadest at the middle and rounded and narrower at the two equal ends (Fig. 3).

Entire - margins without teeth, notches, or lobes (Fig. 3).

Evergreen - a plant with green leaves remaining present through winter.

Exotic - foreign; originating on a continent other than North America.

Exserted - extending beyond some enclosed part.

Fascicle - a tight bundle or cluster, typically refers to leaves or pine needles.

Filament - the long, slender stalk of a stamen that supports the anther (Fig. 2).

Flatwoods - a major forest type of the southeastern Coastal Plain, characterized by flat topography and usually poorly drained areas.

Floricane - the second-year flowering and fruiting cane of *Rubus*.

Forb - a broad leaved herbaceous (non-woody) plant.

Frond - a large, once or twice divided leaf, here referring to fern leaves.

Fruitcose lichen - a growth form of some lichens that is short shrubby or hair-like.

Funnelform - funnel-shaped; gradually widening from the base to the apex; typically refers to the shape of the corolla in various flower types (Fig. 2).

Gland - a structure which contains or secretes a sticky, shiny, or oily substance.

Glandular - having secretory cells, such as nectar glands, gland-tipped hairs, or glandular dots.

Glandular hairy - hairs tipped with a glandular secretion, the tip usually blunt and swollen like the head of a match.

Grain - a grass seed.

Grass - plants of the family Poaceae; herbaceous plants typically with narrow leaves and jointed stems (Fig.4).

Grass-like - having a growth form resembling a grass or members of the Poaceae.

Hammock - an elevated area on flat topography surrounded by a floodplain, marsh or swamp.

Hastate - arrow-shaped leaves, with narrow, outward-projecting basal lobes such as with *Rumex hastatulus* (Fig. 3).

Head - a dense cluster of sessile or stalkless flowers, such as with Asteraceae (Fig. 2).

Herb or Herbaceous - a plant with no persistent above-ground woody stem, dying back to ground-level at the end of the growing season.

Hull - the dry outer covering of a fruit, seed, or nut.

Husk - the outer scale-like coverings of a grass seed, often having hair-like appendages to assist in wind and animal dispersal.

Hybrid - a cross between two closely related species, yielding a plant with a mix of traits from both parents.

Inflated - puffed or blown up; swollen; bladdery.

Inflorescence - the flowering portion of a plant; the flower cluster; the arrangement of flowers on the stem (Fig. 2).

Internode - the space on a herb or grass stem between points of leaf attachment.

Involucre or Involucral - a whorl or collection of bracts subtending or enclosing a flower or flower cluster (Fig. 2).

Irregular flower - a bilaterally symmetric flower; a flower with petals that are not all of the same size or arrangement, such as a pea-type flower.

Keel - a prominent lengthwise ridge.

Lanceolate - lance shaped; widest at or near the base and tapering to the apex (Fig. 3).
Lateral - on or at the sides, as opposed to terminal or basal.
Leaflet - an individual or single division of a compound leaf.
Leaf scar - the scar or marking left on a twig after leaf fall.
Leafstalk - the main stem of a compound leaf of herbs, palms, and ferns.
Legume - a plant in the family Fabaceae; a dry, splitting fruit, one to many seeded, derived from a single carpel and usually opening along two sutures, confined to the Fabaceae.
Lenticel - a raised dot or short line, usually corky in color, on twigs and stems.
Lichen - a plant consisting of a symbiotic relationship between a fungus and a green or blue-green algae, typically forming crusty or scaly growths on soil surfaces, rocks, limbs, or tree trunks.
Ligule - a tiny membranous projection, often fringed with hairs, from the summit of the sheath (top of the throat) in many grasses and some sedges (Fig. 4).
Linear - long and narrow shaped with roughly parallel sides (Fig. 3).
Lobed leaf - margins having deep indentations resulting in rounded portions (Fig. 3).
Logging landing - an intensively impacted portion of a forest harvested area where logs are skidded, delimbed, and loaded for transport.
Lustrous - shiny or glossy.

Margin - the edge of a leaf blade (leaf margins, Fig. 3); the edge of a forest.
Marsh - a poorly-drained portion of the landscape with shallow standing water most of the year, most extensive around intertidal zones.
Mast - nuts, fruits, or other fruiting structures that serve as forage for various wildlife species.
Membranous - thin, filmy, and semi-transparent.
Midvein - the central vein of a leaf or leaflet.
Milky sap - sap being opaque-white and often of a thick consistency.
Mint - refers to a member of the family Lamiaceae; typically aromatic herbs or shrubs with stems square in cross-section and with simple, opposite leaves.
Monoecious - plants having unisexual flowers with the separate male and female flowers occurring on the same plant.
Mottled - spotted or blotched in color.
Mucronate tip - leaf tip having a short, sharp tip, such as a point or spur.

Naturalized - plants of foreign origin, but now well established in their new range.
Nectar - a sugar secretion from special organs on a plant, usually located around flowers.
Nectar disc - a flat nectar-producing organ, in or near the flower, for attracting insects.
Node - the point of leaf or stem attachment, sometimes swollen on grass stems where the sheath is attached (Fig. 4).
Nutlet - a small, dry, non-splitting fruit with a hard cover, usually containing a single seed.

Oblanceolate - lance-shaped with the widest portion terminal; inversely lanceolate (Fig. 3).
Oblong - a shape two to four times longer than wide with nearly parallel sides (Fig. 3).
Obovate - two-dimensional egg-shaped, with the attachment at the narrow end; inverted ovate (Fig. 3).
Obovoid - three-dimensional egg-shaped, with the attachment at the narrower end.
Obtuse tip - blunt or rounded at the apex, sides coming together at an angle greater than 90 degrees but less than 180 degrees.
Odd pinnately compound - pinnately compound leaves with a terminal leaflet rather than a terminal pair of leaflets or a terminal tendril (see illustration as pinnately compound, Fig. 3).

Odorous - with a distinct odor or aroma.
Opposite - leaves born in pairs at each node on opposite sides of the stem (Fig. 3).
Orbicular - round or circular in shape (orbicular leaf, Fig. 3).
Ornamental - a plant cultivated for aesthetic purposes.
Oval - broadly elliptic in shape, with the width greater than half of the length.
Ovate - two-dimensional egg shaped, with the attachment at the wider end (Fig. 3).
Ovoid - three-dimensional egg shaped, with the attachment at the wider end.

Palmately compound - a compound leaf with leaflets arising from one point; leaflets like the spreading fingers of a hand (Fig. 3).
Panicle - an irregularly branched inflorescence with the flowers maturing from the bottom upward (Fig. 2).
Pappus - bristles, hairs, or scales atop the nutlet (achene) in many members of the Asteraceae that assist in wind and animal dispersal.
Parasitic plant - a plant that obtains food or mineral nutrition from another plant.
Pea-type flower - irregular flower characteristic of sweet peas, in Fabaceae (Fig. 2)
Pedicel - the stalk of a single flower (Fig. 2).
Peduncle - the main stalk supporting a flower cluster (inflorescence) or a solitary flower.
Perennial - a plant that persists for three or more growing seasons, even though it may die back to rhizomes or rootstock during the dormant period.
Perigynium - the thin scale-like sheath enclosing the achene of *Carex*.
Petiole - a stalk that attaches the leaf blade to the stem.
Physiographic province - an extensive zone on the landscape with a common geologic origin, having a similar spectrum of topography, soil series, and ecology (Fig. 1).
Piedmont (Pd) - physiographic province extending from the foothills of the Appalachian Mountains to the Fall Line and the beginning of the Coastal Plain, characterized by thick, highly weathered soils, often clayey in texture, generally derived from metamorphic rock (Fig. 1).
Pinnately compound - a compound leaf with leaflets arising at intervals along each side of an axis or leafstalk (Fig. 3).
Pistil - the female reproductive portion of a flower, usually consisting of an ovary, style, and stigma (Fig. 2).
Pith - the soft or spongy central tissue in some twigs and stems, sometimes absent making the stem hollow.
Plume - a tuft of simple or branched bristles.
Pocosin - a type of swamp in an upland coastal region.
Pod - an elongated dry fruit that usually splits open upon maturity, such as a legume.
Pollen - the minute grains produced in the anthers, usually yellowish, and containing the male reproductive cells.
Pome - a fleshy fruit having seeds (no stone), such as an apple; the fleshy portion is derived from the floral tube.
Prickle - a small, sharp projection from a stem or petiole surface, usually slender and straight.
Primocane - the first-year, flowerless canes of *Rubus*.
Prostrate - prone; lying on the ground or nearly so.

Raceme - an elongated, unbranched inflorescence with stalked flowers generally maturing from the bottom upward (Fig. 2).
Range - the general area on the landscape where a species of plant naturally resides.
Ray petals - the strap-like portion of a ray flower of the Asteraceae (Fig. 2).
Recurved - gradually curved backward or downward.
Reflexed - abruptly bent or curved downward.

Resin-dotted - having minute resinous deposits from glands on the surface.
Resinous - with the appearance or feel of resin, usually sticky feeling.
Rhizome - an underground stem, usually horizontal and rooting at nodes (Fig. 4).
Ribbed - tiny raised ridges.
Right-of-way - a narrow corridor of land in straight sections across the landscape, repeatedly cleared and kept in low vegetation, to accommodate roadway structures, poles, and wire for electrical and telephone transmissions, and pipelines.
Rootcrown - the portion of a perennial plant where the stem and roots join, usually located just below the soil surface, often swollen.
Rootlets - small roots below ground or those along the stem of aerial vines used for climbing, such as with *Campsis*.
Rootsprout - plant originating from a root or rhizome that takes root at nodes.
Rootstock - synonymous with rootcrown; the part of a perennial plant near the soil surface where roots and shoots originate.
Rosette (basal rosette) - a circular cluster of leaves on or near the soil surface radiating from a rootcrown, as in dandelions (Fig. 3).
Runner - a general term for an above-ground or below-ground horizontal stem or special root where new plants arise at nodes or at the tip.

Sandhills - areas of the Coastal Plain where old marine and riverine deposits are very high in sand content; dry habitat, often supporting unique plant communities (Fig. 1).
Scaly - covered with minute flattened, plate-like structures.
Scurfy - a plant surface with many coarse hairs or scales resulting in a rough feel.
Semi-evergreen - tardily deciduous or maintaining green foliage during winter only in sheltered locations.
Semiwoody plants - species that have mostly woody stems and deciduous leaves, usually shorter than shrubs, such as species of *Rubus* and *Toxicodendron*.
Sepal - a single unit of the calyx; bract segments of the lower-most whorl of flower parts (Fig. 2).
Serrate - margin with sharp forward-pointing teeth (Fig. 3).
Sessile - attached without a stalk, such as a leaf attached without a petiole (Fig. 3).
Sheath - a more or less tubular portion of a structure surrounding another structure, such as the tubular portion of leaf bases of grasses that surround the stem (Fig. 4).
Shrub - a wood plant, typically multi-stemmed and shorter than a tree.
Simple - not compound; single; undivided; unbranched.
Sinus - the space or recess between lobes on leaves or petals.
Slough - an old stream channel not presently connected or connected only at one end to the main flowing channel on a floodplain, containing stagnant water or mud.
Smooth - not rough to the touch, usually hairless (or only finely hairy) and scaleless.
Sod - dense colony of a spreading grass.
Spathe - a large, leafy bract or bract pair that encloses or partially encloses a flower or an inflorescence (Fig. 2).
Spherical - round in three-dimensions, like a ball; synonymous with globose.
Spike - an elongated, unbranched inflorescence with sessile or unstalked flowers along its length, the flowers generally maturing from the bottom upward (Fig. 2).
Spikelet - a grass flower; tiny spikes.
Spine - a slender, stiff, sharply-pointed structure representing a modified leaf, stipule, or branch, sometimes branched, as with *Crataegus*.
Sporangia - a tiny case bearing spores on ferns.
Spore - a minute (almost not visible), one-celled reproductive body of ferns and lichens, asexual.
Sprawling - spreading irregularly, usually somewhat horizontally.

Stamen - the male reproductive portion of a flower, usually consisting of an anther and filament (Fig. 2).
Stigma - the top part of the pistil that receives the pollen (Fig. 2).
Stipels - the pair of small leaf-like structures at the base of leaflets of compound leaves, such as some legumes.
Stipules - the pair of leaf-like structures at the base of a leaf petiole in some species.
Stolon - a slender, elongate, horizontal stem along the ground surface and rooting at nodes or tip to produce new plants (Fig. 4).
Stoloniferous - tending to reproduce asexually by stolons.
Stone - a hard woody structure enclosing the seed of a drupe.
Striate - having fine, usually parallel grooves.
Style - the part of the pistil, usually narrow, that connects the ovary and stigma (Fig 2).
Subshrub - a very short woody plant, such as *Chimaphila.*
Subtend - a structure just below another, such as flowers subtended by bracts.
Succulent - fleshy or soft tissued.
Suture - a seam or line, usually where a fruit splits at maturity; any lengthwise groove that forms a junction between two parts.
Swamp - a wooded or brushy area usually having surface water.
Synonym - a discarded scientific name for a plant; another common name.

Taproot - the main root axis; a long vertical, central root.
Tardily deciduous - maintaining at least some green leaves into winter or early spring.
Tendril - a slender twining or clasping appendage for support or climbing; a modified stem, leaf, or stipule used by some plants for climbing, may be simple or branched.
Terminal - at the end.
Thorn - a stiff, curved, sharply-pointed modified stem, sometimes branched.
collar where the blade meets the sheath.
Throat - the area inside a flower tube formed from fused petals; the upper side of a grass
Tillering - vegetative shoots from the base of perennial grasses.
Toothed - margin with outward pointed lobes; coarsely dentate.
Trailing - running along the soil or leaf litter surface.
Tuber - a thickened portion of a root or rhizome modified for food storage and vegetative propagation, such as a sweet potato.
Tubular - a cylindric structure, such as formed from fused petals or sepals (flower type, Fig. 2).

Umbel - an inflorescence with pedicels arising from a common point, radiating like the struts of an umbrella (Fig. 2).

Valve - one of the segments into which a capsule splits.
Variegated - marked with stripes or patches of different colors.
Vine - a long trailing or climbing plant.

Whorled - three or more leaves in a circular arrangement from a single node (Fig. 3).
Wing - a flat, thin expansion bordering or surrounding a plant part, such as on seeds and usually assists in wind dispersal.
Winter annual - an annual plant that persists with a basal rosette during winter, flowering and fruiting in late winter or early spring.
Wiry - thin, flexible, and tough.
Woolly - covered with long, curly and matted hairs.

FLOWER PARTS

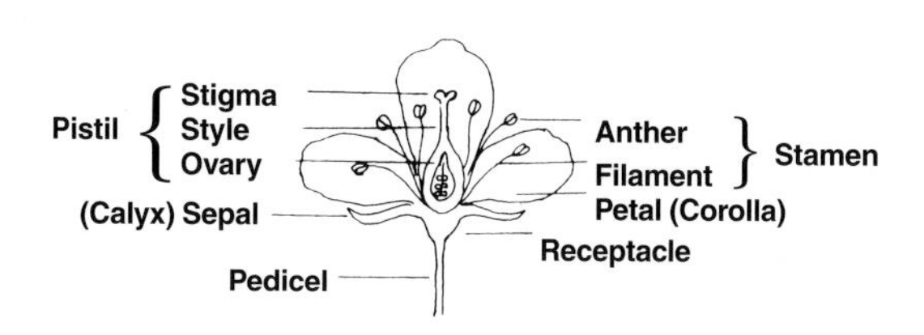

FLOWER TYPES

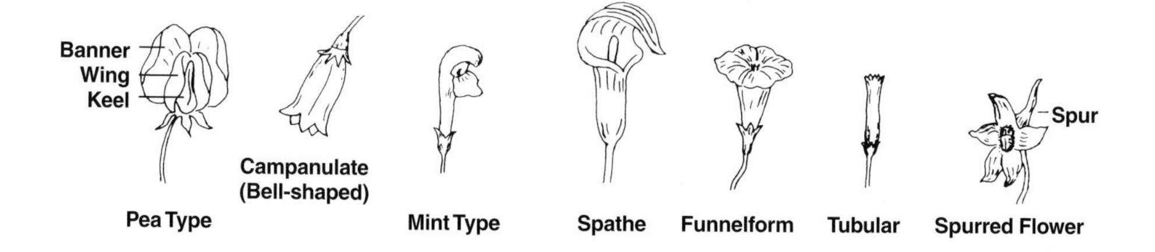

INFLORESCENCES

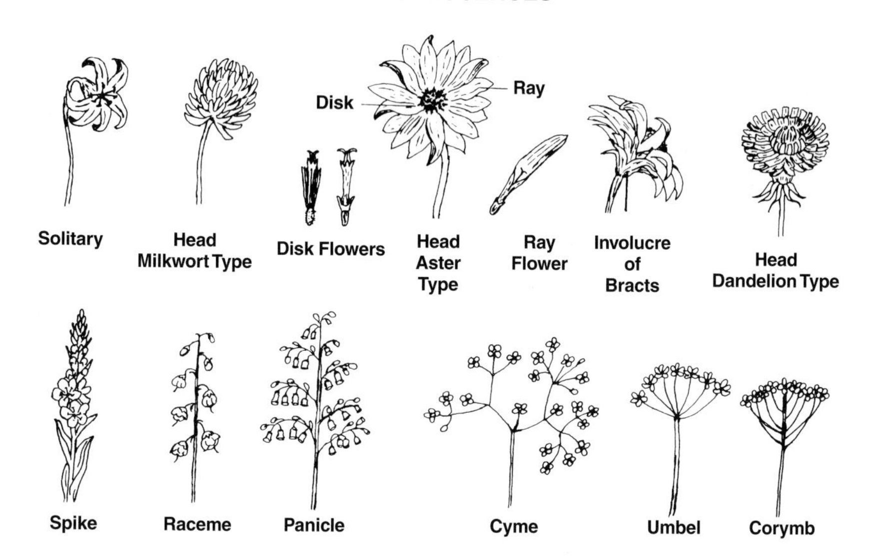

Figure 2. Flower parts, flower types, and inflorescences.

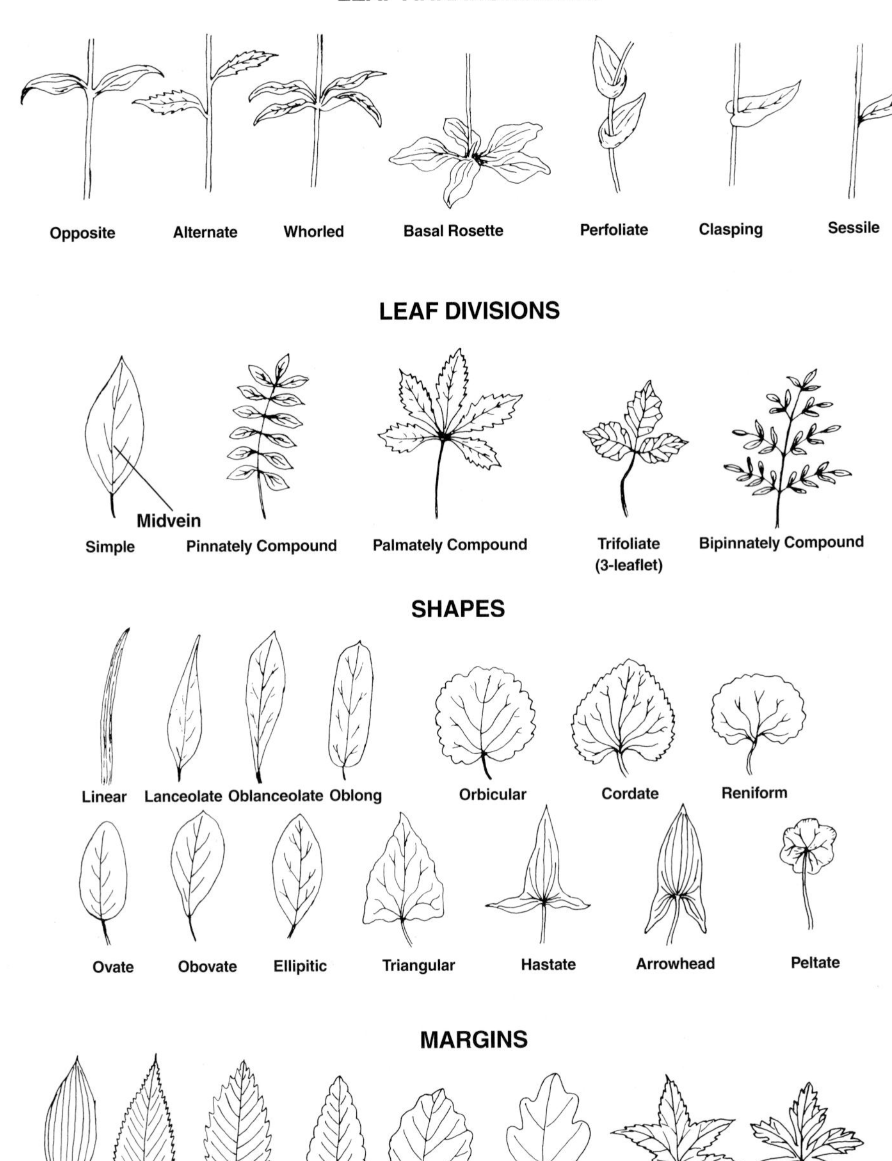

Figure 3. Leaf arrangements, leaf divisions, shapes, and margins.

Used by permission of the University of Alabama Press

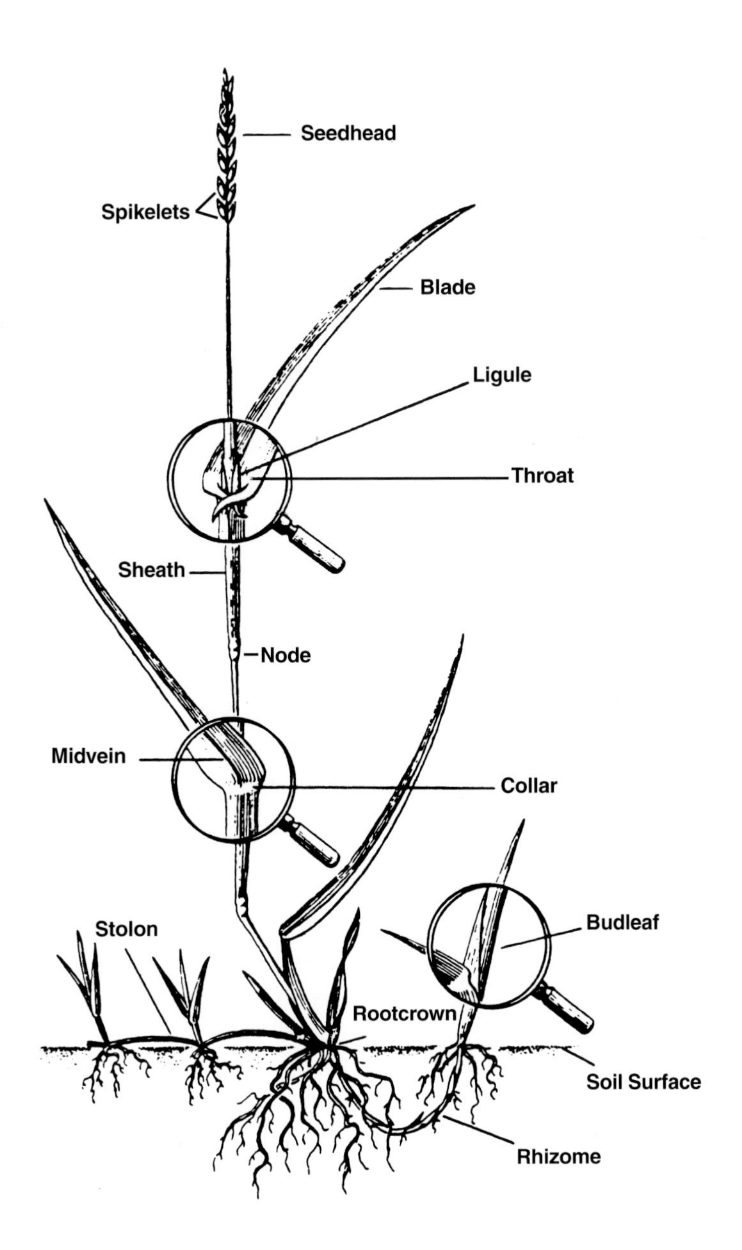

Figure 4. Parts of a grass plant.

Used by permission of The Scotts Company

References

Ajilvsgi, G. 1990. Butterfly gardening for the South. Taylor Publishing Company, Dallas. 348 pp.

Allen, C.M. 1980. Grasses of Louisiana. Ph.D. Thesis, University of Southwestern Louisiana, Lafayette. 281 pp.

Ball, D.M., C.S. Hoveland, and G.D. Lacefield. 1991. Southern forages. Potash and Phosphate Institute, Atlanta. 256 pp.

Bell, R.C., and B.J. Taylor. 1982. Florida wild flowers and roadside plants. Laurel Hill Press, Chapel Hill, North Carolina. 308 pp.

Boyd, R.S, J.D. Freeman, J.H. Miller, and M.B. Edwards. 1995. Forest herbicide influences on floristic diversity seven years after broadcast pine release treatments in central Georgia, USA. 1995. New Forests 10:17-37.

Brown, C.L., and L.K. Kirkman. 1990. Trees of Georgia and adjacent states. Timber Press, Portland, Oregon. 292 pp.

Bryson, C.T., and R. Carter. 1993. Cogongrass, *Imperata cylindrica*, in the United States. Weed Technology 7(4):1005-1009.

Byrd, J.D., J.W. Everest, T.A. Powe, J.D. Freeman. 1996. Poisonous plants of the Southeastern United States. Mississippi Cooperative Extension Service, Publication No. 2166. 51 pp.

Castleberry, S.B., W.M. Ford, K.V. Miller, and W.P. Smith. 1999. White-tailed deer browse preferences in a southern bottomland forest. Southern Journal of Applied Forestry. 23(2):78-82.

Clewell, A.F. 1966. Identification of the lespedezas in North America. Tall Timbers Research Station, Tallahassee, Florida. Bulletin Number 7. 29 pp.

Clewell, A.F. 1985. Guide to the vascular plants of the Florida panhandle. University Presses of Florida, Florida State University Press, Gainesville. 605 pp.

Colvin, D.L., R. Dickens, J.W. Everest, D. Hall, and L.B. McCarty. 1993. Weeds of Southern turfgrasses. Alabama Cooperative Extension Service, Auburn University, Circular - ANR 616. 208 pp.

Crawford, H.S., C.L. Kucera, and J.H. Ehrenreich. 1969. Ozark range and wildlife plants. U.S. Department of Agriculture, Forest Service, Agriculture Handbook No. 356. 236 pp.

Cronquist, A. 1980. Vascular flora of the Southeastern United States. Vol. 1: Asteraceae. University of North Carolina Press, Chapel Hill. 261 pp.

Dean, B.E. 1988. Trees and shrubs of the Southeast. Birmingham Audubon Society Press, Alabama. 264 pp.

Dean, B.E., A. Mason, and J.L. Thomas. 1973. Wildflowers of Alabama and adjoining states. The University Press of Alabama, Tuscaloosa. 230 pp.

Dirr, M.A. 1990. Manual of woody landscape plants: their identification, ornamental characteristics, culture, propagation and uses. Stipes Publishing Company, Champaign, Illinois. 1007 pp.

Duncan, W.H. 1975. Woody vines of the Southeastern United States. The University of Georgia Press, Athens. 75 pp.

Duncan, W.H., and J.T. Kartesz. 1981. Vascular flora of Georgia; an annotated checklist. The University of Georgia Press, Athens. 143 pp.

Duncan, W.H., and M.B. Duncan. 1988. Trees of the Southeastern United States. The University of Georgia Press, Athens. 322 pp.

Edsall, M.S. 1985. Roadside plants and flowers. The University of Wisconsin Press, Madison. 143 pp.

Fernald, M.L. 1950. Gray's manual of botany (8th ed.). American Book Company, New York. 1632 pp.

Foote, L.E., and S.B. Jones, Jr. 1989. Native shrubs and woody vines of the Southeast. Timber Press, Portland, Oregon. 199 pp.

Gee, K.L., M.D. Porter, S. Demarais, F.C. Bryant, and G. Van Vreede. 1991. White-tailed deer: their foods and management in the Cross Timbers. Samuel Roberts Noble Foundation, Ardmore, Oklahoma. 36 pp + appendices.

Gill, J.D., and W.M. Healy. 1974. Shrubs and vines for Northeastern wildlife. U.S. Department of Agriculture, Forest Service, General Technical Report NE-9. 180 pp.

Gleason, H.A. 1952. The new Britton and Brown illustrated flora of the Northeastern United States and adjacent Canada, Vol. 1. New York Botanical Gardens, Bronx. 482 pp.

Gleason, H.A. 1952. The new Britton and Brown illustrated flora of the Northeastern United States and adjacent Canada, Vol. 2. New York Botanical Gardens, Bronx. 655 pp.

Gleason, H.A. 1952. The new Britton and Brown illustrated flora of the Northeastern United States and adjacent Canada, Vol. 3. New York Botanical Gardens, Bronx. 589 pp.

Gleason, H.A., and A. Cronquist. 1991. Manual of vascular plants of the Northeast United States and adjacent Canada (2nd ed.). New York Botanical Gardens, Bronx. 910 pp.

Godfrey, R.K. 1988. Trees, shrubs, and woody vines of Northern Florida and adjacent Georgia and Alabama. The University of Georgia Press, Athens. 734 pp.

Godfrey, R.K., and J.W. Wooten. 1981. Aquatic and wetland plants of the Southeastern United States, Dicotyledons. The University of Georgia Press, Athens. 933 pp.

Gould, F.W., and C.A. Clark. 1978. Dichanthelium (Poaceae) in the United States and Canada. Annals of the Missouri Botanical Gardens, St. Louis. 65:1088-1132.

Grelen, H.E., and V.L. Duvall. 1966. Common plants of longleaf pine-bluestem range. U.S. Department of Agriculture, Forest Service, Research Paper SO-23. 96 pp.

Grelen, H.E., and R.H. Hughes. 1984. Common herbaceous plants of southern forest range. U.S. Department of Agriculture, Forest Service, Research Paper SO-210. 147 pp.

Grim, W.C. 1993. The illustrated book of wildflowers and shrubs. Stackpole Books, Harrisburg, Virginia. 637 pp.

Gupton, O.W., and F.C. Swope. 1987. Fall wildflowers of the Blue Ridge and Great Smoky Mountains. The University Press of Virginia, Charlottesville. 208 pp.

Hale, M.E. 1969. How to know the lichens. William C. Brown, Dubuque, Iowa. 226 pp.

Halls, L.K. (ed.). 1977. Southern fruit-producing woody plants used by wildlife. U.S. Department of Agriculture, Forest Service, General Technical Report SO-16. 235 pp.

Halls, L.K., F.E. Knox, and V.A. Lazar. 1957. Common browse plants of the Georgia Coastal Plain. U.S. Department of Agriculture, Forest Service, Research Paper SE-75. 18 pp.

Halls, L.K., and T.H. Ripley (eds.). 1961. Deer browse plants of southern forests. U.S. Department of Agriculture, Forest Service, Southern and Southeastern Forest Experiment Stations. 78 pp.

Harlow, R.F., and R.G. Hooper. 1971. Forages eaten by deer in the Southeast. Proceedings of the Annual Conference, Southeast Association of Game and Fish Commissioners. 25:18-46.

Harris, J.G., and M.W. Harris. 1994. Plant identification terminology: an illustrated glossary. Spring Lake Publishing, Payson, Mississippi. 198 pp.

Henderson, C.L. 1981. Landscaping for wildlife. Minnesota Department of Natural Resources, Nongame Wildlife Program, St. Paul. 145 pp.

Hunter, G.H. 1989. Trees, shrubs, and vines of Arkansas. The Ozark Society Foundation, Little Rock. 207 pp.

Isely, D. 1990. Vascular flora of the Southeastern United States, Vol. 3, Part 2, Leguminosae (Fabaceae). University of North Carolina Press, Chapel Hill and London. 258 pp.

Johnson, A.S. 1970. Biology of the raccoon (*Procyon lotor varius*) in Alabama. Auburn University Agricultural Experiment Station, Alabama, Bulletin 402. 148 pp.

Justice, W.S., and C.R. Bell. 1968. Wild flowers of North Carolina. The University of North Carolina Press, Chapel Hill. 217 pp.

Kartesz, J.T. 1994. A synonymized checklist of the vascular flora of the United States, Canada, and Greenland, Vol. 1 - Checklist. Timber Press, Portland, Oregon. 622 pp.

Kartesz, J.T. 1994. A synonymized checklist of the vascular flora of the United States, Canada, and Greenland, Vol. 2 - Thesaurus. Timber Press, Portland, Oregon. 816 pp.

Landers, J.L., and A.S. Johnson. 1976. Bobwhite quail food habits in the Southeastern United States with a seed key to important foods. Tall Timbers Research Station, Tallahassee, Florida. Miscellaneous Publication 4. 90 pp.

Leck, M.A., V.T. Parker, and R.L. Simpson (eds.). 1989. Ecology of soil seed banks. Academic Press, San Diego. 462 pp.

Luteyn, J.L. [and others]. 1996. Ericaceae of the Southeastern United States. Castena 61(2): 101-144.

Martin, A.C., H.S. Zig, and A.L. Nelson. 1951. American wildlife and plants: a guide to wildlife food habits. Dover Publications, New York. 500 pp.

McDonald, J.S., and K.V. Miller. 1995. An evaluation of supplemental plantings for white-tailed deer in the Georgia Piedmont. Proceedings of the Southeastern Association of Fish and Wildlife Agencies. 49:401-415.

Miller, J.H. 1996. Exotic plants in southern forests: their nature and control. Proceedings of the Southern Weed Science Society. 48:120-127.

Miller, J.H., B.R. Zutter, S.M. Zedaker, [and others]. 1995. Early plant succession in loblolly pine plantations as affected by vegetation management. Southern Journal of Applied Forestry 19(3):109-126.

Miller, J.H., and K.S. Robinson. 1995. A regional perspective of the physiographic provinces of the Southeastern United States. Proceedings of the Eighth Biennial Southern Silvicultural Research Conference. U.S. Department of Agriculture, Forest Service, General Technical Report SRS-1. p. 581-591.

Miller, K.V., and B.R. Chapman. 1995 Responses of vegetation, birds, and small mammals to chemical and mechanical site preparation. Proceedings of the Second International Conference of Forest Vegetation Management. New Zealand Forest Research Institute Bulletin 192:146-148.

Mitch L.W. 1995. Poison-ivy/poison oak/poison sumac—the virulent weeds. Weed Technology 9(3):653-656.

Nelson, G. 1996. The shrubs and woody vines of Florida. Pineapple Press, Sarasota. 391 pp.

Peet, R.K. 1993. A taxonomic study of *Aristida stricta* and *A. beyrichiana*. Rhodora 95(881): 25-37.

Pope, T., N. Odenwald, and C. Fryling, Jr. 1993. Attracting birds to southern gardens. Taylor Publishing, Dallas. 164 pp.

Radford, A.E., H.E. Ahles, and C.R. Bell. 1983. Manual of the vascular flora of the Carolinas. The University of North Carolina Press, Chapel Hill. 1183 pp.

Randall, J.M., and J. Marinelli. 1996. Invasive plants; weeds of the global garden. Brooklyn Botanical Gardens, New York. 111 pp.

Small, J.K. 1933. Manual of the Southeastern flora. Science Press Printing, Lancaster, Pennsylvania. 1554 pp.

Smith, A.I. 1979. A guide to the wild flowers of the Mid-south. Memphis State University Press, Memphis, Tennessee. 281 pp.

Southern Weed Science Society. 1993. Weed identification guide. Sets I-VI. Edited by C.D. Elmore. Southern Weed Science Society, Champaign, Illinois. 300 pp.

Stucky, J.M., T.J. Monaco, and A.P. Worsham. 1980. Identifying seedling and mature weeds common in the Southeastern United States. The North Carolina Agricultural Research Service and The North Carolina Agricultural Extension Service, North Carolina State University, Raleigh. 196 pp.

Timme, S.L. 1989 Wildflowers of Mississippi. University Press of Mississippi, Jackson. 278 pp.

Tondera, B.P., L.F. Clark, and M.M. Gibson. 1987. Wildflowers of North Alabama. DeskTop Publishing, Huntsville, Alabama. 250 pp.

Turner, N.J., and A. F. Szczawinski. 1991. Common poisonous plants and mushrooms of North America. Timber Press, Portland, Oregon. 311 pp.

U.S. Department of Agriculture. 1971. Common weeds of the United States. Dover Publications, New York. 463 pp.

Uva, R.H., J.C. Neal, and J.M. DiTomaso. 1997. Weeds of the Northeast. Cornell University Press, Ithaca, New York. 397 pp.

Ward, D.B. 1998. *Pueraria montana*: the correct scientific name of the kudzu. Castanea 63:76-77.

Warren, R.C., and G.A. Hurst. 1981. Ratings of plants in pine plantations as white-tailed deer food. Mississippi Agricultural and Forestry Experiment Station, Starkville, Information Bulletin 18. 14 pp.

Wasowski, S., and A. Wasowski. 1994. Gardening with native plants of the South. Taylor Publishing, Dallas. 201 pp.

Watson J., and M.F. Dallwitz. 1992. The grass genera of the world. C.A.B. International, Wallingford, Oxon, United Kingdom. 1038 pp.

Weed Science Society of America. 1984. Composite list of weeds. Weed Science 32 (Supplement 2): 1-137

Weed Science Society of America. 1989. Composite list of weeds. Weed Science Society of America, Champaign, Illinois. 112 pp.

Wofford, B.E. 1989. Guide to the vascular plants of the Blue Ridge. The University of Georgia Press, Athens. 384 pp.

Young, J.A., and C.G. Young. 1992. Seeds of woody plants in North America. Dioscorides Press, Portland, Oregon. 407 pp.

Younghance S.L., and J.D. Freeman. 1996. Annotated checklist of trees and shrubs of Alabama. Sida 17(2):367-384.

Index of Genera by Family

Index of Wildlife Species

Index of Scientific and Common Names

Previously recognized scientific names indicated in italics